Carbon Dioxide Recovery and Utilization

Carbon Dioxide Recovery and Utilization

Edited by

Michele Aresta
University of Bari,
Bari, Italy

KLUWER ACADEMIC PUBLISHERS
DORDRECHT / BOSTON / LONDON

A C.I.P. Catalogue record for this book is available from the Library of Congress.

ISBN 1-4020-1409-0

Published by Kluwer Academic Publishers,
P.O. Box 17, 3300 AA Dordrecht, The Netherlands.

Sold and distributed in North, Central and South America
by Kluwer Academic Publishers,
101 Philip Drive, Norwell, MA 02061, U.S.A.

In all other countries, sold and distributed
by Kluwer Academic Publishers,
P.O. Box 322, 3300 AH Dordrecht, The Netherlands.

Printed on acid-free paper

Printed in the Netherlands.

This book is dedicated to

my Family

Preface

This Book is a summary report of the EU Project BRITE-EURAM 1998 BRRT-CT98-5089 “RUCADI-Recovery and utilization of carbon dioxide”. The Book presents an outline of the emissions in the EU (and a reference to the world emission) followed by a series of papers that are a roadmap to carbon dioxide utilization and cover the scientific and technical aspects of: CO_2 separation, CO_2 technological use, utilization of supercritical carbon dioxide, and actual and perspective use of carbon dioxide in chemical and biological processes. An assessment study of carbon dioxide utilization for the synthesis of methanol is also presented.

The Book has been written for consultation not only by scientists, experts and students, but also by the public. Therefore, an effort has been made by the Authors to use an open and plain language, while presenting accurate scientific and technical data and facts. The result is a collection of reports on technologies, scientific facts and data enriched with updated essential lists of references that represent the route to a deeper insight into the Science and Technology of carbon dioxide.

The work done gives a taste of the usefulness of carbon dioxide and may contribute to strengthen the paradigm shift from the concept of carbon dioxide as a waste to that of carbon dioxide as a resource.

Michele Aresta,
Editor,
Coordinator of the RUCADI Project.

Foreword

The recovery and utilization technologies can contribute to the reduction of both the emission of carbon dioxide and its immission into the atmosphere, two issues our Society has to cope with. The evaluation of the contribution that the utilization option can give to the control of the accumulation of carbon dioxide into the atmosphere is not an easy and straight task: in fact one should consider both the direct and indirect influence of the reuse of carbon dioxide. The fixation of CO_2 for the synthesis of commodities and fine chemicals is an example of the direct reduction of the emission of carbon dioxide, that is used as a source of carbon. The development of innovative technologies that use carbon dioxide and/or are simply less carbon-intensive (lower energy input, more direct syntheses, waste solvent reduction, high yield and high-selectivity syntheses with no- or low-waste production) is an indirect diminuition of the carbon dioxide emission. Both approaches are quite new and require a detailed analysis for understanding the potential of carbon dioxide utilization as a technology for reducing the CO_2 emission and immission in the atmosphere. In fact, a new synthetic methodology based on carbon dioxide does not contribute to its reduction only with the amount of carbon dioxide incorporated into a chemical, but also through the energy-mass-waste-organic solvents reduction. In some cases the latter can represent a much more important contribution than the former. Only the application of the LCA methodology for comparing the existing and the innovative technology can really consent the evaluation of the avoided carbon dioxide.

The utilization of carbon dioxide is often considered as an impossible task to achieve because of the fact that carbon dioxide lays in a potential energy well. Actually, carbon dioxide can be fixed into organic compounds through many different ways, some of which demand a very low energy input, if any.

In this Book a comprehensive analysis is made of the sources of carbon dioxide, and their purity, of the separation technologies and of the uses of carbon dioxide, as a technological fluid or a reagent, or as a reactive solvent. The fixation of carbon dioxide in biomass that can be used as source of energy or chemicals is also considered.

The Book is organized in Sections that gather Chapters dealing with homogeneous contents.

Section I presents a detailed panorama of the carbon dioxide emission by the EU Industry and a comparison with the world data. The purity of carbon dioxide is discussed as a limiting factor to its use. In fact, for some specific purposes, as in the food-industry, contaminants that are often present in carbon dioxide streams must be absent. The elimination of such species to the required level, would rise the price of carbon dioxide and make not economic its use. Therefore, the knowledge of the purity of CO_2 is a key issue for deciding its most economic use and fate. This Chapter also presents the "merchant uses" of carbon dioxide, i.e. those uses in which there is not a chemical conversion of carbon dioxide.

Section II gives a deep insight into separation technologies, presenting advantages and disadvantages of each of them and their cost. An exhaustive discussion on technical plants is also presented and most innovative tecnologies detailed.

Section III describes the use of "supercritical carbon dioxide", and the innovative technologies based on it. It is divided in six Chapters dealing each with the: Particle formation in $scCO_2$, Supercritical fluid chromatography, Homogeneous catalysis in $scCO_2$, Heterogeneous catalysis in $scCO_2$, Polymers synthesis in $scCO_2$, and Modification of polymers in $scCO_2$.

Section IV presents the chemical utilization of carbon dioxide and is organized in four Chapters that address issues such as: Carbon dioxide fixation into organic compounds, with an analysis of the synthesis of compounds that require a low energy input; Linear organic carbonates, essentially focussed on DMC production; Electrochemical conversion of carbon dioxide, that details the opportunities for electrochemical suyntheses based on CO_2; and Carbon dioxide reduction to C1 or Cn molecules, with an analysis of the high-energy options, like fuels synthesis from carbon dioxide and dry-reforming of methane.

Section V introduces the fixation of carbon dioxide into biomass. This topic is divided in two Chapters dealing with the biological utilization of carbon dioxide, as an option for generation of renewable energy. The first Chapter details the use of macro- and micro-algae as source of fuels, the second highlights the use of terrestrial and residual biomass as source of energy, detailing the existing technologies.

The sixth and last Section of the Book presents an assessment study of the carbon dioxide utilization for the synthesis of methanol and an industrial perspective of the use of carbon dioxide..

All together the Book furnishes a comprehensive prospect of the potential of the utilization option and gives information on the state of the art of a number of uses of carbon dioxide, clarifying if they are, or will be in the short term, exploited at the industrial level. This information, coupled with the presentation of the data relevant to the emission and the detailed report on the recovery technologies, gives a perspective of the usefulness of the Recovery and Utilization options and may inspire new scientific research.

The overall picture is that if the recovery of carbon dioxide will be exploited on a large scale as a technology for limiting the immission of carbon dioxide into the atmosphere, then the utilization option may have a key role as it contributes to reduce both the emission and the immission while generating a profit, that may pay back the recovery costs.

The reader will appreciate the effort made by the Authors to use a plain language, while giving both a sound scientific and technological basis to the presentation.

Acknowledgements

As editor of this Book and co-ordinator of the RUCADI Project I have to thank all RUCADI-Partners, whose names are listed at the beginning of each Chapter: with their enthusiasm and competence they made the Project a success. I am also grateful to those who accepted to be authors of the Chapters for the time they have dedicated to the preparation of the papers. We all are grateful to the EU Commisssion for having funded the RUCADI Project and made possible the preparation of this Book. I wish to thank the publisher, Vaska Krabbe, at Kluwer Academic/Plenum Publishers for her continuous assistance during the editing of the book. My warmest thanks go to Dr. Angela Dibenedetto for the time and the attention she has dedicated to the preparation of this Book.

Contents

SECTION I

CARBON DIOXIDE EMISSION AND MERCHANT MARKET

This Section summarizes the activity performed within WP1 and WP3 of the BRRT-CT98-5089 RUCADI Project, "Recovery and Utilization of Carbon Dioxide" by the following Partners:

Rodney J. Allam, *Air Products*, UK
Pantelis Capros, *NTUA*, Athens, Greece
Stephanie Houyou, *University of Liège and ACP*, Belgium
Javier Sanchez Molino, *Carburos Metalicos*, Spain
Andreas Papamichalis, *DMT*, Germany
Frank Sowa, *DMT*, Germany
Jan Vansant, *ACP*, Belgium

Chapter 1

CARBON DIOXIDE EMISSION AND MERCHANT MARKET IN THE EUROPEAN UNION

Jan Vansant
ACP–Dellestraat 55, B-3550 Heusden Zolder – Belgium JANVANSANT@Compuserve.com

1. INTRODUCTION

These last years, carbon dioxide emissions have drawn a lot of attention because of the greenhouse effect. According to the Kyoto protocol, the 15 European countries have committed themselves to reduce their carbon dioxide emissions by an average of 8% with respect to the 1990 level by the period 2008-2012. In this context, the goal of this chapter is to analyse the carbon dioxide emissions in the European Union, as well as the carbon dioxide merchant market.

2. CARBON DIOXIDE EMISSION IN THE EUROPEAN UNION

This paragraph gives an overview of the CO_2 sources in the European Union countries. The goal is first to assess the total emission of each country, which would help evaluate the different CO_2 emission reduction scenarios that would be necessary to respect the Kyoto protocol. Secondly, the objective is to map the CO_2 sources in some identified sectors, in order to have a good mapping of the CO_2 sources in sectors contributing the most to CO_2 emissions and/or on which we can act to decrease these emissions. Indeed, the CO_2 emissions decrease potential all depends on the total CO_2

M. Aresta (ed.), Carbon Dioxide Recovery and Utilization, 3-50.

emission, as well as on the number of plants. In that context, the identification and definition of each plant producing CO_2 is necessary.

CO_2 emission data related to different industrial sectors can be divided in two categories.

For sectors characterized with highly concentrated sources and hence considered as good CO_2 sources, we have collected for each EU country very well documented data for cement, lime, ammonia, and ethylene oxide production. Power plants are also very well documented, due to their importance in the total CO_2 emission. Moreover, significant data is also available for aluminium production in EU and glass production (only for Germany).

Other sectors, such as chemical industries, steel & cokes, refineries, district heating, food-drink-tobacco industry and paper industry, are considered globally. Indeed, sectors such as refineries are in constant evolution: more and more plants are built to produce light hydrocarbons by reaction between heavy hydrocarbon and hydrogen. Consequently, it is hardly possible to have an up to date set of data.

Finally, the following sectors won't be further considered, for different reasons:

- Fermentation: most large breweries now start recovering the CO_2 produced during their fermentation process. They all become self-sufficient, and therefore, do not buy any CO_2 from the merchant market, but also are no longer point sources of CO_2 emissions.
- H_2 plants: not enough data.
- Landfills: too scattered and too short lived to be of economic importance.
- Natural gas offshore: not enough data.

Despite the collection effort that has been made, it is important that the reader be aware of some difficulties encountered and some limitations of the presented set of data, due to the disparity of the structure (different basic assumptions and age of data of the information), the problem of accessing to regional data, confidentiality, and overlapping of information.

2.1 Global carbon dioxide emissions per sector in the EU countries

The following tables (1.1a, 1.1b) present the global carbon dioxide emissions for sector in the EU Countries.

Table1.1a - CO_2 emission(Mt/y) per sector per EU country, 1999[1]

	AU	BE	DK	FI	FR	GE	GR	IR	IT	LX	NL	PO	SP	SV	UK	TOT
01 Bunkers		13.9	4.1	1.7	9.2	6.6	9.8	0.5	7.6	0.0	40.1	1.8	18.5	4.8	7.2	125.8
02 Conventional Thermal Power	11.5	20.0	26.8	19.7	38.2	288.3	38.0	15.5	123.9	0.1	49.8	23.3	93.1	5.1	161.4	914.6
03 Central heating	0.9	0.1	0.3	0.9		5.0								1.4		8.6
04 Energy branch																
Extraction and briquetting of solid fuels						0.1					0.0	0.0	0.2			0.4
Coke ovens and gasworks gas	0.2	0.6			2.4	3.4			0.9		0.4	0.1	0.7	0.2	2.4	11.2
Extraction of petroleum and natural gas	0.2	0.2	1.3		0.6	1.2			0.8		1.7	0.0	0.1	0.0	12.7	18.7
Oil pipeline and gas pipelines	0.4										0.0	0.0	0.0	0.0	0.0	0.5
Oil refining	2.2	4.6	0.9	1.7	16.4	20.8	2.8	0.3	10.5		11.2	2.3	12.6	1.6	19.7	107.4
05 Industry																
Iron and steel industry	6.5	15.2	0.1	5.6	20.4	44.6	0.3	0.1	17.0	0.6	7.3	0.8	9.2	4.7	21.8	154.1
Non ferrous metal industry	0.2	0.4		0.1	1.7	2.8	1.5	0.8	1.3		0.2	0.1	0.9	0.3	2.2	12.5
Chemical industry	0.7	5.6	0.3	1.0	7.4	14.2	0.5	0.7	11.0		6.6	0.4	6.0	0.5	12.5	67.4
Glass pottery & building mat. industry	1.4	3.0	2.0	1.4	10.0	18.2	3.8	0.3	20.4	0.3	1.8	5.0	13.8	1.1	5.6	88.2
Ore extraction industry	0.2		0.2	0.1	0.3	0.9	0.3	0.1	0.2		0.3	0.2	0.6	0.4	1.1	4.9
Food, drink & tobacco industry	0.7	1.0	1.5	0.5	7.3	8.3	0.9	0.9	6.4		3.3	0.8	3.6	0.9	6.9	43.1
Textile, leather & clothing industry	0.4	0.2	0.1	0.1	1.4	1.5	0.3	0.1	4.2		0.3	0.9	1.8	0.1	1.7	13.1
Paper and printing	1.1	0.4	0.1	3.9	4.9	6.9	0.3		4.2		0.9	1.1	2.9	1.7	3.5	31.9
Engineering & other metal industry	0.7	0.5	0.5	0.3	4.8	8.7	0.1	0.2	7.3		1.3	0.2	2.6	0.6	5.4	33.1
Other industries	1.0	5.0	0.8	0.9	8.5	5.5	1.4	0.4	4.5	0.7	0.9	1.2	1.4	0.6	19.9	52.7

Table 1.1b - CO_2 emission (Mt/y) per sector per EU country, 1999 [1]

	AU	BE	DK	FI	FR	GE	GR	IR	IT	LX	NL	PO	SP	SV	UK	TOT
06 Transports																
Railways		0.2	0.3	0.2	1.1	1.8	0.1	0.4	0.4	0.0	0.1	0.2	1.5	0.0	1.4	7.7
Road transport	16.3	22.8	11.2	11.0	125.2	170.3	15.5	8.9	108.0	4.1	28.1	15.5	75.8	18.8	112.0	743.3
Air transport	1.6	4.6	2.6	1.5	19.2	20.8	3.8	1.6	9.7	1.0	10.1	2.2	12.5	2.8	30.3	124.2
Inland navigation		0.8	0.4	0.3	2.3	0.9	2.7	0.1	0.7		2.0	0.1	4.9	0.4	3.1	18.7
07 Household, Commerce, etc.																
Household	8.0	20.1	4.6	3.8	57.8	122.9	6.9	5.7	71.8	1.0	18.9	2.0	15.9	3.4	83.1	426.0
Fishering									0.7							0.7
Agriculture	1.8	1.7	2.5	1.6	8.3	6.0	2.6	0.8	7.5		8.6	1.8	5.4	1.1	2.4	52.0
Other	3.9	8.4	0.9	1.2	35.9	47.8	0.7	2.9	1.9	0.5	10.8	1.0	5.7	4.0	28.0	153.6
Total per country	59.9	129.3	61.5	57.5	383.3	807.5	92.3	40.3	420.9	8.3	204.7	61.0	289.7	54.4	544.3	

Out of this synthesis, we can observe that power plants emit by far the most CO_2, with a total of more than 900 Mt (=megaton) CO_2 per year. Although CO_2 recovery from power plants is of prime importance due to the emission reduction potential, it should be pointed out that the CO_2 concentration in the flue gas is very low (3-12%v), making CO_2 recovery very expensive.

Cement production, refineries, chemical industry and steel & cokes are sectors creating also very important CO_2 emissions.

Even if contributing less to CO_2 emission, ammonia and ethylene oxide production must not be neglected, because of the high CO_2 concentration in the raw gas. That makes CO_2 recovery a lot easier than when it is for example diluted in the flue gas of a power plant.

Other sectors such as transport, commerce and agriculture, contribute also very significantly to CO_2 emission. The only way to decrease the CO_2 emission is to prevent CO_2 formation, for example by means of fuel switching, increased energy efficiency, decreased energy demand, ... As we have no short-term possibility to decrease those emissions, we must consider drastic CO_2 emission reduction in the industry sector.

Consequently, detailed data on CO_2 emission sources has been collected for sectors such as cement, lime, ethylene oxide, ammonia, power plants, non-ferrous metal industry (and especially aluminum) and glass industry. These sectors are considered as "important" due either to their important CO_2 emission or to the high CO_2 concentration in the raw gas, which makes CO_2 recovery easy. These sources, such as ammonia and ethylene oxide production, are at some locations already exploited and the CO_2 is recovered and sold in the merchant market.

2.2 Carbon dioxide emissions point sources

2.2.1 Cement

In 1997 Italy, Germany and Spain produced by far the most cement, thus it can be stated that their CO_2 emissions from cement industry beat that of the other countries as well. In 1997 the share of the EU countries on Europe's cement production amounts to 57,9 %, on the Worlds production to 11,5 %. [2]

The following table presents a compilation of the gathered data per country. Individual plant location as well as plant capacities are available for all EU countries.

Table 1.2 - Data compilation of cement plant per country

Cement production industry	Plant n°	Production (Mt/a)	CO_2-emiss. (Mt/a)	Average production/ plant (kt/a)	Average CO_2 emission per plant (kt/a)
Belgium/Lux.	8	8.2	5.35	1025	669
Netherlands	1	3.4	2.22	3400	2219
Germany	45	34	22.060	756	490
Finland	4	0.9	0.59	225	147
France	60	27.1	17.69	452	295
Greece	8	13.4	8.75	1675	1093
Italy	89	40.9	26.70	460	300
Austria	9	3.8	2.495	422	277
Portugal	7	7.3	4.76	1043	681
Sweden	3	2.5	1.63	833	544
Spain	49	28.7	18.73	586	382
UK	24	13.9	9.07	579	378
Ireland	2	1.8	1.17	900	587
Denmark	1	1.2	0.78	1200	783
Total	310	187.1	122.00	604	394

Plant n° of 1985 (Germany/Austria: 1998/96). Production 1990. (Germany/Austria: 1998/96, Finland 1994, Sweden 1995) Germany/Austria: CO2 emissions of 1997/96; Other countries: CO_2 emissions estimated by applying average specific CO2 factor. [2, 3, 4, 5]

2.2.2 *Lime industry*

In 1995 Germany produced by far more lime than any other EU country, thus it can be stated that its CO_2 emissions from lime industry beat that of the other countries as well. In 1995 the share of the EU countries on Europe's lime production amounts to 78,1 %, on the World production to 18,6 %. [6]

In Germany, in 1996 the lime industry contributed 4,3 % to all industrial direct CO_2 emissions and 20,7 % to all industrial not energy related direct emissions. However, the share of the not energy related direct industrial emissions in all direct industrial emissions was only 14,3 %.

The following table presents a compilation of the gathered data per country. Individual plant location as well as plant capacities are available for Germany, Austria and Sweden.

Table 1.3 - Plant n° of 1999. Production 1995 (former FRG 1997, Switzerland 2000). CO_2-emissons: former FRG/Austria 1997/1995. Other countries: estimated by average specific CO_2-factor [7,8]

Lime production industry	Plant n°	Production (Mt/a)	CO_2 emiss. (Mt/a)	Average production/ plant (kt/a)	Average CO_2 emission per plant (kt/a)
Germany (f. FRG)	116	6.103	7.6	53	66
Austria	5	0.7	0.8	140	160
Sweden	8	0.5	0.60	63	75
Switzerland	1	0.08	0.10	80	96
Belgium/Lux		1.8	2.15		
Netherlands		0	0.00		
Finland		0.3	0.36		
France		3.059	3.65		
Greece		0.5	0.60		
Italy		3.5	4.18		
Portugal		0.2	0.24		
Spain		1	1.19		
UK		2.5	2.99		
Ireland		0.1	0.12		
Denmark		0.125	0.15		

2.2.3 *Ammonia*

Ammonia production plays an important role for the CO_2 merchant market. Indeed, the CO_2 waste stream from the ammonia manufacturing process has a purity (of the dry gas) of > 95%, and CO_2 purification and liquefaction plants are installed at some ammonia production sites, which reduce the effective emissions of CO_2 by these highly concentrated point sources. The following table compiles the production capacities of ammonia production plants (kT/yr), for all EU countries These quantities of ammonia (in kT/yr) are virtually equal to the amount of CO_2 produced by them. Detailed data of plant location and capacities is available.

2.2.4 *Ethylene oxide*

Ethylene oxide production units are generally speaking also good sources of CO_2-rich off-gases. Ethylene oxide production plays an important role as well for the CO_2 merchant market, just as ammonia production. Due to this commercial importance of this sector, a list of the producers in the EU has been established. The following table compiles the production capacities of ethylene oxide (kT/yr), for all EU countries. Not enough data is available at this time in order to determine the quantities of CO_2 released by all of these. Also on some of these plants, CO_2 liquefaction plants have already been attached. Detailed data is available.

Table 1.4 - European ammonia capacity ('000 tonne/year)

	Number of plants	Total capacity kt/year	Average plant capacity kt/year
Austria	1	498	498
Belgium	2	795	397.5
France	6	1614	269
Germany	5	2685	537
Greece	2	270	135
Ireland	1	450	450
Italy	2	510	255
Portugal	1	245	245
Spain	2	475	237.5
The Netherlands	3	2950	983.3
UK	4	1295	323.75
Total	29	11787	

Table 1.5 - European ethylene oxide capacity ('000 tonne/year)

	Number of plants	Total capacity kt/year	Average plant capacity kt/year
Belgium	2	680	340
France	1	197	197
Germany	4	730	182.5
Italy	2	70	35
Spain	1	100	100
Sweden	1	75	75
The Netherlands	2	400	200
UK	1	300	300
Total	14	2552	

Despite the huge amount of collected data, we were not able to determine the CO_2 emission of the ethylene oxide production sector. However, the emitted CO_2 per ton of ethylene oxide is lower than the emitted CO_2 per ton of ammonia. Consequently, according to the production capacities, the emitted CO_2 in ethylene oxide plants does not exceed one fifth of the ammonia sector emission.

2.2.5 Power plants

From the power generation of the EU-countries and their specific CO_2 emissions, the quantities of released CO_2 can be calculated. Although data of 1994 and 1996 are combined, it leads to a usable estimation.

Table 1.6 - Power generation, specific CO_2 emissions and estimated quantities of CO_2 emissions in the EU-countries

Country	Power generation, 1994 [mill. MWh], [9]	Specific emission, [g/kWh], [10]	CO_2 1996	CO_2 emissions (estimated 1994/96), [Mt]
Belgium	69	260		18
Denmark	38	514		20
Germany *	483 (1996: 508)	543		262 (1996: 276)
Finland	62	273		17
France	454	70		32
Greece	37	n.a.		-
UK	304	485		147
Ireland	16	760		12
Italy	220	546		120
Luxembourg	1	n.a.		-
Netherlands	76	459		35
Austria	52	218		11
Portugal	26	431		11
Sweden	138	50		7
Spain	151	341		51
Switzerland (not EU)	n.a.	16		-
Total	2128	349		743

* 1994, 483 mill MWh: only West Germany (f. FRG). 1996, 508 mill MWh: only public electricity supply

These CO_2 emissions depend not only on power generation but also and mainly on the specific CO_2 emissions. The big differences in specific CO_2 emission are based on the use of different primary energies. Only the use of fossil fuels is regarded to cause CO_2 emissions (net). Countries who are able to cover a big share of the power generation by hydropower, nuclear energy or wood (Sweden, France, Austria, Finland, Switzerland), have much lower specific CO_2 emission values than other countries.

Table 1.7 - Power generation by energy carriers, EU, * 1994 [18]; 1997, [10]

Country	Nuclear power	Coal, oil, gas, wood [%]	Hydropower [%]	Others [%]
Belgium	38,9	51,5	9,5	0,1
Denmark	-	90,5	-	9,5
Germany	19,5	70,1	7,8	2,6
Finland	16,2	65,5	18,2	0,1
France	55,0	22,9	22,1	-
Greece	-	71,2	28,5	0,3
UK	17,9	75,5	5,9	0,7
Ireland*	-	92,0	7,9	
Italy	-	70,5	28,3	1,2
Luxembourg	-	7,4	91,9	0,7
Netherlands	2,2	95,7	0,1	2,0
Austria	-	33,9	65,9	0,2
Portugal	-	51,9	48,0	0,1
Sweden	29,5	22,4	47,7	0,4
Spain	16,3	47,7	36,6	-
Switzerland (not EU)	19,2	6,4	74,3	-

The following table presents the number of power plants (>100 MW) per country as well as the total installed capacity.

Table 1.8 - Number of plants and total capacity in EU

	Number of power plants (>100 MW)	Total capacity MW	Average plant capacity MW
Austria	17	5049	297
Belgium	22	8608	391
Denmark	13	7594	584
Finland	32	7914	247
France	47	23470	499
Germany	147	73445	500
Greece	16	5950	372
Ireland	10	3211	321
Italy	81	43738	540
Luxembourg	?	197	?
Portugal	8	3750	469
Spain	45	20352	452
Sweden	20	5934	297
The Netherlands	36	18905	525
UK	88	69947	795
Total	582	297867	

Detailed data on individual plant location and capacities is available.
The necessary data to estimate the CO_2 emissions were available only for Germany. Average CO_2-emission per plant depends on the plant capacity, rate of utilisation, energy efficiency, and fuel specific CO_2-factor. In Germany, brown coal power stations have the highest rate of utilisation while fuel oil and natural gas fired power plants are often used only for peak load service. Therefore these data can hardly be applied to the other countries.

Table 1.9 - Power generation in Germany [5, 11]

Power production		Plant n°	Total capacity (MW)	CO_2-emiss. (MW)	Average plant capacity (MW)	Average CO_2 emission per plant (Mt/y)
Germany	all fossil fuels	147	73445		500	
	Hard coal	28	11207	40.07	400	1.43
	Brown coal	22	19737	139.86	897	6.36
	Fuel oil	8	2117	0.99	265	0.12
	Natural gas	12	3132	3.98	261	0.33

Based on power production, fuel consumption and CO_2 factors, some scenarios of CO_2 emission reduction potential can be derived by switching primary energy for some countries. The results are presented in the next table. The three considered scenarios are coal substitution by natural gas (NG) and nuclear power, and nuclear power substitution by natural gas.

Table 1.10 - CO_2 emission decrease potential, based on fuel switching. [5, 11]

	Actual CO_2 emissions Mt/y	Coal => NG Emission reduction (%)	Coal => nuclear Emission reduction (%)	Nuclear => NG Emission reduction (%)
Germany	262	49.8	91.6	-25.7
France	32	31.6	74.8	-722.7
Italy	120	5.8	14.3	0
Spain	51	38.3	85.7	-44.8
Austria	11	15.8	34.6	0

2.2.6 *Non-ferrous metal industry*

The non-ferrous metal production sector is very heterogeneous. The production depends greatly on the nature of the non-ferrous metal or alloy. Figures available for Germany indicate that aluminium production exceeds that of refined copper, followed by refined lead and refined zinc production. Each non-ferrous metal can be produced by different processes, which complicates the task of data collection. In each non-ferrous metal production, the CO_2 emission has two contributions: non-energy related (from the carbon contained in the ores) and energy related. In that case, emissions are related to energy directly (from the ores processing) and indirectly (emissions from electricity production). However, since most ores do not contain carbon, not energy related emissions can be neglected. In all the non-ferrous metal production processes, aluminium production is by far the most energy intensive process. It is produced by melt electrolysis of Al_2O_3 (extracted from bauxite), thus the main part of consumed energy is electricity. Considering these elements, some detailed data is available for Germany (plant type and capacity). For other countries, detailed data are available only from aluminium production. Germany, UK and France are the largest aluminium producers in EU, followed by Spain, the Netherlands and Italy.

Non-ferrous metal inudstry in Germany

Compared with the cement and lime industry the direct CO_2 emissions of the German non-ferrous metal sector are about a fifth of that of the cement industry and half of the lime industry. The number of plants, however, is much higher (4.5 or 8 times, respectively).

Table 1.11 – Production of non ferrous Metals and metal alloys, [kt/y], [12]

	1996	1997
Primary aluminium pig	576	572
Secondary aluminium	417	432
Refined copper.	671	673
Copper casting alloys	54	60
Refined lead	238	329
Lead alloys	0	0
Refined zinc	328	318
Fine zinc casting alloys	47	51
Remelted zinc, zinc alloys	43	40
Tin and tin alloys	14	16
Total	2389	2492

Table 1.12 – Numbers of non ferrous metal production plants [12]

	1996	1997
Production and first processing of		
- precious metals	15	11
- aluminium	102	94
- lead, zinc, tin	29	32
- copper	52	52
- other metals	23	23
- total	221	212
Metal foundries	253	242
Aluminium subsequent processing	90	n.a.
Hot-galvanising	127	124
Total	601	578

In order to illustrate the heterogeneity of this sector, the following table lists the different metal production branches and the produced quantities.

Since the number of processes with their individual direct and indirect energy consumption is quite high, it is hardly possible to collect more detailed data of CO_2 release.

Aluminium production industry

The following table shows the primary aluminium production in Europe (thousands of metric tons), according to the data of the IPAI (International primary Aluminium Institute) [13].

Table 1.13 - Primary aluminium production in Europe

Period	West Europe	East/Central Europe
1996	3192	3185
1997	3297	3316
1998	3549	3419
1999	3720	3584

Primary aluminium production is the weight of liquid aluminium as tapped from the pots excluding alloying elements, returned scrap or remelted products. However, it should be noted that primary aluminium production figures provided to the IPAI by Russian Federation producers contain alloying additives but these do not exceed 1% by weight of reported totals.

The data received by the IPAI have been aggregated by geographical area to meet the requirement that, in general, company and country data be lost within appropriately aggregated totals. These geographical areas and the primary aluminium producing countries which fall in those areas, are as follows:

- West Europe: Austria, France, Germany, Greece, Iceland, Italy, Netherlands, Norway, Spain, Sweden, Switzerland, United Kingdom.
- East and Central Europe: Bosnia-Herzegovina*, Croatia*, Hungary, Poland*, Romania*, Russian Federation, Slovakia, Slovenia, Ukraine, Federal Republic of Yugoslavia (Montenegro).

Countries which are asterisked are not currently reporting primary aluminium production to the IPAI.

2.2.7 Glass industry (in Germany)

Data are available for Germany only. It can be seen from the table below that the different types of glass have different production, with a different contribution to the CO_2 emission. Container glass is by far the most abundant, followed by plate glass. Crystal and special-use glass is produced only in small amount. [5, 14]

Table 1.14 - Aluminium production in Europe

Aluminium production industry	Plant N°	Production (kt/y)	CO_2-emiss. (Mt/y)	Average production/ plant (kt/y)	Average CO_2 emission per plant (kt/y)
Germany (f. FRG)	5	572	0.74	114	148
Austria	-				
Sweden	-				
Switzerland	1				
Belgium/Lux.	-				
Netherlands	2				
Finland	-				
France	5				
Greece	1				
Italy	2				
Portugal	-				
Spain	3				
UK	5				
Ireland	-				
Denmark	-				
Total EU	23				
West Europe*	33	3720		113	

*Western Europe includes EU + Iceland, Norway, Switzerland Plant number 1999. Production 1999/1997. CO_2 emission 1992.

Table 1.15 - Comparison of the industries regarding point sources and CO_2 concentration

Glass production industry	Company N°	Plant N°	Production (Mt/y)	CO_2-emiss. (Mt/y)	Average N° of plant per company	Average production/emission plant (kt/y)	Average CO_2 per plant (kt/y)
Plate glass	4	15	2032	0.39	3.8	135	26.09
Special and industrial use glass	9	80	446	0.09	8.9	6	1.07
Container glass	26	36	4323	0.83	1.4	120	23.12
Crystal and commodity glass	13	42	106	0.02	3.2	3	0.49
Total	52	173	6907	1.33	3.3	40	7.69

2.2.8 *Summary of collected data on CO_2 emission point sources*

It was not possible to present the exhaustive set of gathered data concerning the identification of CO_2 emission point sources. The following table lists the data collected.

Table 1.16 - Summary of collected data on CO_2 emission point sources

Power supply
Location, capacity, fuel type of all thermal power plants >100MW in all EU countries
National CO_2 emission (total and specific, all EU-countries)
Average CO_2-emission per power plant (Germany)
CO_2-emission reduction potential by switching primary energy (Germany, France, Italy, Spain, Austria)
Cement industry
Location of all cement plants in EU (398 plants)
Plant capacities in Germany (8 of 45 plants) and Austria (9).
National cement production (all EU-countries)
National CO_2 emission (all EU-countries)
Average production and CO_2 emission per plant (all EU-countries)
Lime industry
Location of lime production plants (Germany (129 plants), Austria (5), Sweden(10))
National lime production (all EU countries)
National CO_2 emission (all EU-countries)
Average production and CO_2-emission per plant (Germany, Austria, Sweden,Switzerland)
Ammonia Industry
Location of ammonia production plants (all EU countries)
Plant capacities (all EU countries)
Ethylene oxide industry
Location of ethylene oxide production plants (all EU countries)
Plant capacities (all EU countries)
Aluminium industry
Location of 29 (of 31) aluminium smelters in EU
Plant capacity of 17 plants
Primary aluminium production (Germany and EU)
CO_2 emission (Germany)
Average production per plant (Germany and W-Europe)
Average CO_2-emission per plant (Germany)
Glass industry (Germany)
Location of glass production plants (173)
Location of glass production companies (average 3.3 plants per company)
Total glass production
Average production and CO_2 emission per plant

2.3 Projection of carbon dioxide emissions

Scenarios projecting energy use into the future were used to predict CO_2 emissions as well. The scenarios cover a large number of sectors and countries in considerable detail. The objective of this chapter is to provide quantitative information in order to deliver input for the analysis of carbon dioxide technologies and their markets. To this aim, official statistics on CO_2 emissions from all anthropogenic activities were collected and evaluated for the years 1990 through 1999. These data constituted a sound basis for assessing the evolution of CO_2 emissions in the short, medium and longer term, i.e. for the period 2000-2030 in 5-year intervals, under 'business-as-usual' conditions.

This analysis was conducted with the aid of the PRIMES energy systems model. PRIMES is a modeling system that simulates a market equilibrium solution for energy supply and demand in the European Union (EU) member states. The model determines the equilibrium by finding the prices of each energy form such that the quantity producers find best to supply match the quantity consumers wish to use. The equilibrium is static (within each time period) but repeated in a time-forward path, under dynamic relationships. The model is behavioural but also represents in an explicit and detailed way the available energy demand and supply technologies as well as pollution abatement technologies. The system reflects considerations about market economics, industry structure, energy/environmental policies and regulation. These are conceived so as to influence market behaviour of energy system agents. The modular structure of PRIMES reflects a distribution of decision making among agents that decide individually about their supply, demand, combined supply and demand, and prices. Then the market-integrating part of PRIMES simulates market clearing. PRIMES is a general purpose model. It is conceived for forecasting, scenario construction and policy impact analysis. It covers a medium to long-term horizon, is modular and allows either for a unified model use or for partial use of modules to support specific energy studies.

The reference scenario that was used within RUCADI, named 'baseline scenario', is the basis against which different policy scenarios can be compared. The baseline scenario uses moderate macroeconomic and demographic growth assumptions and assumes a smooth evolution of international energy prices. Emissions are projected under the assumption that no specific policies will be put in place to reduce carbon dioxide emissions. Other policy developments do operate in the context of the baseline, some of them limiting emissions in an indirect way. In addition, technological progress also helps in reducing emissions. However, curbing emissions to comply with Kyoto (for 2010) commitments and the IPCC long-run requirements require additional policies, which are not considered under the baseline assumptions. In that context, the baseline scenario is useful to evaluate the potential sources of CO_2 emissions and their location under the assumption of no policies restraining emission of carbon dioxide from the energy system.

For the quantification of the baseline scenario, a long process of validation has been followed involving the European Commission, national experts and governmental authorities, in particular within the European Union. Earlier results of this effort have led to the 'Shared Analysis' scenario for the period up to 2020 [15]. A major update of the above scenario was effected in subsequent years, which aimed at improving the assumptions and incorporating more recent developments and changes in policy with potentially high relevance for CO_2 emissions. In addition the whole analysis was extended to the horizon of 2030. This report contains the latest of these updates, which was used in the framework of the European Commission's ACROPOLIS project [16].

Table 1.18 presents the CO_2 emissions in each country for the base year 2000 and the year 2015. Emissions are shown for each main economic activity sector (industry, tertiary sector, households, transport, power and steam generation, district heating, fuel production and energy sector). Industry is further subdivided into nine major subsectors.

his baseline scenario, carbon dioxide emissions in Europe are likely to grow, albeit modestly. However, the growth will be such that both the Kyoto commitments and long-run objectives for stabilisation of atmospheric CO_2 concentrations will not be met under baseline assumptions. For the European Union the projections show that in 2010 carbon dioxide emissions are likely to be about 8% above the level of 1990 and will start declining afterwards with a tendency to come down to 1990 levels in the long run (by 2030). Energy intensive industry apart, all sectors and power generation will continue to emit higher levels of carbon dioxide until 2010. Power/steam generation and transport will continue to be the dominating emission sources, accounting for 30-35% of total CO_2 emissions each, both in the present and in the long run.

2.4 Large-scale CO_2 storage option

CO_2 sources identification and CO_2 capture can be an effective way of mitigating climate change only if the captured CO_2 is stored for the long term, i.e. up to several hundreds of thousands of years. There are several ways to store large quantities of CO_2: geological storage, such as depleted

Table 1.17 – CO_2 emissions in European Union (million tonnes) Year 2000

	AU	BE	DK	FI	FR	GE	GR	IR	IT	LX	NL	PO	SP	SV	UK	EU-15
TOTAL	**58.5**	**118.9**	**53.0**	**53.8**	**376.5**	**814.2**	**90.3**	**41.7**	**417.9**	**8.8**	**164.6**	**58.3**	**280.6**	**50.7**	**556.7**	**3144.3**
Industry	12.9	30.0	4.5	12.5	70.8	117.7	9.6	3.3	71.7	1.7	22.3	10.7	44.8	10.5	77.4	500.4
Iron e steel	6.0	15.1	0.1	5.4	23.3	48.9	0.3	0.1	18.3	0.6	6.7	1.0	8.5	5.0	22.5	161.8
Non ferrous metals	0.2	0.3	0.0	0.1	1.7	2.8	1.6	0.7	1.6	0.0	0.2	0.1	0.9	0.3	2.2	12.6
Chemicals	0.9	5.1	0.3	0.7	9.2	15.4	0.5	0.7	9.5	0.0	7.3	1.3	6.2	0.4	12.7	70.3
Building materials	1.7	2.7	1.3	1.3	9.7	19.1	3.8	0.3	17.6	0.4	1.6	3.7	17.7	1.2	5.3	87.4
Paper and pulp	1.3	0.5	0.1	3.3	5.1	7.3	0.4	0.0	4.6	0.0	0.8	1.1	2.7	1.3	3.6	32.3
Food, drink, tobacco	0.8	1.1	1.4	0.4	7.2	8.4	1.1	0.9	5.3	0.0	3.4	0.9	3.7	0.8	7.2	42.5
Engineering	0.6	0.5	0.5	0.2	4.7	8.4	0.1	0.2	6.8	0.0	1.3	0.2	2.2	0.7	5.6	32.1
Textiles	0.4	0.2	0.1	0.1	1.4	1.6	0.4	0.1	4.0	0.0	0.3	0.9	1.8	0.1	1.7	13.2
Other	1.0	4.5	0.7	0.8	8.5	5.7	1.5	0.4	4.0	0.7	0.6	1.4	1.2	0.7	16.6	48.2
Tertiary	6.1	9.9	3.0	2.6	43.7	52.7	3.5	3.7	11.0	0.1	18.9	3.3	11.8	6.0	29.4	205.6
Household	5.9	19.2	4.2	3.4	57.4	122.5	7.2	6.0	73.6	1.5	18.6	2.1	17.0	4.2	81.8	424.5
Transport	18.8	29.0	14.3	13.1	150.6	196.2	23.1	12.2	119.8	5.4	40.2	19.3	100.3	21.8	148.7	912.9
Electricity steam production	10.8	24.6	24.4	19.4	34.9	292.2	44.2	16.0	123.9	0.1	50.1	20.8	93.3	4.9	182.9	942.3
District heating	0.9	0.2	0.3	1.3	0.0	5.3	0.0	0.0	0.0	0.0	0.0	0.0	0.0	1.5	0.0	9.4
New fuels (hydrogen) production	0.0	0.0	0.0	0.0	0.0	0.0	0.0	0.0	0.0	0.0	0.0	0.0	0.0	0.0	0.0	0.0
Energy branch	3.2	6.1	2.3	1.7	19.1	27.6	2.6	0.4	17.9	0.0	14.5	2.2	13.3	1.9	36.5	149.2

Table 1.18 – CO_2 emissions in European Union (million tonnes) Year 2015

	AU	BE	DK	FI	FR	GE	GR	IR	IT	LX	NL	PO	SP	SV	UK	EU-15
TOTAL	**59.9**	**110.0**	**45.6**	**60.4**	**378.3**	**850.7**	**106.7**	**45.4**	**466.1**	**10.5**	**167.0**	**74.5**	**296.0**	**73.5**	**490.9**	**3235.4**
Industry	12.9	23.4	3.3	15.4	55.7	100.5	10.3	6.1	64.2	2.1	17.9	11.8	43.3	8.8	55.6	431.1
Iron e steel	5.0	8.7	0.1	5.5	14.5	39.6	0.3	0.1	10.3	0.5	5.7	0.7	5.9	4.0	13.0	113.8
Non ferrous metals	0.3	0.4	0.0	0.1	2.6	3.5	1.9	1.6	2.4	0.0	0.2	0.2	1.1	0.4	2.3	17.1
Chemicals	0.9	5.2	0.1	2.6	7.3	11.4	0.7	1.1	9.0	0.1	5.9	1.2	6.9	0.1	6.2	58.8
Building materials	1.7	2.4	1.3	1.1	9.9	18.0	3.8	0.4	19.6	0.4	1.6	4.0	15.6	1.1	5.4	86.2
Paper and pulp	1.6	0.1	0.1	3.7	3.8	2.0	0.4	0.4	4.9	0.0	0.1	2.9	5.5	1.4	1.7	28.7
Food, drink, tobacco	0.8	0.6	0.4	0.5	2.3	5.6	0.9	1.3	1.3	0.0	1.6	0.7	2.7	0.1	3.0	21.9
Engineering	1.0	0.7	0.6	0.7	6.3	13.5	0.2	0.7	9.2	0.0	1.7	0.3	3.2	1.0	7.3	46.2
Textiles	0.2	0.1	0.0	0.3	0.8	0.7	0.2	0.1	3.0	0.0	0.2	0.5	1.2	0.0	0.4	7.8
Others	1.3	5.3	0.5	0.9	8.2	6.2	1.9	0.5	4.5	1.0	0.8	1.3	1.3	0.6	16.4	50.7
Tertiary	5.3	11.4	2.9	2.8	40.0	53.8	4.8	4.2	9.0	0.1	18.4	4.5	13.1	7.7	24.6	202.7
Household	6.2	19.4	4.0	4.0	70.9	126.9	10.4	5.8	92.5	1.6	18.9	5.4	25.1	5.5	80.3	476.9
Transport	21.4	31.8	14.9	14.7	161.4	207.1	27.3	14.5	129.2	5.6	45.0	24.0	116.8	23.0	151.1	987.9
Electricity steam production	11.0	19.6	19.4	20.8	34.0	349.0	51.2	14.4	155.3	1.1	59.2	26.4	85.2	22.4	152.1	1021.1
District heating	0.9	0.0	0.2	2.6	1.1	0.5	0.0	0.0	0.0	0.0	0.0	0.1	0.0	5.2	0.0	10.5
New fuels (hydrogen) production	0.0	0.0	0.0	0.0	0.1	0.1	0.0	0.0	0.0	0.0	0.0	0.0	0.0	0.0	0.0	0.3
Energy branch	2.3	4.2	0.9	0.2	15.2	12.8	2.6	0.4	16.0	0.0	7.6	2.2	12.3	0.8	27.2	104.8

oil and gas fields, deep saline aquifers, or unminable coal seams, and deep ocean storage.

Many other ways of storing CO_2 have been suggested, such as conversion into chemicals or storage in a mineral carbonate, generally these are much more expensive and/or have limited capacity compared with natural reservoirs [17]. These possibilities won't be considered here.

2.4.1 Storage in depleted oil and gas fields

Depleted oil and natural gas fields provide a very attractive CO_2 storage option. These reservoirs offer as advantages a proven fluid retention, a well-known geology, and the possibility to reuse the production equipment.

In most of the oil fields already exploited, there is some oil left in place underground. Indeed, conventional primary and secondary oil recovery do not allow to recover all the oil due to rock heterogeneity and to the existence of oil clusters that are hold in place due to the oil-water superficial tension. Oil can be miscible with CO_2 under the pressure and temperature conditions in the reservoir. Consequently, CO_2 injected into suitable depleted oil reservoirs can enhance oil recovery by typically 10-15% [18]. This technique is called Enhanced Oil Recovery (EOR). Most part of the injected CO_2 will hence be stored for the long term, while the part recovered with oil can be captured and re-injected in the depleted oil field [19].

Enhanced oil recovery by CO_2 injection is already used at more than 74 sites in the USA, and 33 Mt/year of CO_2 is being delivered to these fields for injection [18]. Some other projects are under development in the USA and in the North Sea [20]. A large EOR project has recently been commissioned in Canada (Weyburn). EOR is a proven technology, and the additional quantities of recovered oil generates income partly offsetting the cost of CO_2 capture and storage.

Underground storage in natural reservoirs has been practiced by the natural gas industry for many decades. Natural gas is routinely injected into, stored and withdrawn from hundreds of underground storage fields.

2.4.2 Storage in aquifers

Aquifers are underground water-filled strata, and these sites can be used to store CO_2 underground for the long term. The aquifers that could be used are deep underground, contain saline water and are unsuitable for supplying potable water.

However, unlike depleted oil and gas fields, these reservoirs are not automatically good candidates for CO_2 storage. Indeed, the retention capacity is not proven, and additional geologic survey must be conducted to determine whether the aquifer is suitable for CO_2 storage. The main restriction consists in the presence of a geological cap rock with low permeability that prevent any CO_2 leakage to the upper strata and finally to the atmosphere.

All the injected CO_2 would be stored, and no profitable use can be made out of the injected CO_2. Techniques similar to those for depleted oil and gas fields could be used for CO_2 injection.

Sleipner Vest is a natural gas field under the Norwegian sector of the North Sea operated by Statoil. CO_2 must be removed from the gas before delivery to customer. From 1996, nearly one million ton CO_2 per year is captured from natural gas and injected in a deep saline aquifer at a depth of 800-1000 m. The flows of CO_2 injected at Sleipner are being monitored and modelled as part of an international project established by Statoil with the IEA Greenhouse Gas R&D Programme. This work should help resolve many of the uncertainties about CO_2 injection into water-filled strata [18, 21]. CO_2 storage at Sleipner was the first instance of CO_2 being stored in a geological formation because of climate change considerations. This technique is hence proved feasible and no CO_2 leakage has been observed /22/.

2.4.3 Storage in unminable coal seams

Another possible option for CO_2 storage is CO_2 injection in unminable coal seams. As CO_2 is injected in the coal seam, methane is desorbed from the coal surface and can be recovered at the production well. Unlike in enhanced oil recovery, CO_2 remains stored underground, adsorbed on the coal surface. Moreover, coal can adsorb twice as much CO_2 by volume than methane, which means that the storage capacity exceeds the quantity of methane recovered [23]. This technique is called Enhanced Recovery of Coal Bed Methane (ERCBM). The recovered methane can generate income partly offsetting the cost of CO_2 capture and storage.

The technology is proven feasible due to a demonstration plant in New Mexico, USA, where 100.000 ton CO_2 has been injected in 3 years, none of which has found its way to the production wells. A field test of ERCBM using CO_2 and N_2 mixtures is being carried out by the Alberta Research Council under an international project facilitated by the IEA Greenhouse Gas R&D Program [18, 24].

2.4.4 *Storage in deep ocean*

Storage in the deep ocean may offer the largest potential capacity of all. Storing CO_2 in the oceans consists in injecting CO_2 either at a depth exceeding 1000 meters so that the CO_2 is mixed in the medium layers of the ocean, or at a depth higher than 3000 meters, so that the CO_2 sinks and forms a pool at the bottom of the ocean. However, there are major uncertainties about the environmental impact and the retention time of CO_2. Moreover, in addition to technical hurdles, CO_2 storage in the deep ocean could still face legal hurdles and public opinion acceptance [18].

2.4.5 *Storage capacities and cost*

The global potentials for CO_2 storage, estimated by the IEA Greenhouse Gas R&D Programme [18] are shown in the following table.

Table I.19 - CO_2 storage capacity

Storage Option	Global Capacity (Gt CO_2)
Depleted oil and gas fields	920
Deep saline reservoirs	400 – 10.000
Unminable coal seams	> 15
Deep ocean	> 4000

The estimates for deep saline reservoirs were made in the early 1990s. More recent estimates suggest the capacity for storage in geological reservoirs in North West Europe alone could be as much as 800 Gt CO_2 (most of this is deep saline reservoirs). Further research is required to assess the potential storage capacity of deep saline reservoirs.

The cost of CO_2 transport and injection in the storage site is lower than the cost of CO_2 capture out of flue gases (for example by chemical absorption) and compression up to the transport pressure. CO_2 transport cost can be estimated at 1-3 €/t CO_2 per 100 km, whereas injection of compressed CO_2 into deep saline reservoirs or depleted oil or gas fields is expected to cost upwards of 1-3 €/t CO_2. If enhanced oil recovery or enhanced coal bed methane recovery can be achieved, this may generate income partially offsetting the cost of capture and storage [18].

2.5 Conclusions

This chapter gives an overview of the CO_2 sources in the European Union countries. The goal is first to assess the total emission of each country, which would help evaluate the different CO_2 emission reduction scenarios that would be necessary to respect the Kyoto protocol. Secondly, the objective is to map the CO_2 sources in some identified sectors, in order to have a good mapping of the CO_2 sources in sectors contributing the most to CO_2 emissions and on which we can act to decrease these emissions. Indeed, the CO_2 decrease potential all depends on the total CO_2 emission, as well as on the number of plants. In that context, the identification and definition of each plant producing CO_2 is necessary.

CO_2 emission data related to different industrial sectors can be divided in two categories. For sectors characterized with highly concentrated sources and hence considered as a good CO_2 source, we have collected for each EU country very well documented data for cement and lime, ammonia, and ethylene oxide production. Power plants are also very well documented, due to their importance in the total CO_2 emission. Moreover, significant data are also available for aluminium production (in EU) and glass production (only for Germany). Other sectors, such as chemical industries, steel & cokes, refineries, district heating, food-drink-tobacco industry and paper industry, are considered globally.

We observed that power plants emit by far the largest quantities of CO_2, with a total of more than 900 Mt CO_2 per year. Although CO_2 recovery from power plants is of prime importance due to the emission reduction potential, it should be pointed out that the CO_2 concentration in the flue gas is very low (3-12%v), making CO_2 recovery very expensive, i.e. in the range of 40-70 €/ton.

Cement production, refineries, chemical industry and steel & cokes are sectors creating also very important CO_2 emissions.

Even if they contribute substantially lower amounts to CO_2 emission, ammonia and ethylene oxide production must not be neglected, because of the high CO_2 concentration in the raw gas, that makes CO_2 recovery much easier than when it is for example diluted in the flue gas.

CO_2 on these identified sectors offers the most important CO_2 emission reduction potential. However, published cost data for a number of existing CO_2 recovery sites, mostly from flue gas recovery, show that CO_2 recovery is expensive, and amounts to 30 –100 €/t CO_2, including transportation. Although these figures show that CO_2 recovery out of flue gas is expensive and not economically attractive, the different CO_2 capture technologies have considerably improved in the last few years, and are continuously under development.

The projection of future CO_2 emissions showed that according to the baseline scenario, carbon dioxide emissions in Europe are likely to grow, albeit modestly. However, the growth will be such that both the Kyoto commitments and long-run objectives for stabilisation of atmospheric CO_2

concentrations will not be met under baseline assumptions. For the European Union the projections show that in 2010 carbon dioxide emissions are likely to be about 8% above the level of 1990 and will start declining afterwards with a tendency to come down to 1990 levels in the long run (by 2030). Energy intensive industry apart, all sectors and power generation will continue to emit higher levels of carbon dioxide until 2010. Power/steam generation and transport will continue to be the dominating emission sources, accounting for 30-35% of total CO_2 emissions each, at present and in the long run.

According to the CO_2 purity and the volume, several utilizations are possible. Small quantities of pure CO_2 are best suited for the merchant market. However, the total volume of the merchant market is much smaller than the yearly-emitted CO_2, which implies that other solutions must be found. The only way to avoid large-scale CO_2 emission in the atmosphere would be to store it, after capture if the purity is not sufficient. In that case, several options can be found depending on the source localization: underground storage (injection in deep saline aquifers, injection in unminable coal seams coupled with methane recovery, injection in oil fields coupled with improved oil recovery) or storage in the see.

Each of these options has advantages and drawbacks, but they are feasible and offer the possibility to avoid releasing large quantities of CO_2 in the atmosphere.

3 CARBON DIOXIDE MERCHANT MARKET IN THE EUROPEAN UNION

CO_2 finds many "technological" industrial applications. Access to good quality CO_2 is the key issue for the development of CO_2 utilization. CO_2 can be used under different forms: liquid bulk, cylinders and dry ice. The applications can be divided in two categories: those that temporarily use CO_2 (bubbles, cooling, *etc*) and those that permanently remove CO_2 from the atmosphere (pH control of waste water, *etc*).

Today, the merchant market utilization of CO_2 represents less than 0.1% of the CO_2 released into the atmosphere. It is clear, in order to make the utilisation option relevant for carbon dioxide mitigation, that the amount of carbon dioxide used should be increased by at least one order of magnitude. Nevertheless, even if this limit is not reached in the short term, environmental benefits would be obtained by the implementation of benign syntheses and innovative technologies. This specific issue needs to be evaluated carefully as it may be very easy to underestimate its potential.

3.1 CO_2 production and product quality for the merchant market

3.1.1 Production

Carbon dioxide is a naturally occurring gas that exists in small amounts in the atmosphere. It also exists underground, either in a relatively pure state, or blended with water, natural gas or petroleum. With the exception of production from natural wells, manufacturing of carbon dioxide is a secondary operation integrated with the production of other products. Purification and liquefaction of carbon dioxide is capital- and energy-intensive.

For the industrial merchant market, the major sources of by-product carbon dioxide are ammonia production, natural wells, fermentation during the production of alcohol, and in chemical/petrochemical manufacturing (e.g. H_2 production). Together, these sources supply more than 85 percent of the carbon dioxide commercialised in the merchant market.

In ammonia and hydrogen plants, both hydrogen and carbon dioxide are generated from the steam cracking of hydrocarbons. In both cases, the CO_2 must be separated from the final product. In many NH_3 plants, a (large) fraction of the CO_2 output from this source is consumed captively in the production of urea. A lower quantity is purified and liquefied for sale on the merchant market.

In manufacture by fermentation, carbon dioxide is a by-product of ethyl alcohol production, e.g. during the production of beer. In contrast to the ammonia process, the fermentation process is carried out at relatively low temperatures so that cooling is not required.

3.1.2 Product quality

The CO_2 source determines the possible contaminants from very harmless such as N_2 to sulfur components and hydrocarbons that need to be removed for most applications.

ISBT, CGA, EIGA defined standards for purity in order to guarantee that for food-grade CO_2 all possible contaminants have been removed to a level where they will not influence taste or odour and where they will be harmless for the health of consumers. Other elements such as water (humidity) and oxygen are also removed to a level, which is acceptable to the process requirements at the plants that subsequently process the L-CO_2.

Quality control and quality assurance measures currently implemented by the producers of L-CO_2 for food and beverage applications are a lot more stringent than even 10 years ago. Detection levels for some of the possible contaminants now go as low as the ppb (parts per billion) threshold.

3.2 Merchant market overview [25, 26, 27]

CO_2 finds applications in the industry in nearly all sectors of activities. Indeed, CO_2 has a lot of properties (fizz, inert, chemical, acid, pressure, coolant, high density, electrically non-conducting, solvent, bacteriostatic, photosynthesis, pesticide leaving no residue, sublimation (not melting)) that makes it important for industrial application. We defined the following market segmentation:

food processing: refrigeration, chilling, MAP/CAP, freezing, slaughter of swine, stunning of poultry, etc...

carbonated beverages: soft drinks, mineral waters, beer brewing,..

chemical industry: synthesis, polymers, inorganic carbonates, PCC, inertization,...

metal fabrication: welding, steel mills, foundries, aluminium, metallurgy,...

agriculture: green houses, fumigation of silo's, additive to irrigation water, tobacco processing, soil decontamination,...

rubber and plastics processing: blow molding, foaming, deflashing, grinding & shredding, cryo-separation,...

other uses as solvent: dry cleaning, SC-FE, supercritical chromatography,...

water treatment: pH control, carbonatation of tap water, swimming pools, water well rehabilitation,...

nuclear: power plant cooling

well re-injection: secondary oil recovery, methane recovery from coal mines, re-injection in natural gas fields,...

dry ice production

cylinder filling: all types of cylinders and mini-bulk tanks filled with either pure carbon dioxide or blends with other gases

3.2.1 Food processing

Six main applications constitute this market segment, which in the USA accounts for more than 44 % of the merchant CO_2 market (excluding captive on-site productions such as urea production and re-injection in oil-wells):

- Refrigeration: Chilling food products during transportation
- Modified Atmosphere Packaging (MAP/CAP)
- Freezing: Chilling food products during processing
- Supercritical Fluid Extraction

- Stunning: Slaughter of Swine, Stunning of Poultry.
- Decontamination of food stuff

Food safety is an area of increasing importance for L-CO_2 customers. Within the last two years, food processors in the U.S. and other countries have implemented procedures for hazard analysis of critical control points, or HACCP, to identify and correct steps in the processing chain where foods can go off spec. And innovative CO_2-based cooling technologies are, literally, taking the heat off these trouble spots.

In refrigeration, freezing and packaging applications, CO_2 competes with nitrogen. Both gases are non-toxic and leave no residue on the product, so the choice between the two is based upon cost and availability. Lower-cost nitrogen produced via membrane separation and pressure swing adsorption technology competes with carbon dioxide for this application.

Refrigeration: maintaining the "cold chain" for food products during transportation

Liquid carbon dioxide is used to refrigerate food both prior to shipment and during the trip. It is sometimes used to pre-cool trailers equipped with conventional refrigeration systems. It is also used in complete carbon dioxide portable refrigeration systems. These can be fed with either dry ice or contain special boxes that will accumulate CO_2 snow, injected into them from a liquid supply tank. Since carbon dioxide has a high density, it effectively flushes out the warm air, which greatly reduces the load on the truck's mechanical system and eliminates the time lag typically required to cool the truck's interior to a suitable temperature for shipment.

When L-CO_2 is directly injected, some of the liquid carbon dioxide also deposits as layers of snow directly on the food. This snow gradually sublimes during shipment to provide additional refrigeration. The modest equipment costs associated with the use of carbon dioxide chilling and its rapid effect are the main factors behind carbon dioxide's popularity in this use.

Although continuing to see use in tandem with conventional refrigeration systems, complete carbon dioxide refrigeration systems are gaining market share at the expense of mechanical refrigeration systems due to both environmental and cost considerations. Carbon dioxide systems do not employ chlorofluorocarbon (CFC) refrigerants, which are being phased out due to environmental considerations. These systems also eliminate the emissions associated with the diesel fuel used to operate mechanical systems. Additionally, carbon dioxide systems are silent and require little maintenance. Finally, carbon dioxide systems reach the target temperature more rapidly and hold it longer.

Although the general trend over the last decade has been toward the use of liquid carbon dioxide in food refrigeration, solid carbon dioxide (dry ice) has retained some applications in the food industry. Dry ice is used extensively by commercial airlines to chill their food and drinks prior to service to passengers. Advantages of dry ice refrigeration include the fact that it leaves no residue upon evaporation, is non-toxic (which permits direct contact with food), and is relatively light.

Modified Atmosphere Packaging (MAP/CAP)

In packaging uses, carbon dioxide gas, generated from liquid, is used as an inerting medium to prevent flavour loss, deterioration by oxidation, and bacterial growth in food. Carbon dioxide is used for packaging coffee and as an inert atmosphere in the processing and transporting of fruits, vegetables and cereals. Carbon dioxide is completely non-toxic and leaves no residue on the product. Carbon dioxide competes with nitrogen in this use, with the choice of gas determined by costs and regional availability.

One area of potentially strong growth for carbon dioxide in the food processing industry is its use in modified or controlled atmosphere packaging (MAP/CAP), which centers on the absolute control over processing and packaging conditions, and employs breathable flexible film in combination with gas flushing (carbon dioxide, nitrogen, or blends of the two) to significantly extend the shelf life of packaged foods. This reduces the need for preservatives and additives, as well as waste, extends the range of distribution for food growers and processors, and satisfies consumer demands for high quality, minimally processed, easy-to-prepare foods. This technology is still developing, but has already been applied to baked goods, meat and poultry, produce, and prepared foods (e.g., lasagna and fresh pasta).

Freezing: chilling food products during processing

In freezing uses, large quantities of liquid carbon dioxide are used to "flash freeze" or "tunnel freeze" a variety of food products including fresh vegetables, poultry, sea food and bakery products. Foods are flash frozen by transporting them on a conveyor belt through a tunnel where they are exposed to an atmosphere of "flashing" liquid carbon dioxide, which produces the low temperature required. By expanding liquid CO_2 to atmospheric pressure, it will turn into snow (-78°C) and gas in equal portions. The snow provides about 85% of the chilling effect by sublimation. Examples of application of this technology can be found in (Raleigh) and (Murray Hill, NJ.). Even more CO_2 technologies may move from field testing to food processing plants if governments mandate changes to avoid E. coli and other bacteria.

Supercritical Fluid Extractions (SFE)

The newest application for carbon dioxide in the food industry is for supercritical fluid extraction (SFE) processes, which can be used in the decaffeination of coffee and the production of flavours, natural colours and essential oils. This method is gaining favour in coffee decaffeination because it is non-toxic, does not employ solvents such as methylene chloride, and leaves no trace in the final product. Carbon dioxide SFE extraction is also gaining favour for other food extraction applications.

A substance such as carbon dioxide is in a supercritical state when it exists at a temperature and pressure greater than its "critical point." In this state, carbon dioxide is not properly a gas or a liquid; rather, it has some qualities of both. This supercritical product has unique solvent properties, dissolving substances in a manner similar to a liquid solvent, although the process is more rapid due to its gas-like qualities. Unlike liquids (which are non-compressible), supercritical carbon dioxide becomes more dense as pressure is increased. Although other chemicals can be used for SFE applications, carbon dioxide is ideal because it is non-toxic, non-flammable, inexpensive and readily available. Additionally, carbon dioxide reaches a supercritical state at a temperature of only 37°C.

Stunning: slaughter of swine, stunning of poultry

Since many years, CO_2 has been used in slaughterhouses for stunning of pigs instead of using conventional electrical stunning. The use of CO_2 offers a wide variety of advantages, such as reduced animal stress, less injuries, etc., with a better meat quality and yield as result. For poultry, gas mixtures are used.

Decontamination of foodstuff

In a new process called PoroCrit^{SM}, liquid foods and liquid medicines are sterilized and preserved by contact with compressed CO_2 at room temperature. The process uses membrane technology, which is the most effective way to bring CO_2 into contact with the liquids to be sterilized. Microbes such as infectious bacteria, spores and viruses (e.g. E. coli, salmonella, and polio virus) are killed as effectively as with heat pasteurisation. For example, apple juice and orange juice contacted with liquid CO_2 at room temperature are as sterile as if they had been heat pasteurised, and there is no settling out of the pulp on storage. In addition to retaining their full nutritional value, juices and other liquids sterilized in this manner are preserved against spoilage. The PoroCrit^{SM} process is especially attractive for juices where flavour is easily degraded, and for biological therapeutic fluids, which are inactivated by heat processing.

3.2.2 Carbonated beverages

By far the dominant use is in carbonated soft drinks, although carbon dioxide is also used in the production of some types of sparkling wines and other carbonated beverages. Growth in carbon dioxide basically follows soft drink consumption, although the introduction of new carbonated beverages, such as carbonated fruit juices and artificially carbonated sparkling wine, also contributes to market growth. Carbonated soft drinks continue to benefit from a general shift in consumer preference toward non-alcoholic beverages. Consumers currently favour beverages that are low in calories, cold, convenient and sweet. These trends explain the stagnation and/or decline in demand for certain traditional beverages, including tea, milk, coffee, beer and distilled spirits. Soft drink consumption is expected to moderate somewhat from strong growth over the last decade. Bottled water demand is expected to increase favourably. Moreover, carbon dioxide acts as a preservative by preventing fermentation and inhibiting the growth of mold and bacteria.

Beverages are carbonated by use of a carbonator or saturator. Pressurized gas supplied to the carbonation machine is evaporated from L-CO_2. Water cooled to about 5°C, pumped with gas to the top of the carbonator, flows over baffles under pressure where it is saturated with carbon dioxide. It is then ready for mixing with additives and bottling.

Carbon dioxide requirements for soft drinks range from about 2.5 volumes percent of gas per volume of beverage to 4.5 volumes of gas per volume of liquid for highly carbonated beverages such as ginger ale. The total CO_2 volumes consumed by this industry is higher than this because of losses at the point of use by bottlers, use of carbon dioxide for non-carbonation operations such as cleaning and water treatment (see pH control), and the use of carbon dioxide for carbonating non-soft drink beverages such as wines and fruit juices not included in soft drink consumption figures. Furthermore, CO_2 is used in this industry as a propellant gas for emptying tanks and as a shielding gas for preserving the drinks quality (exclusion of oxygen).

Production of beers and ales, does not require external carbon dioxide injection because, especially in large breweries, the brewing process generates adequate quantities of carbon dioxide and this is recovered and purified on-site for further use.

Liquid carbon dioxide used in beverages must be odourless and as pure as possible, as impurities may affect the taste of the beverage. Licensors of proprietary carbonated beverages attempt to maintain a uniform taste worldwide, and many licenses for carbonated beverages contain a clause stating that the license may be revoked without recourse by the licensee if taste satisfactory to the licensor is not maintained. Supply of consistently high quality carbon dioxide to beverage producers is therefore critical.

A growing market is here certainly the delivery of liquid CO_2 in so called mini-bulk systems. These fixed containers are placed inside the premises of a pub, restaurant of gas station shop and are filled directly from cryogenic tanks on distribution trucks. This CO_2 is then used to either pressurize beverage dispensing tools or to mix in situ the soft drinks starting from water and syrup.

3.2.3 Chemical industry

CO_2 as feedstock

As a feedstock, carbon dioxide is used for the production of basic lead carbonate (white lead), as well as sodium, potassium, ammonium and hydrogen carbonates and bicarbonates. Sodium salicylate (an intermediate in the manufacture of aspirin) is produced by the reaction of carbon dioxide with sodium phenolate.

Some sectors of the chemical industry are already trying to develop new chemical and biochemical processes, which use carbon dioxide in the production of high-value chemicals. Such processes have the added benefit of avoiding potential environmental penalties for carbon dioxide emissions. If new or improved processes could be developed, carbon dioxide usage by the chemical industry could be increased by a factor of 10 or more. New approaches presently under development for using carbon dioxide include: production of algae-based biomass from power plant stack gas (see below); pigment production with reduced toxic by-products; polycarbonate production from supercritical carbon dioxide; urethane production using carbon dioxide instead of phosgene; and the direct synthesis of gasoline from carbon dioxide."

Supercritical fluid extractions

As in the food industry, carbon dioxide has an emerging use in the chemical industry for supercritical fluid distillation and the extraction of heat sensitive products. Carbon dioxide SFE is especially attractive for pharmaceutical production because of the lack of solvent residues in the product and the low temperature of the process. It can also be used to remove contaminants from chemical process streams. Supercritical carbon dioxide can be used to vaporize non-volatile substances at moderate temperatures, at which they are normally non-distillable. The ability of supercritical carbon dioxide to vaporize non-volatile compounds at moderate temperatures reduces the energy requirements relative to conventional extraction and distillation.

Other applications

Liquid and solid carbon dioxide are also used for direct injection into chemical reaction systems to control temperature. As an inerting gas, carbon dioxide is used in the chemical industry to purge and fill reaction vessels, storage tanks and other equipment to prevent the formation of explosive gas mixtures and to protect easily oxidized chemicals from contact with air.

Within the last two years, U.S. pulp and paper mills have begun using a CO_2-based technology for washing brownstock. This is patented by AGA AB (Lindingö, Sweden).

3.2.4 Metal fabrication

Shielding gas for welding

The major use for carbon dioxide in the metals industry is for welding operations in private ships, on construction sites and in numerous types of manufacturing operations. Used either alone or with other gases, carbon dioxide is used in the gas shielded arc as a shielding gas to protect the welding zone from the deleterious effects of oxygen, nitrogen and hydrogen.

Carbon dioxide is one of only a few gases suitable for this purpose, and competes primarily with argon and helium. Helium provides the hottest arc and argon allows greater stability in AC (alternating current) arcs and provides a cleansing function on the welding surface. Although slightly less effective in this application than helium and argon, carbon dioxide provides major cost savings, and thus also finds wide use. When used as a shielding gas with semi-automatic microwire welding equipment, carbon dioxide provides welding speeds up to ten times faster than those obtainable with conventional equipment and eliminates the cleaning or wire brushing of the welds.

Carbon dioxide used for welding must be in gaseous form; thus this is largely a cylinder market. However, some large welding operations may purchase bulk liquid and convert it to gas at the point of use.

Foundries

For the production of sand moulds and cores for casting iron and other metals, CO_2 can be used in combination with other additives to form the binding agent between the sand grains.

Separating waste

Messer Griesheim GmbH (Krefeld, Germany) plans to commercialise a CO_2 process for recycling sludge generated during metalworking. The waste consists of lubricant oil used for machining and metal fillings. When mixed with CO_2, the oil dissolves into the supercritical gas leaving behind clean metal particles. Both oil and metal are suitable for reuse, and the CO_2 itself is recaptured. Also contaminated soils can be cleaned this way.

3.2.5 Agriculture

Fumigate grain silos

Carbon dioxide is used as a non-toxic pesticide in grain silos. As a fumigant or insecticide, carbon dioxide kills insects by desiccation, and offers significant advantages over competitive chemical pesticides such as phosphine or methyl bromide. For instance, insects cannot develop immunity to it. Additionally, carbon dioxide leaves no residue, and thus calls for no special aeration procedures. Carbon dioxide is registered with the EPA as a non-restricted use pesticide.

Within this application, carbon dioxide finds most of its use in the non-toxic fumigation of grain silos. In this use, liquid carbon dioxide is converted to a gas onsite, and then injected into the silo, where the gas displaces the existing atmosphere with one fatal to all stages of insect life. Because an atmosphere of 60 percent carbon dioxide is lethal to insects, the silo need not be airtight for the method to be effective.

Additive to irrigation water

Carbon dioxide also may be added to irrigation water to enhance the absorption of nutrients by plants. In the latter use, carbon dioxide is effective because, although it has weak nutrient properties of its own, it acts as an adjunct to other nutrients by changing the pH of the soil, which can increase the nutrient absorption of plants.

Additive to greenhouse atmosphere for additional plant productivity and consistent quality

Plants need water, light, warmth, nutrition and CO_2 to grow. By increasing the CO_2 level in the greenhouse atmosphere (typical to 600 ppm instead of normal 400 ppm value), the growth for some plants can be stimulated in an important way, with often yeald increases up to 20 %, especially for tomato, cucumber, strawberry, etc. but also for potted plants and cut flowers.

Carbon dioxide is added to the atmosphere in greenhouses for accelerating growth rates of vegetables and flowers. When used in concentrations of about 2-3 times that normally present in air, carbon dioxide increases fruit and vegetable yields as much as 15 percent.

This is an important application in some parts of Europe (e.g. Benelux).

Production and processing of algae for industrial applications

The food, pharmaceutical and other sectors show an increasing demand for fine chemicals that have a renewable origin and natural character. At the same time, the agricultural sector in the EU faces small margins on production and tightened rules to reduce the environmental impact. Based on a feasibility study, algae and especially the Spirulina species (a natural source for fine chemicals like phycocyanin, carotene, etc) appear to be a very promising link between fine chemical demand and changes for the agricultural sector. Products to be extracted are among other pigments (applications in food, pharmaceuticals, polymers) and vitamins (applications in food, cosmetics and pharmaceuticals). Additionally, algae cultivation has a positive environmental impact when waste heat of a power plant is used and because of CO_2 fixation.

However, in the moderate climate areas of the EU large-scale production of high quality algae derivatives is hardly developed. When using standard methods, cultivation is not feasible because of climatic reasons. Therefore, a new concept is outlined, in which technologies are developed for algae cultivation, integrated in a greenhouse with supply of waste streams, such as waste heat from power plants and waste nutrients from green houses. Simultaneously, processes and procedures will be developed to extract various high selling additives from algae, particularly from Spirulina. The quality of the products needs to be investigated and if necessary need to be improved. Also the economical feasibility will be investigated. In order to do so, a consortium is formed that consists of partners providing know how on algal cultivation, fine chemical extraction and application development, as well as end user organisation.

3.2.6 Rubber and plastics processing

Blowing agent

The primary challenge for the 100 kTon/yr foam blowing industry has been to find suitable alternatives to chlorofluorocarbon (CFC) and hydrocarbon blowing agents, which are being phased out due to their allegedly detrimental effect on the Earth's ozone layer. Therefore, an emerging application for carbon dioxide is its use as a blowing agent.

Carbon dioxide gas under pressure is introduced into rubber and plastic mixes, and foams the material upon pressure release.

The foam blowing industry is divided into two markets, thermoplastics and thermosets. Thermoplastics, which are used in residential housing as sheeting and roofing, as well as in commercial roofing applications, are the smaller of the two. Although polystyrene is the biggest section of this market, there are sizable segments in polypropylene and polyvinyl chloride as well as several smaller segments.

Thermosets are a much larger market, taking about four times as much business as thermoplastics. This industry closely follows the plastics market. Polyurethane agents go into residential sheeting, refrigerator doors, insulating foam for pipes, spray foam for electrical outlets, and sandwich panels, the thin layer of foam which is placed between layers of metal in walk-in refrigerators, storage trucks and other larger refrigerating devices.

CO_2 systems nowadays match the majority of existing blowing machines (Cannon, Hennecke, Beamech, etc.)

Deflashing of molded rubber and plastic articles

"Flashing (i.e., thin protrusions of material at mold joints) can be quickly removed by tumbling these articles with liquid carbon dioxide."

Blow molding

By injecting liquid CO_2 into the molded product immediately after the blowing, the cooling time can be reduced in an important way. Cycle times of existing machines can therefore be reduced and capacity increased. Due to increase in performance of todays blowing techniques, the use of CO_2 has reduced in popularity, but still shows significant advantages when used with plastic moulds having an important wall thickness.

3.2.7 ***Other uses as solvent***

Spray painting

In *spray painting*, supercritical carbon dioxide is emerging as a functional alternative to VOCs. This technology has been developed by Union Carbide and is marketed under the tradename UNICARB. Although other gases could be used for this application, carbon dioxide is used because it mixes well with many paint polymers.

Aerosols

CO_2 has since long been used as a propellant in many aerosol applications, as a replacement of conventional propellants, which are allegedly known to damage the earth's ozone layer.

Cleaning parts using ultrapure CO_2

Supercritical carbon dioxide is being used for cleaning precision parts, electronic components, as well as for the removal of hydrocarbon machine coolants from metal parts. Specifically, supercritical fluid technology is currently being investigated for possible application in a vast range of cleaning, extraction and thin metal film deposition applications for semiconductors and electronics. In particular, the future of supercritical fluid technology to precision clean semiconductor substrates and electronic boards and parts is promising with potential commercial applications in the areas of information storage devices, semiconductor devices, electrical and electronic components, inertial guidance systems, precision optical devices, medical equipment and devices, and metal finishings. While competition from established cleaning methods may seem severe, there are a number of attractive features which singles out the supercritical carbon dioxide process as a unique method for precision cleaning. These include environmentally-compatible cleaning, low carbon dioxide cost and overall cost, rapid processing, ability to clean large areas and complex shapes, ability to form homogeneous cleaning supercritical fluids for general cleans as well as the ability to incorporate additives to tailor more specialised and specific cleaning.

In all cleaning applications, the increasing demand is for greater efficiency and effectiveness in contaminant removal. There is a market drive for environmentally conscious semiconductor and electronic parts cleaning, in particular due to stringent deadlines and aggressive scheduling imposed by legislative bodies to phase out the production and use of ozone-depleting substances such as CFCs and a reduction in VOC emissions from of organic solvents use. In addition, reducing worker exposure to volatile organic compounds has also become increasingly important in the electronic industry. Existing cleaning processes available to the industry e.g. RCA based cleans for semiconductors, flux cleaning for oxides in electronics, solvent cleaning of flux residues in electronics, present severe limitations including adequate level of cleanliness, volumes of waste solvents generated, worker exposure to solvents, costly chemicals and expensive equipment. As such, there is a requirement for the development and integration of a closed loop, environmentally and substrate compatible, cost effective technology resulting in significant savings in solvent and associated energy costs. Recent work on supercritical fluids indicate that as well as successively removing organic contaminants, the technology may also be used to remove or extract metals by using metal complexing ligands. These discoveries open the potential for a multi-purpose (organic and metallic or ionic contaminant removal) clean compatible with environmentally conscious manufacturing and the process requirements of contaminant-sensitive materials.

Dry cleaning

Dry cleaning of clothes is a large industrial activity in Europe conducted in 60,000 dry cleaning shops, employing 180,000 workers and cleaning 2.2 million tonnes of textiles each year. The dry cleaning sector currently has 75,000 cleaning machines using perchloroethylene (perc) or CFC 113 as cleaning solvents.

CFC 113 contributes to the destruction of the ozone layer and has been phased out already.

The major dry cleaning solvent PERC has risks to human health and shows toxic effects. The severity of those effects, such as neurotoxicological impairment, kidney damage and carcinogenity, is discussed in many European countries and in Asia and America.

In the EU a new VOC directive with stringent restrictions on the use of PERC will become active in the near future. In the Netherlands a reassessment on PERC has been carried out by the Government and a much lower level than at present has been set for the year 2007, which level likely cannot be met by the dry cleaning industry. At least 50% and probably more of this industry has to switch over to another cleaning method than the usual one with PERC.

Dry cleaning operations in Europe release more than 70,000 tons of PERC per year into the environment. Therefore dry cleaning shops and plants already had and still have to invest in expensive environmental provisions such as diffusion barriers, deep cooling machines with absorption units, measures to prevent soil and groundwater contamination, treatment of contact water and waste, in order to prevent human and environmental exposure to PERC.

As the industry has to meet very stringent low environmental levels for PERC in the near future, the industry needs to look now much stronger and very hard for new and cleaner technologies with less environmental and economical risks in order not to endanger the workplaces of 180,000 workers in the dry cleaning shops and factories as well as 40,000 workers in the supply and equipment manufacturing sector. Some companies have raised the idea of cleaning with CO_2 (carbon dioxide) under sub- or supercritical dry cleaning (SCDC) conditions.

A major portion of the PERC and CFC 113 might be replaced by SCDC with CO_2 if the right technology, apparatuses and detergents are developed. Important aspects for the cleaning effect are the removing of soil and stains on the one hand without redeposition and garment damages on the other hand. Therefore parameters such as mechanical action, the fluid/garment ratio, the fluid flow, the type of detergent, the detergent concentration, application of additives, relative humidity, time, temperature and pressure have to be investigated, separately and in combination, in order to obtain an optimal process which meets the criteria for cleaning effect, greying, colourfastness, shrinkage, appearance and surface structure of the textiles/leathers. Also the behaviour of different textiles, leathers and other materials on the garments in relation to different types of soil and stains will be a part of the investigation.

The results of the all the research activities reported so far were promising but variable; the cleaning efficiency was strongly dependent on process conditions and the results showed large variability. The cause of these variable results is the fact that the mechanisms of cleaning in supercritical CO_2 were only empirically examined in practical applications. No major insight was developed regarding mechanisms and rate limiting steps in the cleaning process.

In the USA, Nu-Life cleaners [28] have recently started the introduction of Micell's DryWash cleaning process using a CO_2-based cleaning liquid for dry cleaning clothes.

Production of fine particles

Several companies, such as Separex in France and Messer have also developed a number of technologies that use supercritical CO_2 in the production of extremely fine particles. One recently commercialised process dissolves compounds ranging from pharmaceuticals to paint pigments in supercritical CO_2. When the pressure is suddenly released, the material precipitates as particles smaller than 10μm with a very homogeneous size distribution.

3.2.8 Water treatment

Re-carbonation of sweet surface waters to be used as tap water

Modern water works use CO_2 together with lime to raise the hardness of drinking water, especially with soft surface waters. At the right pH-level, the mentioned treatment allows the formation of a protective surface layer in water-mains pipes, thus avoiding corrosion and improving water quality.

Water treatment market: pH control

The chemical industry also uses carbon dioxide for water treatment. Carbon dioxide is non-toxic and leaves no residues upon evaporation.

Indeed, lowering the pH of alkaline water with CO_2 is on the rise. Thanks to the natural buffering capacity of bicarbonate, which, along with carbonic acid and carbonate, is produced when CO_2 and water mix, neutralizing with the gas lowers corrosivity and makes overshooting the pH range difficult. The technology has been around for 25 or 30 years, but in the last 3 to 5 years we've seen a significant increase in its use as people become more aware of the dangers of storing and handling sulphuric acid.

This method offers thus many advantages towards the use of classical acids, such as less reagent, less equipment and monitoring devices, less downtime and increased safety, equipment life, etc. Especially for this growing application, ACP offers the NEUTRAC pH technology: a turnkey system for treatment of such waters. A pilot-testing unit can also be put at disposal of the customer to carry out preliminary trials and gathering valuable process data. See appendix 5.

CO_2 used in H_2O system is known to also dissolve $CaCO_3$ deposit from the pipes. This is commercially not used yet.

Water well rehabilitation

The water well rehabilitation technique [29] consists in injecting gaseous carbon dioxide at the desired depth in the treatment area. This produces a highly abrasive carbonic acid solution that penetrates far into the surrounding formation. Liquefied carbon dioxide is then injected at various temperatures and pressures. When it comes in contact with the water, it expands rapidly, producing tremendous agitation. The continued, controlled injection, of the liquefied carbon dioxide assures the freezing of water within the formation around the well, resulting in superior disinfection and dislodging of mineral encrustation.

After treatment the well is mechanically developed using surge/airlift methods to remove the newly dislodged particulate matter from the well and formation. The well pump is then reinstalled and the well returned to service, providing and increased supply of water for its intended use.

3.2.9 Use in the nuclear sector

CO_2 is used in the nuclear energy sector as a cooling medium for power plant cooling.

3.2.10 Well re-injection [30]

Secondary oil recovery – re-injection

In most of the oil fields already exploited, there is some oil left in place underground. As oil can be miscible with CO_2 under the pressure and temperature conditions in the reservoir, CO_2 is injected into suitable depleted oil reservoirs in order to enhance oil recovery by typically 10-15%. This technique is called Enhanced Oil Recovery (EOR).

Part of the injected CO_2 will be recovered at the production well together with oil, while most of the CO_2 will be trapped in the depleted reservoir.

Enhanced oil recovery by CO_2 injection is already used at more than 74 sites in the USA, and 33 Mt/year of CO_2 is being delivered to these fields for injection. Some other projects are under development in the USA and in the North Sea. A large EOR project has recently been commissioned in Canada (Weyburn).

Coal mining, coupled with methane recovery

Methane is present both in the pores of coal, but is mostly adsorbed on the coal surface. CO_2 injection in the coal seam decreases methane partial pressure, methane is desorbed and can be recovered at the production well. Unlike in enhanced oil recovery, CO_2 remains stored underground, adsorbed on the coal surface instead of methane. Moreover, coal can adsorb twice as much CO_2 by volume than methane, which means that the storage capacity exceeds the quantity of methane recovered. This technique is called Enhanced Recovery of Coal Bed Methane (ERCBM), and is proven feasible due to a demonstration plant in New Mexico, USA, where 100.000 ton CO_2 has been injected in 3 years, none of which has found its way to the production wells. A field test of ERCBM using CO_2 and N_2 mixtures is being carried out by the Alberta Research Council under an international project facilitated by the IEA Greenhouse Gas R&D Programme.

3.2.11 Dry ice production

Production

Dry ice (Carboglace™), or CO_2 in its solid form, is produced by expanding liquid CO_2 at atmospheric pressure. The expansion results in the formation of equal portions of gas and CO_2 snow. The CO_2 snow formed is compressed in extruders or presses at high pressure, resulting in the manufacture of blocks or pellets.

Most producers supply dry ice in blocks, slices (optionally. packed) and sticks with different dimensions in specially designed, well isolated containers for easy use and minimal losses. A quick and efficient distribution system ensures an excellent service to customers.

Properties

A few important features are:
Dry ice temperature of –78,9 ° C
Heat vaporisation: 645 kJ/kg (152 kcal/kg) at 0°C (twice the refrigerating capacity of water ice per unit weight)
Specific density: 1,56 kg/ dm^3
Dry sublimates at atmospheric pressure, leaving no residue
Dry ice creates also a protective, bacteriostatic atmosphere when used with foodstuffs.

Applications

Dry ice is used in practically all type of industries, mainly because of its cooling properties. It is particularly interesting for application where "spot cooling" is needed. The most important are:

Dry ice as a cooling medium

The fact that it sublimes directly into the atmosphere, as well as its higher cooling capacity and lower temperature make dry ice an excellent alternative to water ice for most cooling applications.

Examples are:

- cooling of foodstuffs, meals, ice-products, etc., during transport or when taking frozen food back home from the supermarkets store;
- cooling of catering products in aeroplanes, trains;
- cryogenic transport of pharmaceutical products;
- cooling of moulds, cores, shrink-fitting of metal parts;
- experiments in lab's;

CO_2 pellet blasting for paint stripping and other surface cleaning

This technique, which was originally developed to remove paint from the outer body parts of airplanes, has only recently started to gain industrial acceptance in a much broader field.

The principle of surface cleaning using dry ice is based on the projection of dry ice particles at high speed (usually using a flow of dry compressed air) onto the surface to be cleaned. These particles are usually 3 mm diameter pellets, but can also be small crystals scraped off from large blocks of dry ice.

When hitting the surface to be cleaned, these particles will explode. The thermal shock (dry ice = at –78°C) associated with the sudden expansion of solid CO_2 to gas, effectively remove contaminants of all sorts from the substrate.

The most important advantages of dry ice blasting vs. water cleaning or sand blasting are:

Only removed substance as waste (dry ice sublimates to gas),

Not as abrasive as sand: will not damage softer substrates,

Electrically non-conducting, so even electrical cabinets can be cleaned,

Food-grade abrasive, so food production lines can be cleaned on the spot,

Currently this technique is being used for removing coatings and paints, removing residues from molds, cleaning production machinery, food processing lines, printing equipment and welding spots in automotive assembly lines.

In most cases substantial savings can be obtained as the production equipment can be cleaned without disassembly.

Smoke effects in theatres and bars

When dry ice particles are thrown in a liquid at elevated temperature, the dry ice sublimes rapidly into CO_2 gas. This is heavier than air, and spreads itself slowly as a white misty cloud around the source. This is used in some bars to produce foaming cocktails or in theatres and disco's.

3.2.12 Cylinder filling

CO_2 can be used in cylinder for several applications, such as:

Horeca

Fire fighting: Consumer use and industrial applications. CO_2 smothers fires without damaging or contaminating materials and is used especially when water is ineffective or undesirable.

Consumer use: Cartridges for carbonating. Carbon dioxide also finds use in gas operated firearms.

Medical use: Finally, the medical market is a fast growing consumer of industrial gases, with the primary gases used being oxygen (for respiratory therapy, CAT scans and numerous other uses), carbon dioxide (laser surgery), helium (magnetic resonance imaging), and small volume specialty gases (laser surgery).

3.2.13 Other applications (not using large quantities of L-CO_2 yet)

CO_2 continues to substitute for traditional halogenated-carbon refrigerants that eat up atmospheric ozone or have even more severe greenhouse effects than CO_2. Norsk Hydro ASA (Oslo) and an Norwegian university have developed a CO_2-based cooling system for automobiles. Daimler Chrysler AG (Stuttgart) is currently testing the Mobile air Conditioning 2000, and Norsk claims the technology competes favourably in price, weight, space and energy efficiency. Likewise, Sanyo Electric Co. (Tokyo) recently developed a closed-type rotary compressor that uses CO_2 as the coolant.

CO_2 may soon be cooling the canned drink offering you some relief. Soon Chill Can N.V. (Bloomfield, N.J.) will conduct a pilot launch of self-chilling beverage cans. Looking just like a conventional 500-mL can on the outside, the self-cooling version contains a small pressurized cylinder packed with CO_2-charged activated carbon. Pressing a button on the bottom of the can releases the cylinder's pressure and initiates CO_2 desorption, resulting in a 15°C drop in the temperature of the liquid in about 3 minutes. In addition, the final 10% of the CO_2 takes about 20 minutes to desorb, prolonging the cooling effect.

3.3 Volume of the CO_2 merchant market

Determining the volume of each segment of the CO_2 merchant market is problematic for the simple reason that those CO_2 volumes are confidential. For the carbonated beverage sector we can evaluate the volume of the CO_2 merchant market by collecting information on the production of carbonated beverage.

For the other segments, we could identify some studies where data on the CO_2 merchant market are gathered. They are mainly realized by American societies, and therefore the results are very often limited to the American continent. However some of these studies also include the European CO_2 merchant market. However the results of those studies are not available. Indeed, as this data collection was realized by private companies, the reports are not free of charge and are prohibitively expensive. Three studies where data on the CO_2 merchant market are gathered are presented hereunder:

CEH Report. Carbon Dioxide [31]

Industrial Gases (Freedonia group), Industrial and Specialty Gases [32]

International Society of Beverage Technologists [33]

3.3.1 Carbonated beverage segment

One can estimate the amount of CO_2 used for the production of soft drinks by monitoring the total amount of soft drinks produced in each European countries. For practical purposes we assume that 8 gr CO_2 are added per liter of soft drinks produced. For 1998 we can, thus, come to the following table:

Table 1.20 - Carbonated soft drink production per year in EU

Country	Carbonated soft drinks - Million liters per year*	Min t CO_2 required
Belgium	930	7.440
Italy	2.824	22.592
Spain	3.521	28.168
U.K.	5.000	40.000
Netherlands	1.300	10.400
Germany	6.583	52.664
France	2.164	17.312

* Remark: Using the UNESDA, CISDA statistics, we came to a total requirement of 206.520 t of CO_2 for the production of carbonated soft drinks in the EU member countries. If, however, we use the C&C overall statistics, claiming an average consumption of 118 L per capita for the 370 million inhabitants of the EU members, one gets a total consumption of 43 294 million liters of carbonated soft drinks per year, and thus a need for 346 354 t per year of CO_2.

For the breweries, the situation is even more complicated. Although the beer consumption figures are readily available for all EU countries, deriving CO_2 merchant market volumes from these is impossible unless we have access to the data related to the recovery units for fermentation CO_2.

As far as market tendencies for the beverage markets, we see from the trends of the last five years that the market from soft drinks has grown 3% on average in the EU countries and grown 6-7% in the other European countries. The beer consumption is coming down slowly in most European countries.

For the market segment related to food processing it is clear that the need for CO_2 will sharply increase in the coming years due to an increased demand for processed food and the European HACCP directive. We notice that this market segment accounts for 4% in Belgium to 16% in Italy of the total CO_2 merchant market in Europe. In the USA, on the other hand, the food market consumed 3.100.000 tons of CO_2 in 1996. This represents 58% of the total CO_2 market. If Europe would use CO_2 in the same manner as it is being used in the US food processing market, the volume of Europe's CO_2 merchant market would increase with more 2.000.000 tons per year.

3.3.2 European merchant market global data

Commercial use of CO_2 in the world

Estimated at <10% of the 28 Billion $ industrial gases market.

Table 1.21 - Comparison

	% of Value	% of volume	Consumption/ capita (*)	Cost: $/t
Total:	2 700 M.$	13 500 kt/y	Average 2.4 kg/y	
America's	37 %	57 %	9.6 - 25	130
Asia	31 %	21 %	0.8 - 10	326
Europe	26 %	19 %	5 - 10	248
Others	6 %	3 %	0.15 - 6	400

(*) High consumption per capita is noticed in countries with:
High degree of industrialization; High standard of living; Availability of low cost CO_2; R&D dedication; New applications.

CO_2 Merchant market in Europe

Installed (nominal) Production capacity: **3 280 000 t/y**
Merchant market estimated at **2 120 000 t/y**

Table 1.22 - Application in northern and southern Europe

Application	Northern Europe: (kg/person & year)	Southern Europe: (kg/person & year)
Soft drinks & mineral waters	1.4	1.3
Breweries	1.5	1.1
Chemical Ind. (inerting, reactive)	1.0	0.12
Refrigeration	1.0	0.20
Freezing	0.4	0.1
Nuclear power stations	0.4	N/A
Greenhouses	0.2	N/A
Water neutralization	0.2	0.1
Welding + foundries	0.15	0.37
Slaughterhouses (pigs, poultry)	0.1	0.05
Blow molding + foaming	0.06	0.08
Fire extinguishers	0.05	0.07
Supercritical extraction	0.05	0

3.4 Conclusions

CO_2 finds many "technological" industrial applications. Access to good quality CO_2 is the key issue for the development of CO_2 utilization. CO_2 can be used under different forms: liquid bulk, cylinders and dry ice. The applications can be divided in two categories: those that temporarily use CO_2 (bubbles, cooling) and those that permanently remove CO_2 from the atmosphere (pH control of waste water). However, most merchant market applications lead to CO_2 release after all.

Today, the merchant market utilization of CO_2 represents less than 0.1% of the CO_2 released into the atmosphere. It is clear, in order to make the utilisation option relevant for carbon dioxide mitigation, that the amount of carbon dioxide used should be increased by at least one order of magnitude. Nevertheless, even if this limit were not reached in the short term, environmental benefits would be obtained by the implementation of benign syntheses and innovative technologies. This specific issue needs to be evaluated carefully as it may be very easy to underestimate its potential.

The following segmentation was determined for the merchant market, each of these segments was outlined:

food processing: refrigeration, chilling, MAP/CAP, freezing, slaughter of swine, stunning of poultry, etc...
carbonated beverages: soft drinks, mineral waters, beer brewing,..
chemical industry: synthesis, polymers, inorganic carbonates, PCC, inertization,...
metal fabrication: welding, steel mills, foundries, aluminium, metallurgy,...
agriculture: green houses, fumigation of silo's, additive to irrigation water, tobacco processing, soil decontamination,...
rubber and plastics processing: blow molding, foaming, deflashing, grinding & shredding, cryo-separation,...
other uses as solvent: dry cleaning, SC-FE, supercritical chromatography,...
water treatment: pH control, carbonatation of tap water, swimming pools, water well rehabilitation,...
nuclear: power plant cooling
well re-injection: secondary oil recovery, methane recovery from coal mines, re-injection in natural gas fields,...
dry ice production
cylinder filling: all types of cylinders and mini-bulk tanks filled with either pure carbon dioxide or blends with other gases

Determining and/or evaluate the volume of those CO_2 merchant markets segments was problematic for the simple reason that those CO_2 volumes are confidential. Despite the huge number of contacts, our efforts remained often unsuccessful. For the carbonated beverage sector we can evaluate the volume of the CO_2 merchant market by collecting information on the production of carbonated beverage. We could identify some studies where data on the CO_2 merchant market are gathered. They are mainly realized by American societies, and therefore the results are very often limited to the American continent. However some of these studies also include the European CO_2 merchant market. Indeed, as this data collection was realized by private companies, the reports are not free of charge and are prohibitively expensive.

Some additional future applications for carbon dioxide are considered in another chapter. This completes the market segmentation and some future applications are for example supercritical CO_2 (micro-encapsulation, impregnation and extraction of residual impurities by swelling of the polymer matrix, oligomer processing, dendritic micelles, dyeing process, reactive modifiers, complexing agents), utilisation of CO_2 in carboxylation reactions, synthesis of carbonates and carbamates (carboxylation of organic substrates, insertion of CO_2 into C-C bonds, synthesis of carbonates, synthesis of carbamates), CO_2 reforming of methane and biomass (biomass as source of fuel, biodiesel production, freshwater algae [34])

REFERENCES

1. EUROSTAT database (1999)
2. BDZ (Bundesverband der Deutschen Zementindustrie), (Federal Association of the German Cement Industry): annual report 1998-99 internet pages (www.bdzement.de)
3. Map, 6/1998
4. Cembureau (European Cement Association)
5. First Annual Progress Report of DMT. RUCADI-Project BRRT-CT98-5089. Brite Euram Programme of the European Commission.
6. European minerals yearbook, 2. edition
7. "Beschluß der Bundesregierung zum Klimaschutzprogramm der Bundesrepublik Deutschland auf der Basis des Vierten Berichts der Interministeriellen Arbeitsgruppe "CO_2-Reduktion" (IMA "CO_2-Reduktion")". Bundesministerium für Umwelt, Naturschutz und Reaktorsicherheit (Federal Ministry of Environment), 1997.
8. Own estimation basied on data of electricity consumption provided by /12/ applying the calculation method and emission factors of RWI Essen, 1997-1999
9. Jahrbuch, Bergbau Erdöl und Erdgas Petrochemie Elektrizität Umweltschutz. 104. Jahrgang. Verlag Glückauf GmbH Essen. (Yearbook 1997, Mining Mineral Oil and Natural Gas Petrochemistry Electricity Environment), 1997.
10. VDEW e.V. (Vereinigung Deutsche Elektrizitätswerke e.V.) internet pages (www.strom.de) publication: "Strommarkt Deutschland 1996"
11. BMWi (Bundesministerium für Wirtschaft und Technologie), Publication: „Energie Daten 1999". (Federal Ministry of Economy and Technology)

12. Wirtschaftsvereinigung Metalle e.V. (Economic Association Metals). Internet pages (www.ne-metalnet.de)
13. IPAI (International Primary Aluminium Institute), London
14. BV Glas (Bundesverband Glasindustrie und Mineralfaserindustrie e.V., federal association of glass industry and mineral fibre industry), Düsseldorf, Germany
15. "European Union Energy Outlook to 2020", special issue of the review "Energy in Europe", European Commission, Directorate General for Energy, November 1999.
16. Assessing Climate Response Options: Policy Simulations, Insights from Using National and International Models (ACROPOLIS), Research Project under the EU's 5th Framework Programme.
17. Freund, P. "Progress in understanding the potential role of CO_2 storage". Proc. 5th Intl Conf. On Greenhouse Gas Control Technologies, CSIRO, Sydney, Australia, 272-277, 2001.
18. Freund, P. "Reducing greenhouse gas emissions by capture and storage of CO_2" Intl Conf. Power Generation and Sustainable Development, Liège, October 2001.
19. Bondor "Applications of carbon dioxide in enhanced oil recovery" Energy Conv. Mgmt., 33, 579-586, 1992.
20. Moritis "Enhanced oil recovery increases 24% worldwide; Claims 10% of US production" Oil & Gas Journal, 51-79, 1992.
21. Baklid, O. "Sleipner Vest CO_2 disposal, CO_2 injection in a shallow underground aquifer" Society of Petroleum Engineers, paper 36600, 1996.
22. Torp "The Sleipner CO_2 injection: a large-scale case story" Intl Conf. On Power Generation and Sustainable Development, Liège, October 2001
23. Mathieu, P., Mostade "Use of CO_2 from power generating plants to enhance Coal Bed Methane Recovery" Intl Conf. On Power Generation and Sustainable Development, Liège, October 2001.
24. Stevens, Riemer "CO_2 sequestration in deep coal seams: pilot results and worldwide potential" 4th Intl Conf. On Greenhouse Gas Control Technologies, Interlaken, Swizerland, 1998.
25. www.wittemann.com/co2.htm
26. www.purgit.com
27. www.acpco2.com
28. http://www.nuco2.com/
29. http://www.aquafreed.com/
30. Freund, P. "Reducing greenhouse gas emissions by capture and storage of CO_2" Intl Conf. Power Generation and Sustainable Development, Liège, October 2001
31. http://ceh.sric.sri.com/Public/Reports/743.2000/
32. www.globind.com
33. http://www.bevtech.org/annual.htm
34. Kadam, K.L. "Power plant flue gas as a source of CO_2 for micro-algae cultivation, economic impact of different process options" Energy Conv. Mgmt. 38, S505-S510, 1997.

SECTION II

SEPARATION TECHNOLOGIES

This Section summarizes the activity performed within WP2 of the BRRT-CT98-5089 RUCADI Project, "Recovery andUtilization of Carbon Dioxide" by the following Partners:

Rodney J. Allam, *Air Products & Chemicals*, UK
Sven Geleff, *Union Engineering*, Denmark
Rune Bredesen, *SINTEF*, Norway
Gabriele Clarizia, *CNR-ITM*, Italy
Enrico Drioli, *CNR-ITM*, Italy
Philippe Mathieu, *The University of Liége*, Belgium
Stéphanie Houyou, *ACP*, Belgium

Chapter 2

CARBON DIOXIDE SEPARATION TECHNOLOGIES

Rodney J. Allam[#], Rune Bredesen[§] and Enrico Drioli*
[#] *Air Product&Chemicals England;* [§]***SINTEF Materials Technology*** *- Dept Process Metallurgy and Ceramics, Norway; *Research Institute on Membrane Technology CNR-ITM, c/o University of Calabria, P. Bucci Street 17/C, I-87030 Rende (CS), Italy*
e.drioli@itm.cnr.it

1. INTRODUCTION AND OBJECTIVES

1.1 Introduction

Carbon dioxide is the major atmospheric contaminant leading to temperature increase caused by greenhouse effect. From 1750 until 1998 the concentration of CO_2 in the atmosphere has increased from 280 ppm to almost 380 ppm. Known anthropogenic sources account for 7 billion metric tons per year. The principal anthropogenic source is the combustion of fossil fuels, which accounts for about three-quarters of total anthropogenic emissions of carbon worldwide [1]. It is now quite well established that carbon dioxide generated by anthropogenic activities has four major sources:
1. Production of electrical energy (36%)
2. Industry (24%)
3. Transportation (30%)
4. Heating of homes and industrial work-places (10%).

M. Aresta (ed.), Carbon Dioxide Recovery and Utilization, 53-120.

Fiscal measures and changes in behaviours can influence demand for energy, but in the technical area three distinct option for reducing emissions exist:

•improving energy efficiency
•switching to low carbon fuels
•switching to "no-carbon" fuels

The first two options are cost-effective in many places today and will provide useful reductions in CO_2 emissions. Deep reductions could be provided by the third option, such as switching to renewable energy or nuclear power, but the world is still heavily dependent on fossil fuels. For this reason, it is important to have also technology options, which would allow continued use of fossil fuels without substantial emissions of CO_2. This can be done with capture and storage of CO_2 from flue gases.

1.2 Objectives

This report describes the state-of-the-art of CO_2 separation technologies applied to large point CO_2 emission sources. It will focus on the recovery of CO_2 in three main types of process solutions:

- end of pipe solutions; generally represented by separation after combustion of low pressure flue gases.
- oxy-fuel combustion; combustion of fossil fuel with pure oxygen and recycling of flue gas to produce high concentration CO_2 directly from the process.
- carbon elimination before combustion; processing of fuel before combustion to eliminate carbon either as CO_2, solid or in other ways. In practice this would usually mean to convert the primary fuel to hydrogen.

Many possible combinations of process and separation method can be conceived for CO_2 removal from industrial power production processes. The process solution classification into end-of-pipe, oxy-fuel and carbon removal from the fuel, can be considered with any of the possible separation processes existing today, and therefore applied to either retrofit existing plants or in the design of new systems. The report will provide detailed description of each of the current major CO_2 separation technologies, which are:

- Absorption
- Cryogenic
- Adsorption
- Membranes

These separation methods may be applied before or after combustion depending on the process solution. In order to understand the options more completely it is necessary to be aware of the various process routes for production of hydrogen. Carbon removal from the fuel before combustion will then involve conversion of the fossil fuel to hydrogen with removal of CO_2 or elemental carbon, etc. A summary of hydrogen production may be as follows:

√ Steam-methane reforming (SMR) of natural gas resulting in two emission points for CO_2, one from the furnace flue gas, the other from the process gas following water-gas shift.

√ Partial oxidation (POX) with pure oxygen in which the CO_2 produced within the process by fuel combustion to provide the high temperature required for complete conversion of the fuel to CO and H_2 is removed from the high pressure gas stream together with CO_2 from the shift reaction at one point-emission point.

√ A variant on partial oxidation is the autothermal reforming process (ATR), which consists of a POX burner located above a fixed catalyst bed. The high temperature gas from the burner at about 1300°C enters the catalyst bed where steam-methane reforming and shift reactions occur. The outlet temperature is about 900°C.

√ Increase in efficiency can be achieved by utilising the high outlet temperature of POX, ATR and SMR in a gas heated reformer where further steam-methane reforming can take place resulting in up to 30% more H_2 production with the elimination of excess steam production.

√ An alternative to conversion of the fossil fuel carbon to CO_2 is the newly developed process of natural gas pyrolysis in electrical plasma, which results in production of high-grade carbon dust and pure hydrogen. This process is, however, less energy efficient than routes, which produce CO_2 and will not be considered further at this stage. Other future processes for carbon capture may also be envisaged, but will not be discussed further in this report focusing on current technologies.

1.2.1 Oxy-fuel

Oxy-fuel processes consist of the replacement of air in a combustion system with a synthetic air-like mixture of pure oxygen and recycled flue gas so that the fossil fuel can be burned at the same flame temperature. It results in a flue gas containing 85-95% CO_2 content. The process is ideal for the conversion of existing large power boilers for CO_2 removal. A 500 MW electrical output power boiler would require about 8000 Mt/d of O_2 and produce 11000 Mt/d of CO_2. The actual CO_2 concentration depends on the fuel used, the excess oxygen required for combustion, and the air leakage into the boiler. The CO_2 would be compressed typically to greater than 100 bar for pipeline transmission to a disposal site. At an intermediate pressure in the compression train, say at about 40 bar, the CO_2 would be dried in a desiccant drier and as an option subjected to a cryogenic low temperature purification step. In its most elaborate form, the plant would consist of a heat exchange system, which cools the dry 40 bar feed gas to within 2-3°C of the CO_2 freezing temperature where a distillation step is carried out in two linked distillation columns. This separates a non-condensable fraction rich in oxygen, which can be recycled to the boiler, and a high boiling point fraction, which consists of most other emissions from the boiler such as $SO_2/NO_x/HCl$.

Note that the oxy-fuel conversion of a boiler will substantially remove all emission contaminants. It will be necessary to vent a small stream from the light gas return to control the build-up of non-condensable components in the recirculating system.

This process has been demonstrated on a small scale, but needs a relatively large scale test with a difficult fuel such as coal to confirm operating characteristics, performance and reliability.

2. CO_2 SEPARATION TECHNOLOGIES

2.1 Absorption

2.1.1 Introduction

The main principle in all separation processes using gas absorption is the transfer of one or more substances from the gas phase to the liquid phase through the vapor-liquid phase boundary. The substance to be absorbed is dissolved in the liquid by physical dissolution, which may be followed by a chemical reaction. Absorption processes can be divided into chemical and physical absorption processes according to whether or not a chemical reaction takes place after dissolution of the substance into the liquid absorbent.

The absorption stage takes place in a packed or a tray column. The desorption (or regeneration or stripping) stage in involves the transfer of substances from liquid to vapor phase and can be regarded as a special case of absorption. The equipment used for desorption will depend on whether the absorption process is based on chemical reaction or physical dissolution.

In this paper the absorption processes which can be used for CO_2 removal from different sources will be outlined. The principles of chemical and physical absorption will be explained. Some common solvents which can be used for absorption will be described and the process equipment used will be discussed.

2.1.2 The principles of absorption

Physical absorption

Physical absorption processes are based on dissolution of a gas into the liquid phase. The dissolution of a gas into a liquid can be described by Henry's law, which is valid for low concentration systems with only one soluble component.

$$P_i = H_i\, x_i$$

where P_i is the partial pressure of component i; $P_i = y_i / P_{total}$ (y_i is mole fraction i in vapor phase) H_i is the Henry's law constant of component i, x_i is the mole fraction of i in the liquid phase.

Henry's law predicts a linear equilibrium curve. The higher the partial pressure of component i in the gas phase, the higher the solubility of the same component in the liquid phase. Henry's law constant is temperature dependent. There is a general trend of increased solubility at decreased temperature.

The following Table 2.1 compares Henry's law constant at 313 K for a few selected components. From Table 2.1.2.1 it is seen that carbon dioxide and hydrogen sulfide have small Henry's law constant compared to nitrogen and oxygen which means that at a fixed partial pressure the solubilities of CO_2 and H_2S are much larger than those of N_2 and O_2. The values of Table 2.1 are valid for systems where the liquid phase is 100% water.

Table 2.1 – Henry's law constants of selected components in water at 313 K. [2]

Component	H (313 K)
Carbon dioxide, CO_2	2330
Hydrogen sulphide, H_2S	731
Oxygen, O_2	53473
Nitrogen, N_2	102955

The absorbents, which are used for absorption of CO_2, have been selected to offer high CO_2 solubilities. The Rectisol process is a physical absorption process using pure methanol as the solvent. The solubility of CO_2 in methanol (at 263K and P_{CO2} = 1 bar) is 10 L of CO_2 /L of methanol, while the solubility of carbon dioxide in water at the same conditions is only 2.5 L of CO_2 /L of water [2].

Physical absorption is mainly used in high pressure CO_2 separation processes where one can utilize the high CO_2 partial pressure to obtain high solubilities. Regeneration of a physical absorbent is accomplished by lowering the pressure as this will lead to a decrease in CO_2 partial pressure and therefore a decrease in solubility.

Chemical absorption

In chemical absorption the component of interest will be transferred to the liquid phase like in a physical absorption process. Already in the liquid phase the component will react with the absorbent to form a semi-stable complex.

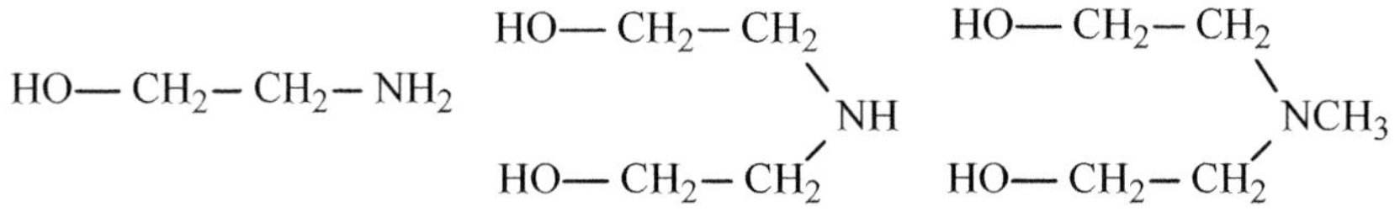

Monoethanolamine Diethanolamine Methyldiethanolamine

Figure 2.1.- Structure of monoethanolamine (MEA), diethanolamine (DEA) and methyldiethanolamine (MDEA)

Alkanol amines are widely used in chemical absorption. Alkanol amines are amines with at least one hydroxyl group. The structural formula of selected alkanol amines is given in Figure 2.1.

The hydroxyl group of alkanol amines serves to facilitate the water solubility of the alkanol amine. The reaction with carbon dioxide takes place at the amino group:

$$\underset{Alkanolamine}{HO\text{-}CH\text{-}R\text{-}CH_2\text{-}NH_2} + CO_2 + H_2O = \underset{Complex}{HO\text{-}CH\text{-}R\text{-}CH_2\text{-}NH_2{*}H_2CO_3} \quad (1)$$

The reaction from alkanol amine to "complex" takes place at low temperatures (20-50°C). To invert the reaction and desorb the carbon dioxide the temperature should be increased to 110-130°C. At these conditions the solubility of carbon dioxide in the liquid phase will be very small due to a high vapor pressure of this component. This means that CO_2

will be transferred to the vapor phase. In this way the alkanol amine is regenerated.

The reaction from left to right in Eq. (1) is exothermal. This means that the absorption process will develop heat, while the desorption reaction from right to left in Eq. (1) need energy to be added to proceed. Table 2.2 compares the heat of reaction for the absorption of CO_2.

Table 2.2 - Heat of reaction for absorption of CO_2 in selected alkanolamines [3]

Alkanolamine	Heat of reaction (kJ / kg CO_2)
Monoethanolamine (MEA)	1919
Diethanolamine (DEA)	1519
Triethanolamine (TEA)	989
Methyldiethanolamine (MDEA)	1105

Monoethanolamine, MEA, is a popular alkanolamine for CO_2 absorption. This is primarily due to the fact that MEA has a high carrying capacity (moles of CO_2 absorbed per moles of MEA) compared to other alkanolamines, which compensates for the high heat of reaction, as less MEA has to be used in the process.

Alkonolamine absorbents are normally used in aqueous solutions. In a MEA-water system the maximum MEA concentration is approximately 10% by weight as more highly concentrated MEA solutions are highly corrosive. In a later section the corrosion aspect will be discussed together with other limitations in absorption systems.

2.1.3 Process equipment for absorption systems [4]

Absorber column

All absorption processes have an absorber column in common, while the rest of the equipment may differ from process to process. The column can be either a packed or a tray column although packed columns are the most widely used. The gas to be treated enters the column from the bottom, while the absorbent liquid is fed to the column at the top (countercurrent flow). In order to minimize the loss of evaporated absorbent through the top of the absorber a demister is installed at the top or even better a washing step is included.

The packings which are used in absorbers are either random or structured packing. These have some advantages over tray columns of which low foam and low pressure drop are the most important. Structured packings offer large contact areas, high efficiencies and a minimal tendency of foaming.

Process equipment for physical absorption

The physical absorption takes place at low temperature, most often at –5 to –10°C. As mentioned before, the gas solubility in the solvent will increase with decreasing temperature, but there is a lower limit for the temperature, which is due to the fact that the viscosity of the absorbent will increase with decreasing temperature resulting in a poor mass transfer. The height of the column will therefore increase with decreasing the temperature.

The gas entering the absorber is cooled in a gas-gas heat exchanger using the treated gas as cooling media.

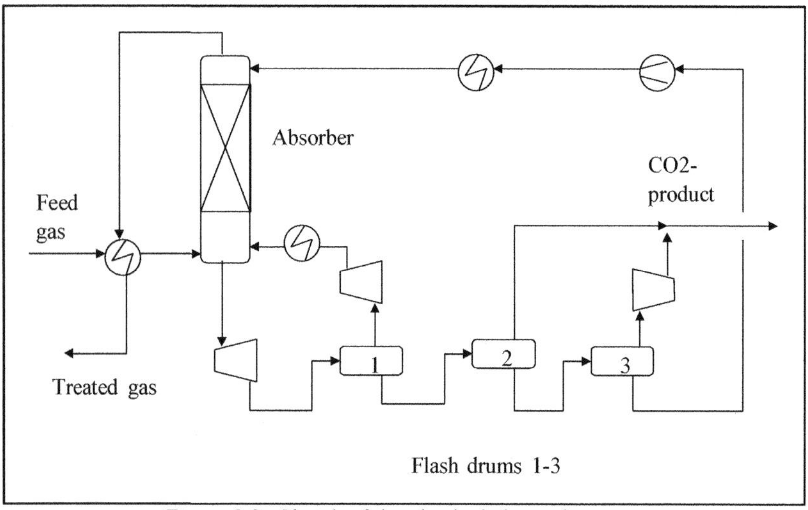

Figure 2.2 - Sketch of the physical absorption process.

As mentioned the regeneration of a physical solvent is accomplished by lowering the pressure, which will cause the gas absorbed to be liberated as a gas. The separation of gas and liquid is accomplished in a flash drum. Often many drums are used for regenerating physical absorbents. Figure 2.2 shows a simplified flow diagram for a physical absorption system with three successive flash drums. After the first expansion the flashed gas, which is rich in CO_2, is recompressed, cooled and led bask to the bottom of the absorber in order to increase the CO_2 concentration and thereby the partial pressure to obtain a higher solubility of CO_2 in the absorbent.

The absorbent from the first flash drum is further expanded and more gas is flashed off. In the third flash drum the pressure is lowered further to strip off more gas. The flash gas from the third drum is recompressed to the pressure of the flash gas from the second flash stage and the two gas phases are mixed. This phase will contain the removed CO_2. The regenerated absorbent leaving the third flash drum from the bottom is compressed, cooled and pumped back to the absorber.

Process equipment for chemical absorption
Chemical absorption takes place in the absorber from 20 to 50°C. As for the physical absorption an even lower temperature would be preferred but due to high viscosities the temperature has a minimum value for the same reasons as in physical absorption.

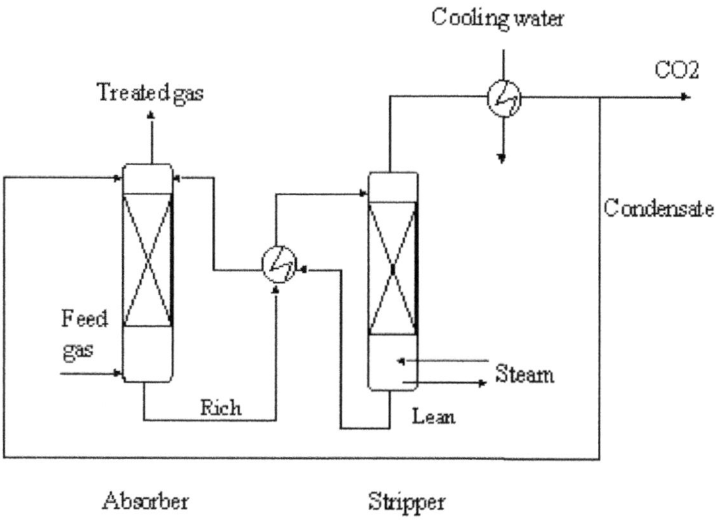

Figure 2.3 - Sketch of the chemical absorption process

Figure 2.3 shows a traditional chemical absorption system with absorber and stripper columns.

After absorption the treated gas exits the absorber at the top, while the rich absorbent exits from the bottom. Due to the exothermal reaction between CO_2 and the absorbent the temperature will be higher than that of the incoming liquid. The rich absorbent is further heated in a liquid-liquid heat exchanger before it enters the stripper column.

The stripper column is typically a packed column with a steam reboiler for heating of the absorbent. In the stripper the absorbent is heated in order to strip off the carbon dioxide. As the temperature is close to 100°C a considerable amount of steam is also produced. The liberated CO_2 leaves the stripper column at the top together with the steam. The CO_2:steam ration in this stream may be as high as 50:50.

The regenerated absorbent leaves the stripper column at the bottom. It is cooled in a the before mentioned liquid – liquid heat exchanger using the lean absorbent as the cooling media. Further cooling may be needed as will be provided by cooling water. The cold regenerated absorbent is pumped back to the absorber. The hot gas leaving the stripper is cooled in a heat exchanger whereby most of the steam is condensed. The condensate is lead back to the absorber.

2.1.4 Limitations of absorption processes.

The use of absorption processes for removal of CO_2 has certain limitations of which the most important will be discussed in this section.

Foaming

The formation of foam in an absorption process is not desirable. Foam will change the hydraulics in the columns and there will be a risk of flooding the column. Foam formation in an absorption system depends on several different aspects. Firstly, a tray column is more likely to have foam problems than is a packed column because the packed column will be able to break the foam emulsion. Structured packing will minimize the foaming tendency of an amine system.

Alkanolamines tend to foam at special conditions, while for instance alcohols do not tend to foam. Foaming of alkanolamine solutions are caused by the presence of for instance heavy hydrocarbons or particulate matters in the raw gas. Degradation products in the amine solution will also increase the risk of foaming.

Particulate matter

Absorption systems will not work properly if the incoming gas has a high content of particulate matter such as dust and soot. These will increase the risk of foaming, they will accumulate in the system and lead to high pressure drops and lowered efficiency. 100-150 mg/m^3 of gas is normally encountered as the upper limit for particulate matter.

Dust, soot etc. can quite easily be removed by inserting a filter on the raw gas line. Ceramic filters will remove these contaminants quite efficient.

Sulphur

The presence of sulphur in the raw gas can be either as sulphur dioxide, SO_2, or as hydrogen sulphide, H_2S. The form of sulphur will depend on where in a process the carbon dioxide removal is done. For power generation an end of pipe CO_2 removal will mean that SO_2 should be dealt with while downstream CO_2 removal will most likely mean that sulfur is present as H_2S.

H_2S will follow carbon dioxide in an absorption process; this means that these two components will be together in the off-gas from the desorption process. If a selective removal of H_2S and CO_2 is needed a more specific solvent system has to be used.

Sulphur dioxide is a serious problem in chemical absorption processes. In the absorber SO_2 will react with the alkanolamine to form stable salts. This will deactivate the amine solution and will also increase the risk of corrosion. For CO_2 removal for instance from coal fired power plants it is necessary to remove the SO_2 before the absorber. This can accomplished in a scrubber, which will also serve to cool the gas down to the required absorption temperature.

Oxygen

The presence of oxygen in the gas to be treated may be critical. Oxygen will tend to form heat stable salts with the amine solution if the partial pressure becomes too high. 5 % (vol.) at 1 bar may be considered the maximum oxygen content for absorption systems using a simple amine-water system.

An elevated level of oxygen in a gas will require the use of specialized solvents. Such solvents are typically an amine-water solution which contains some kind of inhibitor. Using inhibited solvent systems will allow for oxygen concentrations above 20% (vol.), but they will also allow for a higher amine concentration without any risk of corrosion. MEA systems which normally have a upper concentration limit of 10 % (vol.) can run at 30% (vol.) by the addition of a proprietary inhibitor (Flour Daniel's EconAmine process).

2.2 Cryogenic distillation

2.2.1 Introduction

Carbon dioxide separation at low temperature is governed by the relatively high triple point temperature of CO_2 which is -56.6°C at a pressure of 5.18 bar abs., see Figure 2.4. Below these temperature and pressure, no liquid phase forms when CO_2 is cooled and purification based on phase separation or distillation cannot be achieved. Above this temperature and pressure, carbon dioxide can be separated in a pure form by partially liquefying the mixture of CO_2 and other inert gases followed by a distillation step if very pure CO_2 is required.

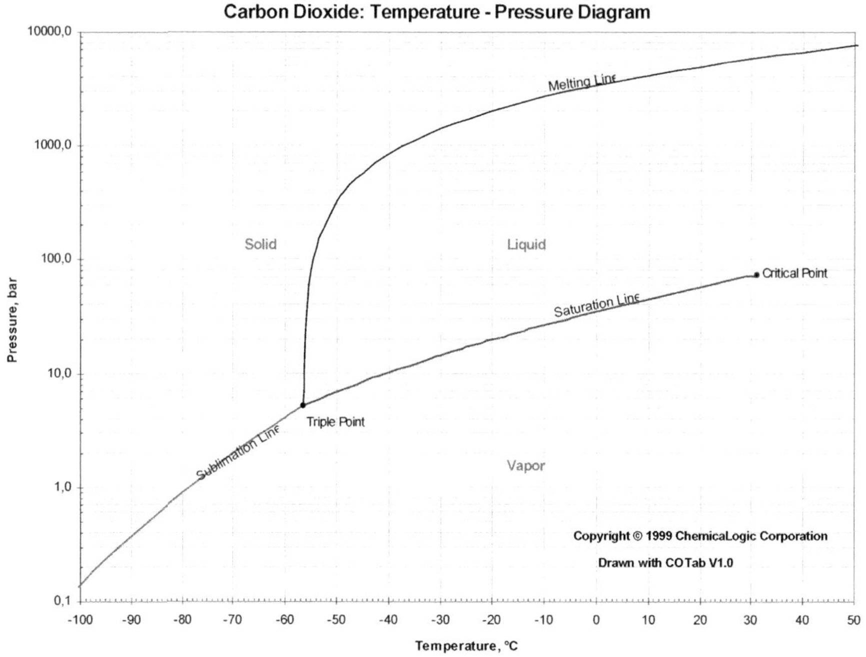

Figure 2.4 - Temperature-pressure diagram for pure carbon dioxide.
The solid, liquid and vapor states are indicated. The saturation line is the phase boundary between vapor and liquid states. By compressing pure CO_2 or by cooling it is possible to cross the phase boundary from vapor to liquid.

For gaseous mixtures of CO_2 and other gases (inerts or impurities) cooling and compression of the gas mixture will not immediately lead to liquefaction of the mixture. This is due to a so-called phase envelope. Within the phase envelope vapor and liquid coexist at varying compositions depending on temperature and pressure. Operation within this phase envelope permits the separation of CO_2 from inerts. The separation can be done in a simple CO_2 condenser, which will allow for a few separation stages. The condenser is normally operated at constant pressure and the flash to vapor and liquid is achieved by lowering the temperature. More separation stages can be obtained with a multistage distillation column. The column will be equipped with a reboiler and a condenser. The column internal can be either trays, random or structured packings. Depending on the level of impurities and the carbon dioxide purity required a condensator or a multistage distillation column can be chosen. For a CO_2 rich stream from a MEA system at a power plant a condenser will be sufficient as the impurities are mainly N_2 and O_2 at low levels.

Gas streams with high levels of higher hydrocarbons (C_2^+) will require the use of a distillation column. In special cases it might be necessary to decrease both pressure and temperature in the condenser in order to create sufficient vapor flow in the distillation column for the separation to occur. The process of liquefaction and purification of carbon dioxide is energy-intensive due to multistage compression and cooling.

2.2.2 Applications

The separation of CO_2 from a gas mixture at low temperature can only be carried out economically when the CO_2 concentration in the feed gas is high, probably greater than 75%.

Consider the problem of stripping an inert gas, say nitrogen, from CO_2 by distillation. In order to minimise CO_2 loss from the top of the column it would be necessary to operate as close to the triple-point temperature as possible, but the minimum partial pressure of CO_2 achievable in the vent gas would be 5.18 bar abs. This means that as the concentration of other components in the CO_2/gas mixture increases the pressure of the stripping column would have to be increased in order to achieve a certain CO_2 recovery from the process. At 75% feed purity and 90% CO_2 recovery, the column pressure would be about 26 bar pressure, for 95% CO_2 recovery it would be 46 bar.

Refrigeration must be provided at a temperature level required by the distillation or phase separation process which depends on the inert gas content and the required CO_2 recovery. This limitation of processing at an initially high CO_2 concentration limits the cryogenic or, to describe it more accurately, the low temperature process to purification of an enriched CO_2 stream. A typical application would be an oxyfuel CO_2 removal system such as a steam boiler in a power station burning coal with recycled flue gas and using pure oxygen rather than air for combustion [5] The absence of bulk nitrogen in the system means that the only inert gas present comes from air leakage into the boiler and the small proportion of inerts in the oxygen. This results in a typical CO_2 concentration on a dry basis of about 85% to 90%. If the CO_2 is required for enhanced oil recovery it is necessary to reduce the inerts content to typically 1% to 5% to avoid miscibility problems in the oil bearing structure. This would be a good application for a low temperature CO_2 separation system. Note that in the case of coal combustion, the inerts would contain a high concentration of oxygen which would be introduced into the boiler in excess of stoichiometric requirements to ensure complete coal combustion. The inert stream would be recycled back to the boiler with a certain fraction being vented to atmosphere depending on the total inerts level which could be tolerated in the recycle system which in turn would depend on the purity of the oxygen used and the air leakage rate into the system.

The separated CO_2 would also contain heavier impurities such as SO_2, SO_3, HCl and NO_2. These could also be separated from the CO_2 by distillation if required. A typical flowsheet for a system as described in given in Fig 2.5. [5]

The refrigeration system used for the process is a recycled CO_2 stream which is compressed, liquified, expanded to a low pressure above 5.18 bar abs and evaporated to produce the required cooling duty. The CO_2 is generally required at high pressure as a liquid for transmission to its point of underground injection which requires a pressure of 60 to 200 bar. The CO_2 low temperature purification plant would generally be operated at an intermediate pressure in the compression train. The impure feedgas would be dried in a switching dessicant drier before entering the low temperature purification process. Drying would invariably be required for any practical CO_2 pipeline transportation system to allow carbon steel to be used for the pipeline without the risk of corrosion.

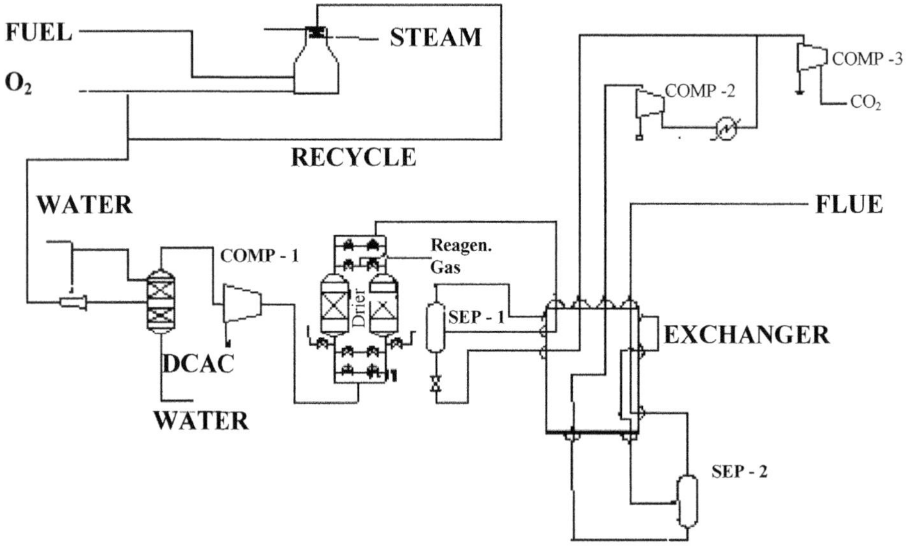

Figure 2.5 – Flowsheet of a plant for CO_2 purification

2.3 Adsorption

2.3.1 Introduction

The process of separation by physical adsorption is based on the ability of porous solids to reversibly adsorb large quantities of certain components in a mixture. It can be used to effect separations in liquid or gaseous mixtures. These separations are of two types. Firstly the removal of trace components to purify a fluid; an example would be dehydration of a gas. Secondly, adsorption can be used for bulk separations such as air separation to produce oxygen or for the separation of straight chain and branched or cyclic isomers.

We are concerned here with the use of adsorbtive processes for the bulk separation of relatively pure CO_2 from gas streams produced in a variety of industrial systems. We will start with a summary of the characteristics of microporous adsorbents particularly those suitable for carbon dioxide separation, followed by an analysis of the principles of adsorption processes and finally, descriptions of specific industrial processes which are major sources of CO_2 emission to show how adsorbtive CO_2 separations can be applied in practice.

2.3.2 Microporous adsorbents

Practical adsorbents are in general microporous solids with pore diameters ranging up to a few tens of Angstrom units. There are two main types. The first comprises adsorbents having a distribution of pore size and a mean pore diameter which are controlled by the manufacturing process. Typical examples of this group are silica gel, alumina and activated carbon. In the second group there are the zeolites where the micropore size is controlled by the crystal structure.

Silica gel
Silica gel is a partially dehydrated form of polymeric colloidal silicic acid with a chemical composition $SiO_2\ nH_2O$. It can be prepared from soluble alkali metal silicates by acid hydrolysis or removal of the cation by ion exchange. The silicic acid polymerises to form linked SiO_4 tetrahedra which aggregate to form particles 20 to 200 Å diameter. Drying agglomerates these particles to a microporous structure. Synthesis conditions control pore size and size distribution. The presence of hydroxyl groups gives surface polarity which favours the adsorption of polar species such as water phenols, etc.

The importance of silica gel is as a drying agent in a layered adsorbent bed in which CO_2 is adsorbed in a subsequent layer by an adsorbent such as a zeolite whose CO_2 capacity would be destroyed by pre-adsorbed water.

The equilibrium water isotherms of common dehydration adsorbents are shown in Figure 2.6. Silica gel has the highest capacity as well as the lowest heat of adsorption which means that the adsorbed water is more easily desorbed. It is generally used in two layers. The first layer is a lower capacity material, resistant to physical damage from liquid water carry over.

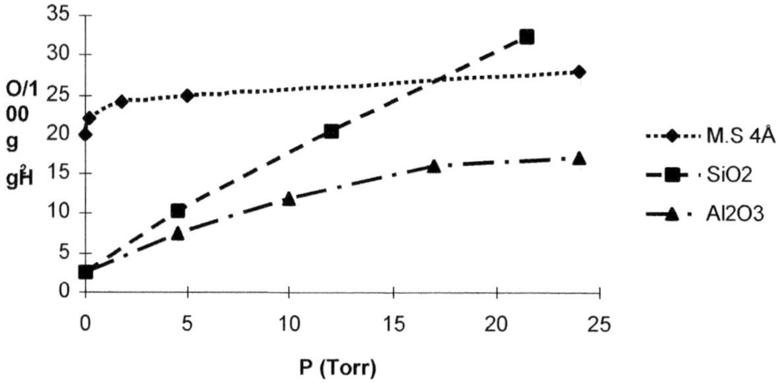

Figure 2.6 - Equilibrium isotherms for sorption of water vapour at 25°C on 4A molecular sieve, activated alumina and silica gel

Activated alumina

This is a porous form of aluminium oxide usually prepared from bauxite (Al_2O_3 $3H_2O$) by dehydration and recrystallisation at elevated temperature. It is more strongly polar than silica gel and has both acidic and basic characteristics. It has a good capacity for water at elevated temperatures and can reach a lower dewpoint than silica gel (-70°C vs. -50°C).

It can adsorb both water and CO_2 together although the water is more strongly adsorbed. The CO_2 adsorption capacity can be enhanced by impregnation of the alumina with potassium carbonate which seems to promote a degree of chemic-sorption with conversion of the carbonate to bicarbonate. While potassium carbonate is useful at ambient temperature, sodium oxide impregnation leads to an adsorbent which can be used for CO_2 adsorption at elevated temperature, say 200°C to 300°C.

Activated carbon

It consists of randomly orientated micro-crystallites of graphite. Active carbon is prepared by the thermal decomposition of organic matter or coal followed by activation with steam or CO_2 at temperatures of 700°C to 1100°C. It has a non polar surface and is hydrophobic. Active carbon is most often used to adsorb hydrocarbons and its use in CO_2 removal is important in purifying hydrogen or hydrogen/carbon monoxide gas mixtures where it forms one of the layers in an adsorbtive separation system usually following the dehydration layer.

It is possible to form active carbon with a very narrow pore size distribution, typically 4 to 9Å. The most important use is for air separation to produce small quantities of pure nitrogen in a pressure swing process which depends not on the equilibrium selectivity of the material but on the kinetics of adsorption with the oxygen adsorption rate being much faster than nitrogen or argon.

Zeolites

Zeolites are porous crystalline alumino silicates consisting of combinations of SiO_4 and AlO_4 tetrahedra which combined to form a very large number of possible configurations. These all have the characteristic of an open crystal lattice with molecular scale pores which can accommodate adsorbed species. The size of the cavities and the windows into the cavities determine which species can be adsorbed. The structure also has metallic cations which balance the excess negative charge of the system.

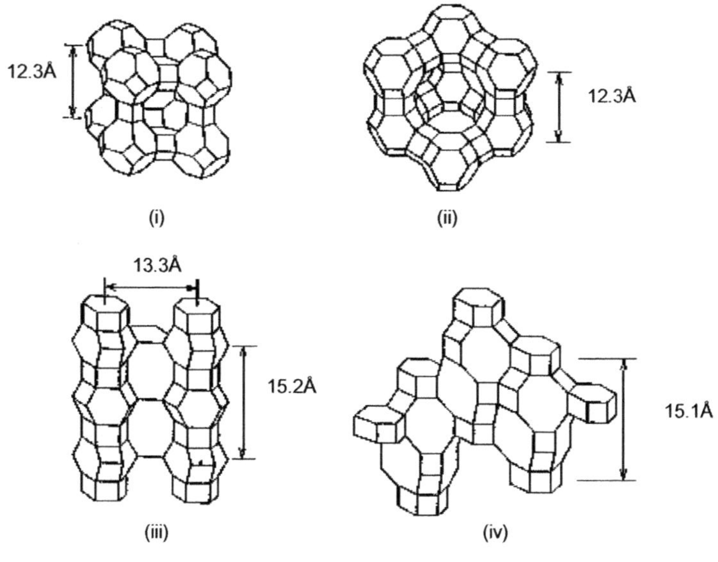

Figure 2.7 - Schematic representation showing framework structures of (i) zeolite A, (ii) zeolites X and Y, (iii) erionite and (iv) chabazite

These cations can be exchanged and this together with the variety of crystal structure gives a wide range of adsorption properties both capacity and selectivity for gas separation and purification processes. The Si/Al ratio is never less than 1 but has no upper limit. Higher alumina ratios (above 8) give progressively more marked hydrophobic properties. Figure 2.7 shows the schematic framework structures of the most common forms of zeolite.

The A and X zeolites are most important for CO_2 removal having good capacity at near ambient temperature. The gas must be dried in a pretreatment layer because pre-adsorbed water will greatly reduces the capacity for CO_2 and other less polar molecules.

The steps in the manufacture of a pelleted molecular sieve adsorbent are shown in Figure 2.8.

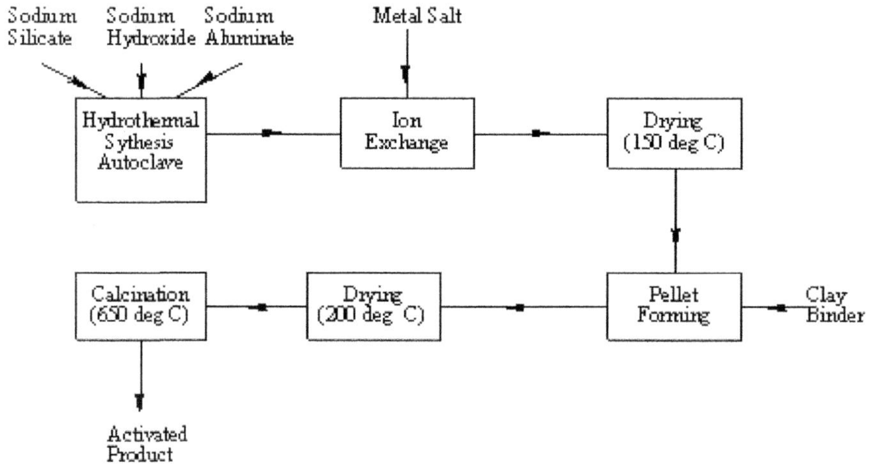

Figure 2.8 - Steps in the manufacture of a pelleted molecular sieve adsorbent

High temperature chemi sorbents

Carbon dioxide is usually released from high temperature processes such as combustion for power generation or hydrogen production. Adsorption separation of CO_2 is now possible at temperatures of 300°C to 500°C using chemi sorbents. A typical example is a potassium carbonate promoted hydrotalcite in pelleted form which reversibly chemi-sorbs CO_2 in the presence of high partial pressures of water vapour. This adsorbent exhibits a high working capacity for CO_2 in a PSA cycle with relatively fast adsorption and desorbtion kinetics. The material is stable in the present of water vapour.

2.3.3 The principles of adsorption

Heat of adsorption

The heat of adsorption provides a direct measure of the bond strength between the adsorbed molecule and the surface. In gaseous systems adsorption is always exothermic. The forces involved are both Van der Waals, which would predominate in active carbon, and electrostatic which would dominate in a zeolite which has an ionic structure. With small polar molecules such as water the heat of adsorption is very large.

Henry's law

For a low concentration system where adsorbed molecules are isolated from their neighbours, the equilibrium between fluid and adsorbed phase concentrations will be linear. This is the basis of Henry's law, which can be expressed in terms of concentration or partial pressure for the gas phase (c or P), and amount adsorbed n (mols/unit volume))

$$n = Kc = K'p \quad \text{where} \quad K = K'RT$$

The temperature dependence of the Henry's law constant obeys the van't Hoff equation:

$$\frac{d\ln K'}{dT} = \frac{\Delta Ho}{RT^2} \qquad \text{and} \qquad \frac{d\ln K}{dT} = \frac{\Delta Uo}{RT^2}$$

where ΔHo and ΔUo are the differences in enthalpy or internal energy between the adsorbed and gaseous states. Neglecting differences in heat capacity between the phases and integrating

$$K' = K'_0 e^{-\Delta Ho/RT} \qquad \text{and}$$

$$K = K_0 e^{-\Delta Uo/RT}$$

This means lnK and 1/T have a linear relationship over a wide temperature range.

Adsorption isotherms

The Brunauer classification divides physical adsorption isotherms into five types as shown in Figure 2.9 [6]. Type I represents a true microporous solid, II and III would correspond to adsorbents with a wide range of pore sizes. Type IV has two types of adsorption, normal micro pore followed by capillary condensation in the larger pores. Type V is occasionally observed when there is a very strong inter-molecular attraction. The simplest model used for treating monolayer adsorption is that of Langmuir. Various other treatments are used to model isotherms in real systems.

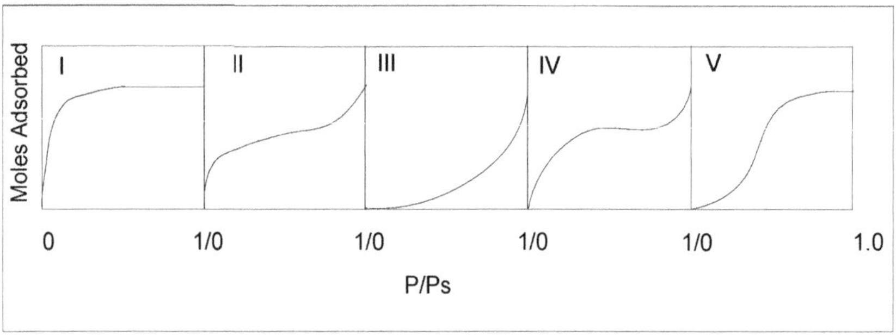

Figure 2.9 - The Brunauer classification of isotherms

Diffusion

Rates of adsorption and desorbtion are controlled by diffusion within the pore network rather than by the kinetics of surface adsorption. The mechanism of pore diffusion depends on the type of adsorbent used. In zeolites where the electrostatic forces are high, the adsorbing molecule is always influenced by these surface forces and transport occurs by surface diffusion often described as micropore or inter-crystalline diffusion. When diffusion occurs in large pores it is known as macropore diffusion. The driving force in a real system is the gradient of chemical potential. Measurement of diffusivities is covered in standard textbooks.

Kinetics of adsorption

The actual rate of physical adsorption in a particle is generally controlled by heat and mass transfer resistances rather than by the intrinsic adsorption kinetics alone. Real adsorbents consist of small microporous crystals which are formed into a macroporous pellet. They have two diffusional resistances in the pellet, the micropore and macropore, together with an additional resistance due to transport through the boundary layer surrounding the pellet. Since adsorption is exothermic there is a continuous transfer of heat during adsorption from the pellet to the surrounding fluid by conduction within the pellet and convection across the fluid boundary layer. The latter resistance is usually controlling.

Flow through packed beds

Most adsorption processes take place in beds of adsorbent pellets where they are contacted by fluids flowing during the adsorption and regeneration phases. The hydrodynamics of fluid flow in packed beds is dealt with in standard chemical engineering text books. Important variables are particle size, fluid velocity and bed dimensions which determine the pressure drop, axial mixing and heat transfer properties which are important factors in the economics of the process. The bed shape is either cylindrical or annular, a shape which in general is gaining increasing favour as a means of limiting bed pressure drop.

2.3.4 *Dynamics of adsorption systems*

General

In a practical adsorption system it is the overall dynamics of the packed bed rather than the kinetics for a single particle which determines the design of the process. In a plug flow system with no heat or mass transfer resistance or heat of adsorption, the outlet concentration profile would be the same as the inlet but with a time delay depending on the adsorption equilibrium for the adsorbed component, bed size and fluid flowrate. In a real system the outlet profile spreads due to axial dispersion and heat and mass transfer resistances. In addition the heat of adsorption will increase the bed temperatures at the adsorption front as it moves through the bed leading to a higher outlet temperature for the fluid.

The form of the concentration front

The general characteristic of the concentration front or mass transfer zone is determined only by the type of equilibrium adsorption isotherm exhibited by the system, although kinetics will modify the shape of the breakthrough curve. The three types of isotherm are shown in Figure 2.10.

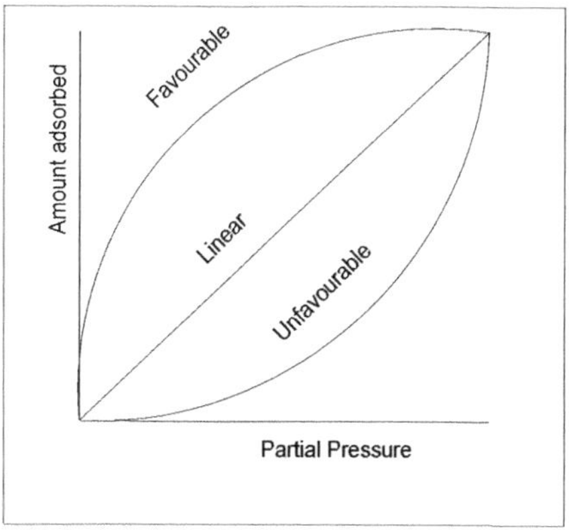

Figure 2.10 - Types of equilibrium adsorption isotherm.

They are the favourable, the linear and the unfavourable shapes. The favourable shape results in an initial broadening of the mass transfer zone at the front of the bed due to axial dispersion but a point is quickly reached where the compressive effect of the isotherm shape balances the axial dispersion and a constant adsorption profile then moves through the bed.

The linear and unfavourable shapes lead to a continuous broadening of the mass transfer zone as it moves through the bed. Examples of favourable isotherms are CO_2 and water on zeolites and alumina.

We are concerned here with CO_2 separation from multicomponent gas systems which exhibit multiple mass transfer zones in beds with several different layers of adsorbents. Modelling of such systems is now possible using dynamic simulation programmes. These programmes consider the entire adsorption and desorbtion cycle including pressurisation and depressurisation of the beds.

2.3.5 Gas separation by cyclic adsorption system

Most practical systems operate in a cyclic manner in which the beds are alternately saturated with the adsorbed component, then regenerated.

Methods of regeneration

Thermal swing regeneration is accomplished by heating the bed counter-currently with a stream of hot gas while in a pressure swing operation the bed pressure is reduced at constant temperature then the bed is counter-currently purged at low pressure. Frequently the two systems are combined with a thermal regeneration using a low pressure heated gas. The operation of pressure swing and thermal swing systems is shown in Figure 2.11 for these three cases.

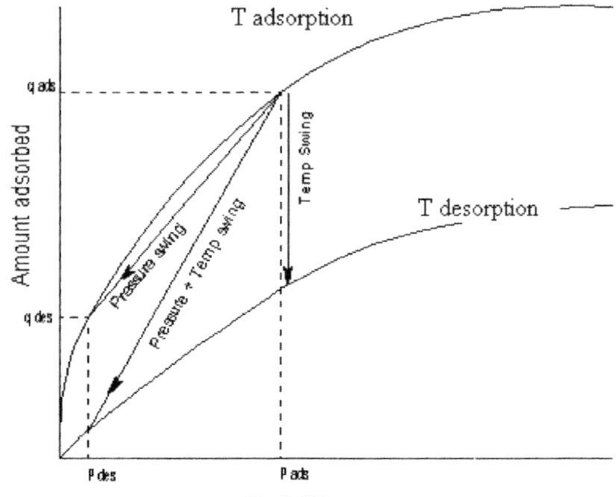

Figure 2.11 - Pressure Swing and Thermal Swing Adsorption Systems

This diagram is however only correct if the regenerating gases are in equilibrium as shown at a partial pressure Pdes. In practice the concentration of the adsorbed component in the inlet regeneration gas may be close to zero and in the case of a thermal swing system, the bed would be cooled down by a counter-current cold gas flow following the heating period to ensure no contamination of the outlet end of the bed in the on-line period. In these cases a regeneration outlet end of the bed Pdes would approach zero and virtually all the adsorbed component would be removed during regeneration.

Thermal swing systems

The temperature dependence of the adsorption isotherm in the van't Hoff relationship gives a large drop in amount adsorbed with a modest temperature rise and particularly favours strongly adsorbed species such as water and zeolite which also have higher heats of adsorption. The non-adsorbed component can be recovered at high purity with very low losses but the adsorbed component will be contaminated by the regeneration gas. One disadvantage of the thermal swing system is the cost of regeneration heat. Another is the relatively long cycle time required to carry out the thermal regeneration of the whole bed, usually from one to eight hours. Because of this minimum cycle time, thermal swing systems are difficult to design for bulk separations such as CO_2 removal. They are generally only used for trace component removal such as drying.

Pressure swing systems

The pressure swing system is well suited to rapid cycling with typical cycle times of from 30 seconds to 10 minutes. This maximises the productivity of the system and bulk separations can be achieved. The non adsorbed component can be recovered at high purity, but the necessity for pressure reduction means that the feed gas present in the voidage of the bed is lost in the waste as blowdown stream. This loss is minimised by using a multiple bed cycle in which depressurisation gas is used to partially pressurise other beds. In this way the recovery of the non-adsorbed components can be maximised.

The displacement cycle

In a CO_2 recovery process using pressure swing adsorption, the CO_2 is the adsorbed component which we wish to recover at reasonable purity (>97%) and recovery (>90%). Normally the adsorbed component is contaminated with regeneration gas. This can be avoided by introducing a cocurrent stream of adsorbed component, -CO_2, at feed gas pressure into the bed and using it to sweep out the feed gas from the bed voidage. The bed can then be depressurised to ambient pressure and this can be followed if necessary by a vacuum step to further recover relatively pure CO_2 before repressurising counter-currently with product gas from the outlet end of the bed which will confine CO_2 to the inlet end of the bed.

It is this combination of pressure swing, cocurrent CO_2 rinse, vacuum regeneration and multiple bed repressurisations which is the key to the design of CO_2 recovery systems.

Practical aspects of adsorption systems

The arrangement for the installation of adsorbent in a typical cylindrical bed adsorber is shown in Figure 2.12. In a pressure or vacuum swing adsorption system for bulk CO_2 removal there might be 3 layers of adsorbent alumina, carbon, 5A zeolite separated by floating wire mesh screens to prevent mixing of the layers. Non-adsorbent spherical high density ceramic balls would be placed in a layer above the support grid for gas distribution and on top of the bed to prevent local fluidisation or adsorbent movement during rapid pressurisation or depressurisation of the bed. A cylindrical filter screen at the top pipe connection also acts as an inlet gas distributor. Internal or external insulation would be used with a temperature swing system. Currently annular bed systems are being introduced for adsorption. They are particularly useful for very large gas flowrates. They combine low pressure drop with a high proportion of the vessel volume occupied by adsorbent.

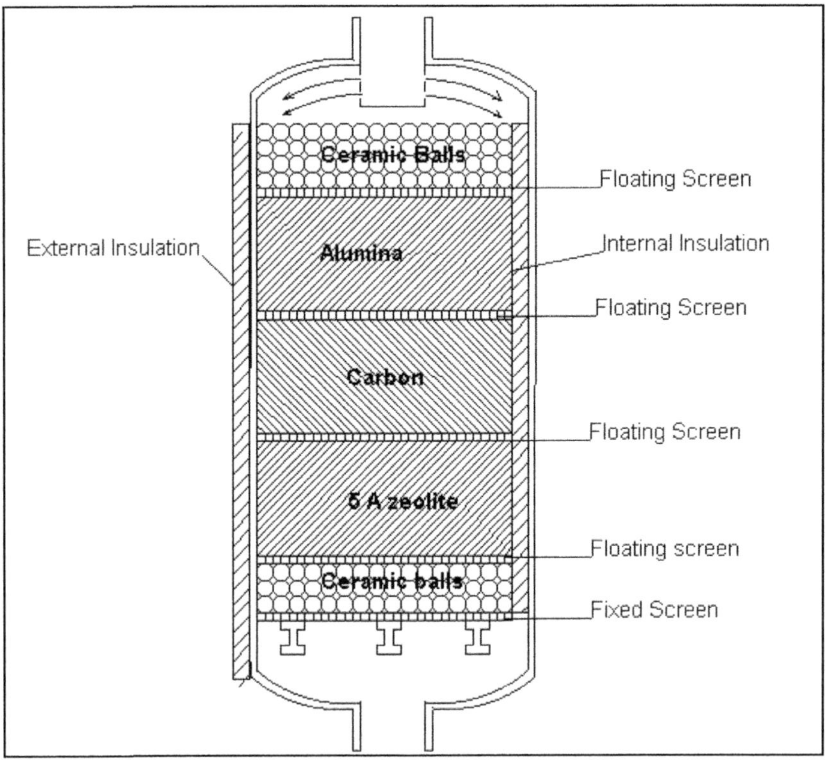

Figure 2.12 - Schematic representation of a typical bed adsorber arrangement

The switching valves are critical items in a PSA or VSA system. With cycle times as low as 30 seconds very fast operation (<1 sec) and very high flowrates usually require butterfly valves with diameters up to 100 cm. Typical operating pressure differences across the closed valve would be up to 25 bar with tight specification on leakage for up to 10^6 cycles between maintenance periods. Very fast pressurisation and depressurisation of the beds means careful attention to local gas velocities to avoid fluidisation.

2.3.6 *Industrial CO_2 removal by adsorption*

Power generation systems

Electric power production by the combustion of fossil fuels is the largest source of CO_2 emission. Considerable research engineering analysis and some demonstration has been devoted to this subject over the last 10 to 15 years. The two major current systems comprise combustion of fossil fuel, mainly pulverised coal in boilers to raise high pressure steam for steam turbines or natural gas fired gas turbine cogeneration systems. Future coal or oil based systems will be either conventional boilers with very high pressure steam systems or gasification of the fuel in an oxygen or air blown gasifier at high pressure (IGCC) to make a fuel gas which can be burned in a gas turbine combined cycle cogeneration system. Sulphur and NO_X removal in these systems are carried out by conventional means. In power boilers flue gas scrubbing is used coupled with low NO_X burners, while the IGCC systems at present use ambient or low temperature absorbtion for H_2S removal with chemical or physical absorbents such as MDEA or cold methanol (Rectisol). Future coal or oil fired stations are more likely to be based on IGCC systems with good prospects for development of hot gas sulphur cleanup systems based on chemisorbtion. Adsorption can be used for CO_2 removal from the flue gas of a conventional power station and also from the fuel gas of an IGCC system. Another approach is to convert the natural gas fuel for a conventional gas turbine combined cycle system to hydrogen and CO_2 and remove the CO_2 by adsorption. In all cases the cycle design of the system is adjusted to allow optimum utilisation of the adsorption process for CO_2 removal.

Pulverized coal fired power station

The CO_2 removal system would be installed following the SO_X scrubber treating the stack gas. The flowsheet shown in Figure 2.13 has been proposed by M.Ishibashi et al. [7] It consists of three sections. In the first, the flue gas at 1.1 to 1.2 ata is dried to a dewpoint of -50°C in a 2 tower PSA drier using silica gel or alumina adsorbent with a vacuum regeneration to achieve the necessary pressure ratio for operation. This dry gas can now be passed through a two stage CO_2 removal process. The first stage uses a PSA cycle with a temperature swing imposed on the inlet end of the bed, probably to remove residual water by introduction of heated purge gas halfway down the bed. The cycle operates with a vacuum blower, Crude CO_2 at 40% - 50% purity enters the second CO_2 removal step which concentrates the CO_2 to 99% purity using a vacuum PSA cycle with a cocurrent CO_2 rinse step to achieve the required CO_2 purity. The CO_2 recovery is claimed to be 90%. Table 2.3 gives performance details.

Table 2.3- Specification of Pilot Plant for PTSA-CO_2

Properties of Exhaust Gas			
CO_2	10.80%	CO_2 Separation Method	4-Adsorber Type-2-Staged PTSA
H_2O	12.90%	Gas Processing Capacity	1000 m^3/hr
O_2	5.10%	CO_2 Recovery Ratio	90%
N_2	71.20%	CO_2 Recovery Rate	190 kg/hr
SO_X	76 ppm	Cenc. of Recovered CO_2	99%
NO_X	60 ppm	Adsorption Temperature	30 - 50%
Dust	20 mg/m^3N	Adsorption Pressure	1.1 - 1.2 ata
		Desorption Pressure	0.1 - 0.3 ata

The figures refer to a pilot plant installed adjacent to TEPCO's Yakusuka thermal power station in Japan operating on stack gas.

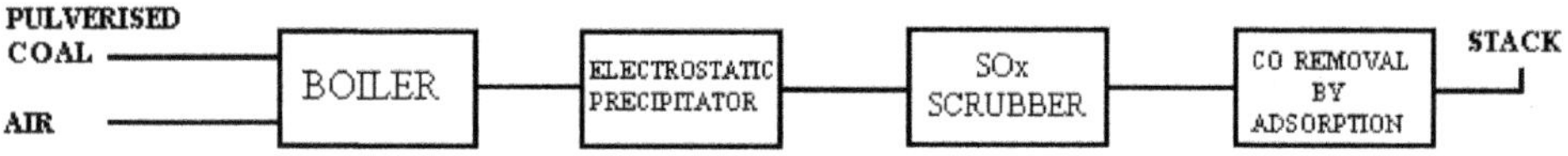

Figure 2.13 CO_2 Removal by adsorption

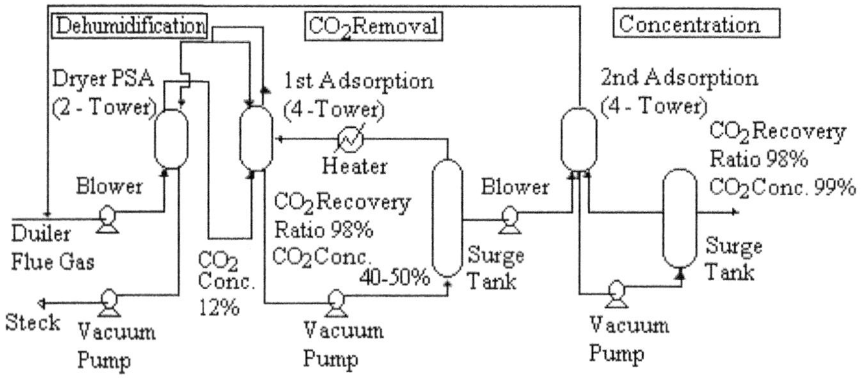

Figure 2.14 Conventional pulverised fuel fired power station

Natural gas fired combined cycle

There are two options for CO_2 removal from a natural gas fired combined cycle generator, from the exhaust gas or by removing the carbon in the fuel gas as CO_2 before combustion. Because of the large amount of excess air used to reduce the combustion gas temperature the CO_2 concentration in the effluent gas is only about 3%. Conversion of the natural gas fuel to hydrogen will give a partial pressure of typically 5 ata based on a fuel pressure of 25 ata or 20% CO_2 concentration on a dry basis. It is thus attractive to consider the fuel conversion option because of the much lower gas volume which is processed (0.5%) and the much higher capacity of the adsorbents for CO_2 at the higher partial pressure. The overall system shown in Figure 2.15 is based on the use of an oxygen blown autothermal reformer with a feed desulphurised natural gas plus steam.

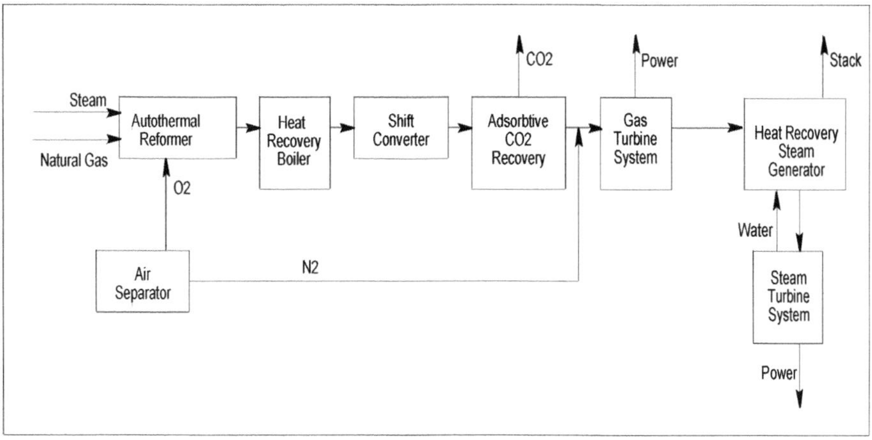

Figure 2.15 - Natural gas fired combined cycle power generation

The reformer consists of a partial oxidation burner firing downwards with the combustion gases passing through a catalyst bed where the heat of combustion is used for further conversion of natural gas by endothermic steam-methane reforming. The hot gases pass successively through a waste heat boiler, a catalytic shift converter and an adsorbtive CO_2 removal system. The reactions (based on methane) are as follows:

$$CH_4 + \frac{1}{2} O_2 \rightarrow CO + 2H_2 \quad \text{POX}$$

$$CH_4 + 2O_2 \rightarrow CO_2 + 2H_2O \quad \text{burner}$$

$$CH_4 + H_2O \leftrightarrow CO + 3H_2 \quad \text{autothermal catalyst bed}$$

$$CO + H_2O \leftrightarrow CO_2 + H_2 \text{ at all points in the process}$$

Note that the oxygen plant supplies all its nitrogen to mix with the hydrogen fuel following CO_2 removal to reduce flame temperature and NO_X formation in the gas turbine and to boost the power output and efficiency of the system.The CO_2 removal system is shown in Figure 2.16 in its most basic form in a two bed configuration. The adsorbent could be potassium carbonate promoted activated alumina. The cycle consists of an adsorption step followed by a cocurrent rinse with pure CO_2 compressed in the rinse compressor to feed gas pressure. This will displace feed gas from the bed and vessel voidage. There is then a countercurrent depressurisation to atmospheric pressure recovering CO_2 followed by a vacuum step at about 0.3 ata. The bed is then repressurised countercurrently with the hydrogen product before a new adsorption cycle commences. The cycle time (absorbtion plus regeneration) would be about 300 seconds. A cycle of this type has been demonstrated in a research pilot plant achieving a CO_2 purity of 98% while the total CO_2 recovery of the system is about 90%.

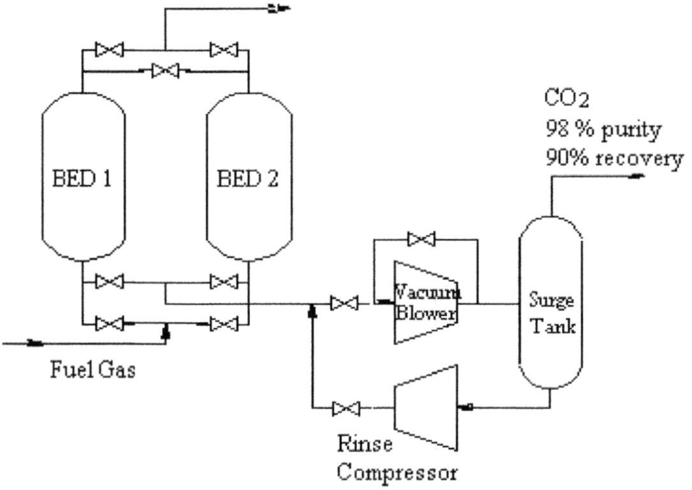

BED 1	ABSORPTION	Cocurrent CO2 rinse	Countercurrent depressurize to surge tank	Vacuum regenerate to surge tank	Countercurrent repressure with product
BED 2		ABSORPTION			

Figure 2.16 - CO_2 removal from CO_2/H_2 gas mixture by adsorption

Coal and oil fired IGCC

Gasification of coal or heavy residual oil takes place in an oxygen blown partial oxidation reactor which produces a high pressure fuel gas in which the sulphur is present mainly as H_2S. In future systems it is likely that sulphur compounds can be removed using hot chemisorption systems. This would be followed by CO shift conversion giving a CO_2 hydrogen mixture at a temperature of 250 - 450°C. A coal gasifier would produce a fuel gas having about 45% CO_2 / 55% H_2 mixture at a pressure of about 24 ata. A hot CO_2 removal by adsorption is possible using newly developed high temperature CO_2 adsorbents such as potassium carbonate promoted hydrotalcite developed by Air Products. This material has been shown to have high usable CO_2 capacity in a PSA system and will operate at temperatures up to 500°C. The adsorption cycle used would be similar to the system described in the previous paragraph. It will be necessary to demonstrate this cycle and this material at a reasonable throughput on a typical IGCC fuel gas composition before the system could be installed in a full size plant.

Hydrogen production

Hydrogen is produced on a very large scale for ammonia and methanol, hydrogenation reactions and in the petroleum refining industry for hydrotreating to remove sulphur from transportation fuels and to treat heavy oil fractions. Hydrogen production is likely to increase considerably in the future when it begins to be used as a clean transport fuel in cars, buses and lorries, particularly in centres of population. The production of hydrogen in large centralised plants will allow the removal of CO_2 at high purity and recovery for sequestration. Most of the hydrogen produced for refining and chemicals use is purified using a PSA adsorption process and adsorption will continue to be attractive in the future for combined pure H_2 and CO_2 production. Hydrogen is currently produced from a variety of fossil fuels, natural gas, light and heavy oil, coal, petroleum coke. These feedstocks are converted to hydrogen by either steam-hydrocarbon reforming or autothermal reforming for light feedstocks or partial oxidation using pure oxygen for heavy feedstocks coal and coke.

H_2 pressure swing adsorption

The purification of a hydrogen containing gas is usually carried out in a pressure swing adsorption system which typically has a three layer bed comprising alumina or silica gel for water and heavy hydrocarbon removal, a carbon layer for light hydrocarbon and some CO_2 and a zeolite layer such as 5A or 13X for CO_2 and light gas removal such as nitrogen and argon. The process steps in a typical 6 bed PSA system are shown in Figure 2.17. The steps comprise adsorption followed by depressurisation cocurrently with the gas being used to provide purge for regeneration and gas for repressurisation of regenerated beds. This is followed by countercurrent blowdown to near atmospheric pressure countercurrent purge and then a two-step countercurrent repressurisation followed finally by countercurrent repressurisation to feed gas pressure with hydrogen product. A typical system fed from a steam/natural gas reformer would be operating at 22 ata 30°C and can achieve 99.999% hydrogen purity, and 86% to 90% hydrogen recovery.

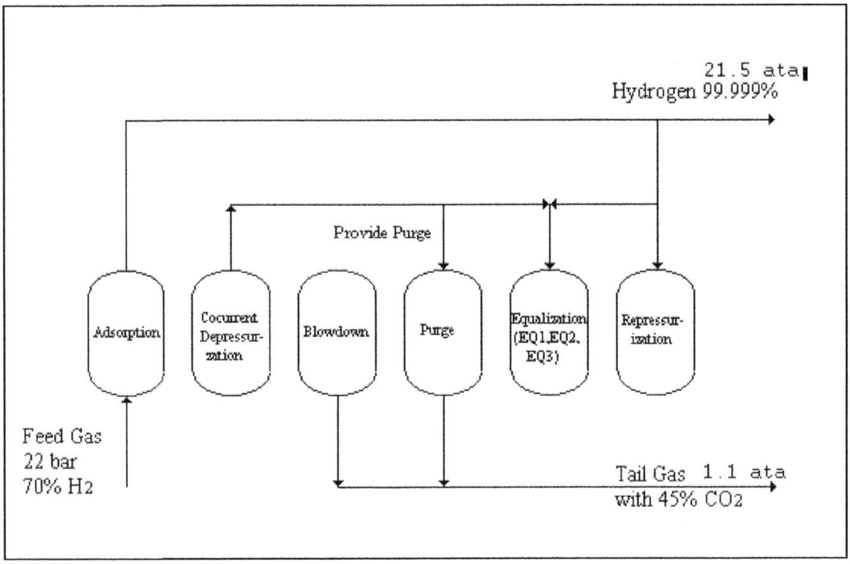

Figure 2.17- Conventional hydrogen pressure swing adsorption cycle

PSA for simultaneous pure H_2 and CO_2 production

The PSA described in the previous section will produce a fuel gas waste stream at about 1.2 ata pressure containing about 45% CO_2. It is possible to separate the CO_2 removal and hydrogen production into a two part linked system which is shown in Figure 2.18. The first part of the system operates to separate the most strongly adsorbed component CO_2 while the second part separates the lighter components such as methane, nitrogen and argon from the pure hydrogen product. In order to produce high purity CO_2 product, the first stage process uses a cocurrent CO_2 rinse step with a CO_2 rinse compressor followed by blowdown of CO_2 and a vacuum regeneration step. This process has been demonstrated on a commercial scale over a period of 15 years in Butler Pennsylvania by Air Products. A CO_2 purity of about 99% is achieved at better than 90% CO_2 recovery.

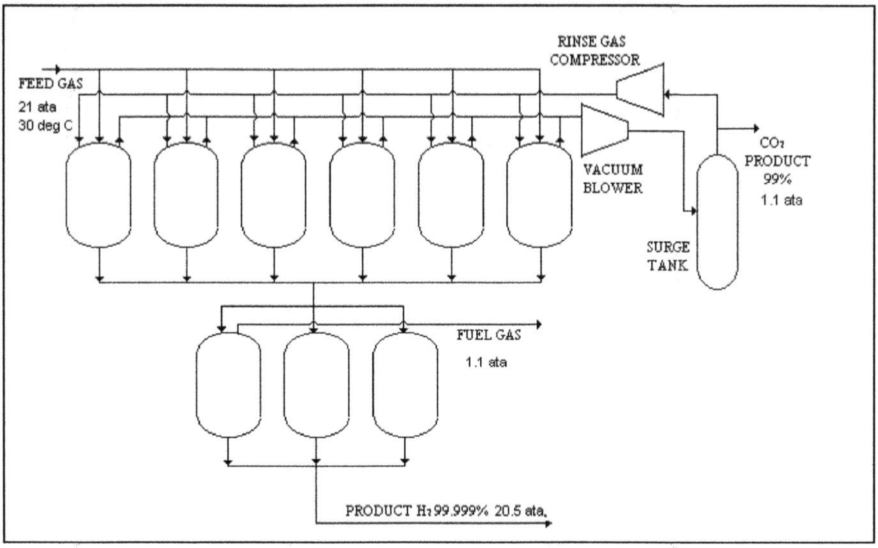

Figure 2.18 -Simultaneous pure H_2 and CO_2 production by pressure swing adsorption

H_2 and CO_2 production by sorption enhanced reaction

The importance of hydrogen production processes in the future has led to the development of a number of novel systems for production and purification. One of the most interesting is a process which combines steam methane reforming with an adsorption process which continuously removes the CO_2 by-product of the reaction by adsorption onto a high temperature CO_2 adsorbent. The system is shown diagramatically in Figure 2.19.

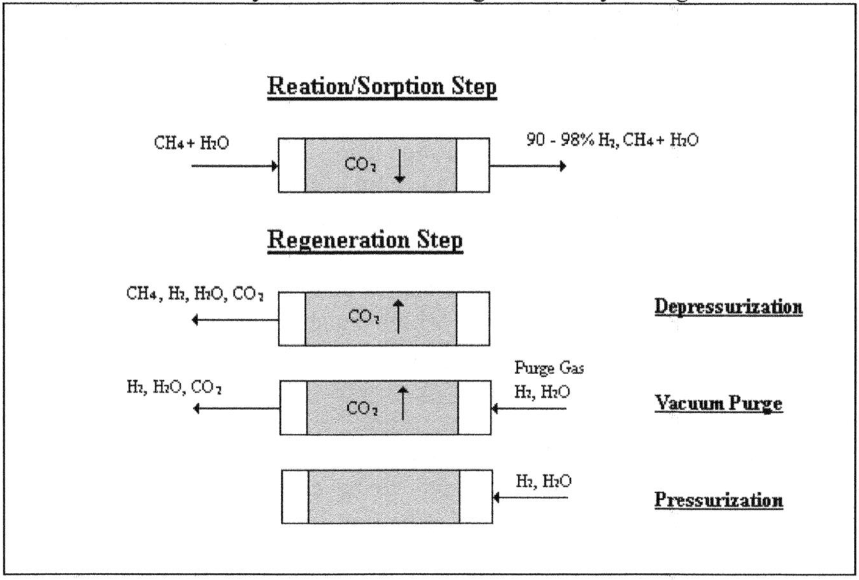

Figure 2.19 -H_2 by adsorption enhanced reforming

The system operates as a high temperature cyclic adsorption system with a number of beds, which are packed with an intimate mixture of a nickel based steam methane reforming catalyst and a high temperature CO_2 adsorbent such as a potassium carbonate promoted hydrotalcite, which is active in the presence of high partial pressures of water vapour. The bed is fed with a preheated mixture of natural gas and steam at a temperature of between 400°C and 600°C and 18 ata pressure. The following reversible reactions take place:

$$CH_4 + H_2O \leftrightarrow CO + 3H_2$$

$$CO + H_2O \leftrightarrow CO_2 + H_2$$

The result of the adsorption of the CO_2 is to drive the hydrogen production to completion by the Le Chatelier's principle. The vessel is then depressurised and regenerated. The system requires heat input during both the adsorption/reaction and the regeneration phases and this heat is supplied by a circulating heat transfer fluid. The system has been demonstrated by Air Products on a laboratory pilot plant scale. It could be used with the CO_2 rinse and vacuum regeneration cycle described previously to allow pure CO_2 production at high CO_2 recovery. A demonstration on a commercial scale is the next step in development of the system. Projected economics are superior to conventional steam/natural gas reforming with shift conversion and a H_2 PSA. It would be an ideal process for simultaneous CO_2/H_2 production.

Iron and steel production

The production of iron in the iron and steel industry involves the reaction of carbon usually in the form of coke with air at high temperature to produce CO which reduces the iron oxide to molten iron. The carbon leaves the blast furnace top as a mixture of CO and CO_2 which is then burned as a fuel gas. It is possible to operate iron making processes at high efficiency with complete removal of CO_2. An important economic and ecological driving force for the adoption of new process routes to iron making is the desire to reduce or eliminate the use of coke which must be produced at high cost from coking coals in a highly polluting coke oven battery.

Corex process with CO_2 removal

The Corex process developed by Korf followed by Voest-Alpine is the first of these new iron making systems to be commercially demonstrated (Figure 2.20). It uses a vertical shaft reduction kiln fed at the top with iron ore and limestone and utilising a hot CO/H_2 containing reducing gas produced from a melter - gasifier which enters the bottom at about 850°C and passes up counter-current to the iron ore/limestone. The melter - gasifier is fed by a stream of reduced iron powder plus limestone from the shaft kiln plus a stream of powdered coal and it is blown with a stream of pure oxygen. It operates at 5 - 6 ata pressure. The heat of reaction between the powdered coal and oxygen which produces the reducing gas for the shaft kiln is used to melt the iron and slag which are tapped from the base of the vessel. The top gas contains about 30% CO_2 and 14% H_20 together with significant quantities of dust particles. The remaining 56% is almost all CO/H_2. The dust is removed by water wash which also cools the gas to near ambient temperature and condenses most of the water vapour. The gas can now be compressed and fed to an adsorption CO_2 removal process similar to that described in Section 2.2.3.6.3 and shown in Figure 2.16. This will remove CO_2 at high purity and high recovery leaving a H_2/CO gas stream which can be preheated and recycled to the shaft kiln to produce excess direct reduced iron powder product.

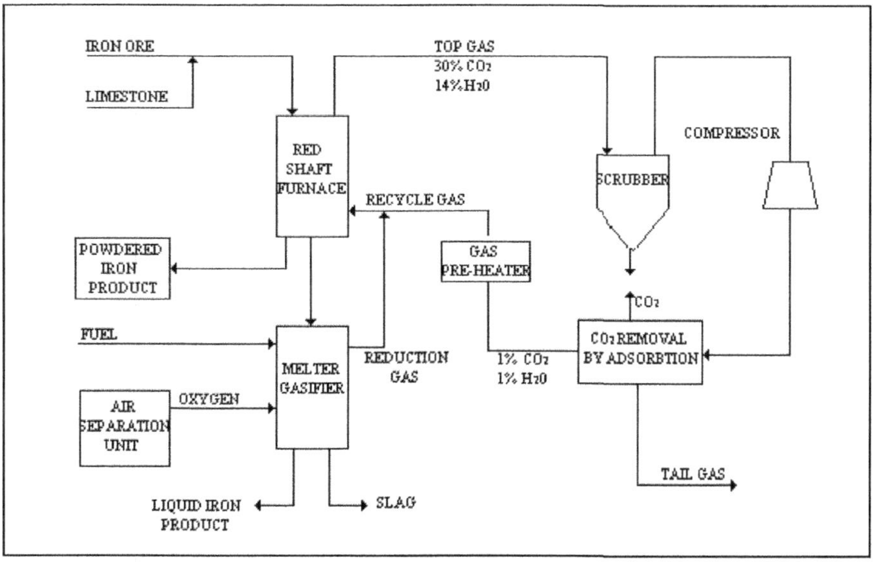

Figure 2.20 -Corex iron making with recycled reducing gas

Blast furnace with recycled top gas

Modelling work has been carried out on the concept of recycling the top gas from a blast furnace to provide fresh CO rich reducing gas for reduction of iron ore and melting of the iron plus slag. The system is shown in Figure 2.21. The top gas is scrubbed to cool it, remove dust and water vapour then compressed for recycling. The CO_2 can be removed at high purity and high recovery in an adsorption separation using a cocurrent CO_2 rinse step and vacuum regeneration similar to the system described in section 2.2.3.6.3. The purified gas is preheated in gas fired recuperative heater then injected into the base of the blast furnace. The pure oxygen required for combustion is separately injected into the base of the furnace.

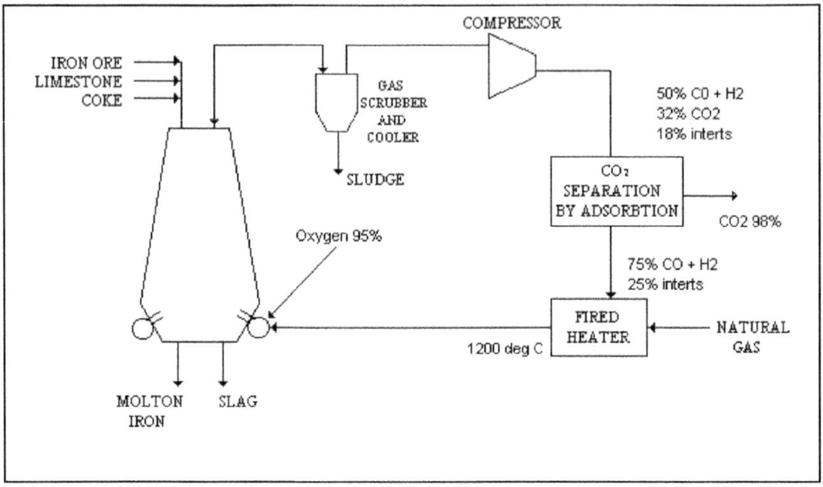

Figure 2.21- Blast furnace with recycled reducing gas

2.4 Membranes

2.4.1 Fundamentals and membrane types

A membrane is commonly regarded as a barrier between two phases capable of being selectively permeated by one or some components from one phase to the other. With reference to the nature, one may further divide membranes into biological and synthetic types; the synthetic type comprises organic and inorganic membranes.

Organic membranes are mainly made by using semicrystalline and amorphous polymers. The latter type can be divided in rubbery and glassy polymers, depending on their glass transition temperature (Tg). Usually, rubbery polymers with Tg < less than the operating temperature (T), show high permeabilities, while glassy polymers with a Tg > T, show high selectivities for gases. The membranes may furthermore be divided into symmetric and asymmetric types. The first type has uniform structure and composition through the membrane, while the second type has a selective layer (skin) with a thickness of 0.1-0.5 micron on a porous support (thickness: 50-150 micron) which gives the mechanical strength. Since the resistance normally is proportional to the thickness of the selective layer, the asymmetric structure significantly improves the overall membrane flow rate. A special type of asymmetric membrane is the composite membrane, in which different materials make the final structure.

Inorganic membranes, as organic types, may firstly be divided into nonporous (dense) and porous, and secondly into supported (asymmetric) and unsupported (symmetric) structures. Nonporous membranes as metal membranes (e.g. Pd alloys) and oxide membranes (mixed conductors and electrolytes), transport matters by bulk and grain boundary diffusion. These membranes, if defect free, generally show infinite separation factors. Inorganic porous membranes typically made from binary metal oxides (e.g. alumina, zirconia, titania, silica), zeolites, glasses and carbon show, have on the other hand, finite selectivities varying in a large range depending on membrane type and operation condition.

Inorganic gas separation membranes are expensive to produce for several reasons, and furthermore, are still lacking the sufficient performance required in most high temperature applications. Therefore, these membranes have not yet intruded the gas separation market to any noticable extent.

2.4.2 Membrane modules

The membrane module separates the feed streams into a permeate and a retentate (concentrate) stream (Figure 2.22). Important critera for design of module housings are i) to give a mechanical stable support for the membranes, ii) an efficient packing density of membrane area per unit volume, iii) good hydrodynamic conditions for the separation process, iv) permit easy cleaning, maintenance and replacement of the membranes.

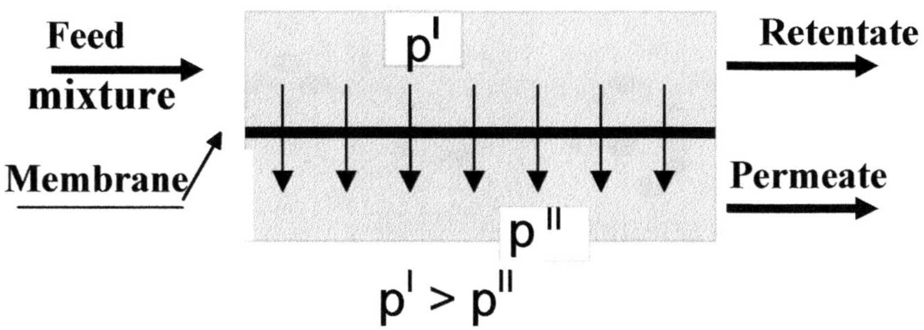

Figure 2.22 -Membrane gas separation driven by difference in partial pressures p^I and p^{II}

"Plate and frame" and "spiral wound" are module designs based on flat membranes, while tubes, multichannel and hollow fibres are assembled in a multitubular heat exchanger design. Tubular and plate and frame modules, typically, have packing densities of 10-400 m^2 of surface area per m^3 module volume very efficient in liquid applications, but not particular in gas separations due to the low membrane area per module volume. The "Spiral wound" modules of 300-1000 m^2/m^3 and "Hollow fibre" modules of 30000 m^2/m^3 are preferred in the latter applications.

A gas separation plant may require several stages of separation to obtained the desired purity of the permeate. In the design of the complete gas separation system, each stage must be considered in terms of pressure drops, flow rates, selectivities, and recycling of gas streams in order to obtain the highest possible cost efficiency. Different types of modules may therefore be used in series and parallel in an optimised system.

2.4.3 Gas transport and separation mechanisms

Gas separation is normally a pressure driven membrane operation, the main exeption being electrically driven reverse fuel cell operation. Dense membranes separate due to differences in solubility and diffusivity, which for organic membranes usually result in finite (Figure 2.23), while for inorganic membranes in infinite separation factors (e.g. Pd-based, oxide electrolytes and mixed conductor membranes).

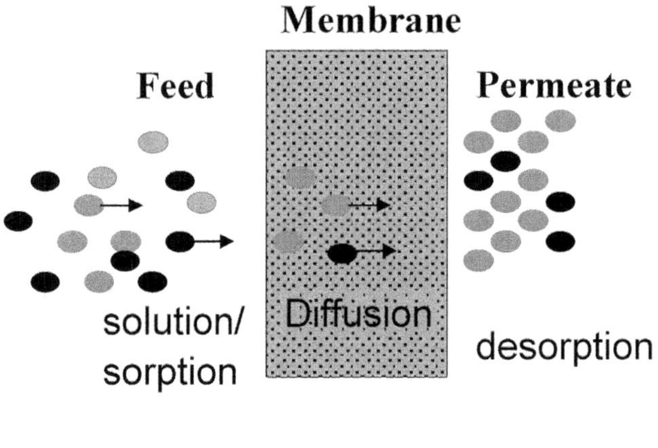

Permeability (P) = Diffusivity (D)•Solubility (S)

Permselectivity $\alpha_{A,B} = P_A/P_B$

Fig. 2.23 - Transport through a dense polymeric membrane by a solution – diffusion mechanism

The flow rate through dense membranes has different partial pressure dependencies depending on the transport mechanism. The gas flow of component i (J_i) through polymeric membranes following the solution-diffusion transport model is given by

$$J_i = D_i \cdot S_i \frac{(p_i^I - p_i^{II})}{l}$$

where D_i, S_i, p_i are the diffusivity, solubility and partial pressure, respectively, of component i, and l is the membrane thickness. Analogue relationships for hydrogen transport and oxygen transport through metal and ion conducting oxide membranes, respectively, are

$$J_{H_2} = D_{H_2} S_{H_2} \cdot \frac{((p_{H_2}^I)^{1/2} - (p_{H_2}^{II})^{1/2})}{l}$$

for metal membranes, and

$$J_{O_2} = \div \frac{RT}{16F^2 L} \int_{\ln p_{O_2}^I}^{\ln p_{O_2}^{II}} \frac{\sigma_{el}\sigma_{ion}}{\sigma_{el} + \sigma_{ion}} d \ln p_{O_2}$$

for oxygen ion conducting oxide membranes (F is the Faraday constant, σ_{ion} and σ_{el} are the ionic and electronic conductivities, respectively).

Porous membranes may separate gases by different mechanisms, see Figure 24 For gas separation the pore size must be small enough so that Knudsen diffusion, size exclusion,, or other selective processes as adsorption, surface diffusion, and pore condensation may occur.

The flow rate through porous membranes will have different pressure dependencies depending on mechanism. Laminar or viscous flow in larger pores is non-selective, but may contribute to the overall flow rate in a Knudsen type membrane. The flow rate of a single gas in a Knudsen type membrane may be written according to (mol m^{-2} s^{-1})

$$J = \cdot \left(\frac{\varepsilon r^2 Pm}{8\eta\tau_L RTL} + \frac{2\varepsilon rv}{3\tau_K RTL} \right) \Delta P$$

where ε is the porosity of the membrane, r is the average pore size, P_m is the mean pressure, η is the viscosity of the gas, τ_L and τ_K are the membrane tortuosity for laminar flow and Knudsen diffusion and ΔP the pressure difference, respectively, and v is the mean molecular velocity given by

$$v = \sqrt{\frac{8RT}{\pi M}}$$

where M is the molecular weight of the gas molecule. Knudsen diffusion, which may take place for pore sizes typically lower than ≈ 20-50 nm, separates only due to differences in molecular mass according to $(M_1/M_2)^{1/2}$. This mechanism thus gives a low selectivity, never higher than about 5. In microporous membranes (r < 1 nm), the interaction with the pore wall may become significant giving large deviation from or complete different expression for the mass transport. If the membrane-gas adsorption properties are very different for the gas components then pore blocking, pore condensation or pore wall surface diffusion, may lead to much higher selectivities than the Knudsen value. Such effects are typical in, for instance, gas separation with zeolite membranes. Due to the strong dependence on adsorption, the selectivity depends strongly on temperature and type of gas mixture.

Finally, liquid membranes should be mentioned, where an immobilised liquid that can dissolve a gas component (e.g. carbonate for CO_2 dissolution) is supported by a porous solid structure. A special case related to CO_2 separation at high temperature is melted carbonate liquid membranes [8].

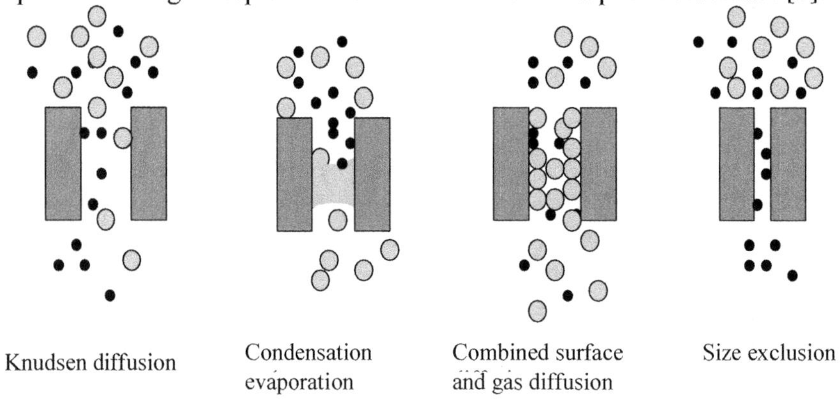

Figure 2.24 - Gas separation mechanisms in porous membranes [9].

2.4.4 *Commercial polymeric membranes for gas separation*

The actual market of gas and vapour separation (2000 year) has been estimed equal to 160•10^{6}U.S.$ with a 8% annual growth for the next 20 years. This datum confirms previous evaluations of other authors [10]. In the following table are reported these information distinguished for the different applications [11].

Table 2.4- Sales estimates and sales predictions for the principal gas and vapor separation applications [11].

Application	Annual Membrane System Sales (10^6 U.S.$)		
	Year 2000	Year 2010	Year 2020
Nitrogen from air	75	100	125
Oxygen from air	<1	10	30
Hydrogen			
• H_2/gas (CO, N_2, C_1, C_2)	25	40	80
• H_2/vapore (C_{3+}, CO_2)	5	20	70
Natural Gas			
• CO_2 Removal	30	60	100
• NGL Removal and Recovery	<1	20	50
• N_2 Removal, dehydration	<1	10	25
Vapour (C_{2+})/Gas (N_2, Ar)	10	30	90
Vapour/Vapour (including dehydration)	5	20	100
Air dehydration/other	10	30	60
TOTAL	160	340	730
Annual growth, %		8	8

Since the introduction of the first commercial gas separation system by Monsanto (Permea, Inc.) in the late seventies, several new players have entered in this area. In the following is furnished an update list of the main companies involved in gas separation by membranes [12].

Table 2.5 - Principal industrial manufactors of membrane systems for gas separations

Company	Polymer	Module type	Applications
Air Products (Permea)	PSF	HF	O_2, N_2, H_2, H_2O, CO_2
Praxair (IMS)	CA	HF	N_2, H_2O
Du Pont/L'air Liquide (Medal)	PI/PA	HF	N_2, H_2, O_2
MG Industries (Generon)	TBPC	HF	N_2
Whatman (Aquillo)	PSF	HF	N_2
Dow (Cynara)	CTA	HF	CO_2
UOP (Separex)	CA	SW	CO_2, H_2
UBE Industries	PI	HF	O_2, N_2, H_2, H_2
Kvaerner (GMS)	CA	SW	CO_2
GKSS licensees	SR	SW, PF	VOCs
MTR	SR	SW	VOCs
Parker-Hannifin	PPO	HF	Air dehydration

CA= *cellulose acetate*, CTA=*cellulose triacetate* PMP= *polymethylpentene*, PA= *polyaramides*, SR= *silicone rubber*, PSF=*polysulfone*, PI= *polyimides*; PEI= *polyetherimides*, TBPC= *tetra bromo polycarbonate*. HF=hollow fiber,

The main successful applications of membranes in the field of gas separation are currently the nitrogen production and the hydrogen recovery. CO_2 reduction in natural gas sweetening is the main application of polymeric CO_2 selective membranes

2.4.5 CO_2 separation by polymeric membranes

The principal polymeric membranes used for CO_2 separation are reported in the Table 2.6 [13], while a more complete description is available in literature [14].

Table 2.6 - Permeability (Barrer) and permselectivity of significant rubbery and glassy polymers at 35°C.

Polymer	He	CO_2	O_2	N_2	CH_4	CO_2/N_2	CO_2/CH_4
Poly[1-(trimethylsilyl)-1-propyne]	5080	28000	7730	4970	13000	5.6	2.2
Poly(dimethylsiloxane)	590	4550	781	351	1430	13	3.2
Poly(dimethylsilmethylene	100	520	91.0	35.9	150	15	3.5
Poly(cis-isoprene)	43.4	191	37.5	14.5	47.4	13	4.0
Poly(butadiene-styrene)	32.9	171	32.9	10.3	34.2	17	5.0
Ethyl cellulose	39.8	75.0	12.4	3.4	6.8	22	11
Polystyrene	22.4	12.4	2.9	0.52	0.78	24	16
Poly(ethyl methacrylate)	23.8	7.01	1.9	0.33	0.35	21	20
Bisphenol-A polycarbonate	13.0	6.8	1.6	0.38	0.36	18	19
Cellulose acetate	16.0	4.75	0.82	0.15	0.15	32	32
Bisphenol-A polysulfone	10.8	4.6	1.2	0.19	0.18	24	26
PMDA-4,4'-ODA polyimide	8.0	2.7	0.61	0.10	0.059	27	46
Poly(methyl methacrylate)	8.4	0.62	0.14	0.02	0.0052	31	119
Poly(etheretherketone)W*	6.99	2.71	0.55				

*at 25°C [15]

Flue gas

The separation of CO_2 from power plant flue gases is one of the most important applications due to the large amount of these emissions in the atmosphere.

In the flue gas, CO_2 is typically present at low concentrations (partial pressures). Whitout considering the presence of moisture, for a combustion with 110% of theoretical air, if coal is used, CO_2 represents about 15 mol% of the flue gas. This percentage is still lower if oil and natural gas are used as fuel, respectively, 13% and 9% [16].

Rautenbach and Struck [17] have identified limiting values of profitable application of membranes to separation of CO_2 and H_2S from flue gases. The principal problem in such membrane applications is the usually low driving force due to low total pressure and small partial pressure diference. Furthermore, commercially available membranes are not selective enough; usually having a separation factor of 20-50 for CO_2/N_2 separation with permeance 0.01-1 $m^3(STP)m^{-2}h^{-1}bar^{-1}$.

An evaluation shows that the membranes need a separation factor higher than 200 and permeability higher 50 Barrer to be competetive to other CO_2 recovery technologies for flue gas treatment [18]. Hybrid systems consisting of a membrane system at the tail of an amine scrubbing system may be a possible option. These systems may offer advantage either to increase efficiency of the amine scrubbers or de-bottleneck various facilities [10].

Cardo-type polyimide asymmetric hollow fibres showed higher CO_2/N_2 permselectivity than commercial available hollow fibres in virtue of a CO_2 affinitive functional group. In order to improve their mechanical strength, spinning using blended dopes of cardo-type polyimide and polyamide was tried. The increase in mechanical resistance is not accompanied by decrease of both selectivity and flux [19].

Dry-wet phase inversion method applied to prepare polyethersulfone (PES) asymmetric membrane, followed by silicone resins coating in order to seal the defects like pinholes on the skin layer of the asymmetric structure, allowed to improve transport properties of membrane for CO_2/N_2 separation. The selectivity for CO_2/N_2 increased but the permeability decreased, as the contents of PES and volatile organic solvent and evaporation time increased. As result of the optimised experiments a CO_2/N_2 selectivity equal to 61 combined with a oxygen permeance of 21 GPU were obtained. These values are very interesting since, if confirmed in long term tests, will permit a convenient separation of CO_2 from flue gases [20].

Finally, composite membranes combining (nano-sized) inorganic particles in a polymeric matrix should be mentioned. These membranes show CO_2 separation performance [21] higher than obtained with the organic matrix only. Such membranes appear currently very promising in the further development of CO_2 separation membranes.

Natural gas processing

A large part of applications in which CO_2 recovery is involved, that are increasing their importance mainly in energy field, are indicated as Natural Gas processing. This term involves three main different process typologies: Landfill gas upgrading CO_2 enhanced oil recovery and Natural Gas treatment. In all these processes, that represent a large industrial source of

CO_2, the same components are present, but with different concentration levels.

Landfill gas upgrading

Landfills generate methane (roughly 50%) that is contaminate mainly by CO_2. Landfill gas is at low pressure (atmospheric), available only at low flow rate (60,000 m^3/day) and is expensive to upgrade. Such gas is usually more valuable than wellhead gas because it is often located near industrialised areas.

In the following table are indicated the average composition and the characteristics of landfill gas.

Table 2.7 - Composition and properties of landfill gas [22].

Component	Composition, %	Property
CH_4	54	Energetic potential
CO_2	40	Green house effect
N_2	4	Reduction of heating value
O_2	1	Explosion danger
H_2O	1	Condensation
H_2S	100 mg/m^3	Smell, corrosion
C_2-Cn	200 mg/m^3	Toxic
CFC	100 mg/m^3	Formation of dioxin

The experience of two years with a pilot plant for 200 m^3_{STP}/h feed gas operating on a landfill dumpsite to produce a gas with natural gas quality has been reported [22].

A scheme of this plant is represented in the following figure 2.25.

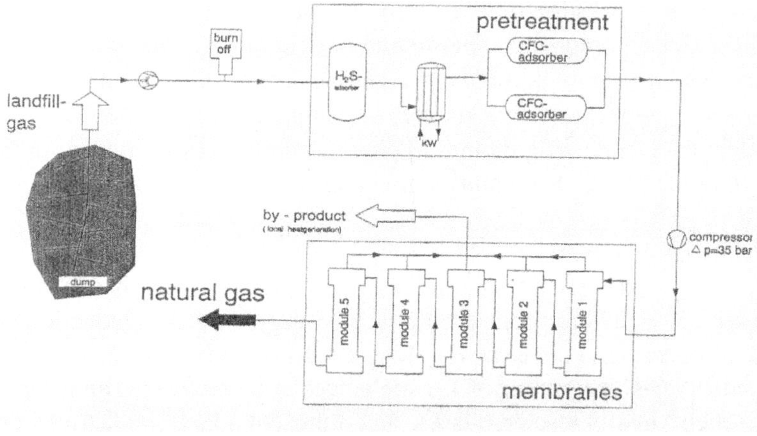

Figure 2.25- Flow chart diagram of the pilot plant for treating landfill gas [22].

The process consists of two stages, an adsorption for the separation of the trace components and a membrane unit for the separation of CO_2. In the first stage of adsorption step H_2S is oxidised to sulphur and deposited on activated carbon. The loaded carbon has to be replaced after the breakthrough of sulphur and can be regenerated. Subsequent to dehydration of the gas, halogenated hydrocarbons are removed by activated carbon. Unlike the removal of H_2S, this stage is designed as a thermal-swing adsorption because of the low adsorption capacity of the carbon for halogenated hydrocarbons. This stage is regenerated by steam. The halogenated hydrocarbons leave the process in the form of an aqueous mixture that can be treated by conventional methods.

The produced gas is almost pure methane containing only minimal residues of toxic traces (< 2 mg/m^3_{STP} of H_2S and CFC); it can be used in local gas distribution systems or stored at high pressure as an engine fuel.

The membrane used (solution-diffusion type) are installed in form of 5 hollow fiber modules (by UBE Industries) with a total surface area of approximately 700 m^2.

The flexible plant design allows feed flow rates up to 200 m^3(STP)/h, feed pressures up to 35 bars, feed temperatures up to 90°C and different module arrangements like one- or two-stage cascades.

The permeabilities of the involved gases increase with temperature following the Arrhenius law, whereas the selectivity for the mixture decreases. It determines a saving in membrane area at high temperatures accompanied by losses in methane recovery.

Cost calculations indicate that the product losses are more important than savings in membrane area. Applied feed pressure is inversely proportional to the membrane area: for this reason the savings in membrane area has to be weighed against the higher energy consumption at elevated pressures. Calculations indicate that savings in membrane area are more influential on process economics than savings in power consumption. From this analysis, gas permeation seems favourable to other processes for dumpsites with a production of less than 1000 m^3(STP)/h landfill gas.

CO_2 enhanced oil recovery

CO_2 is pumped into dying oil reservoirs at high pressures to extend the life of an oil field.

The CO_2 then diffuses through the formation to drive the residual oil toward existing oil wells. Over time, the CO_2 also arrives at the well and begins to dilute the natural gas associated with the well.

The aim of the gas separation is to recover both the natural gas and the CO_2 (for re-injection). Because both the volume and composition of the gas change with time, it is difficult to economically design and to operate a conventional amine treatment plant for this application.

The CO_2 is re-injected at high pressures back into the reservoir: therefore CO_2 recovered as permeate must be recompressed. The choice of higher permeate pressure reduces membrane efficiency and an optimisation study is required to find the best operating conditions.

When carbon dioxide is injected into the field, it mixes with the oil and promotes underground flow. It must be removed at the surface and, as the flooding procedure is carried out, the CO_2 content of oil grows accordingly.

Cost comparisons for producing CO_2 for enhanced oil recovery as a function of membrane selectivity highlighted the importance of selectivity on system cost [23]. In this example an 80% CO_2 stream at 500 psig was processed to recover 95% pure CO_2 permeate at a fixed CO_2 recovery of 91%. As the selectivity of the membrane was raised, the permeate pressure could be increased while still meeting the objectives of the separation. Higher permeate pressure led to reduced recompression costs and hence lower overall capital cost. Estimated capital costs declined by 31% when the CO_2/CH_4 selectivity was raised from 15 to 45.

The above study also found that capital costs to treat acid gases with 20% CO_2 were 60% lower for membranes as compared to the Selexol absorption process. In the range of 10 to 20 % CO_2, membranes were not competitive with DEA treatment, although a combined membrane/DEA system had a 9% cost advantage over DEA alone.

A comparison of CO_2 removal by membranes alone versus removal by either a combined membrane/Benfield potassium carbonate process or a standard TEA treatment showed the sensitivity of the economics to the particular details of each separation. The desired objective was to reduce the CO_2 level of a 44% CO_2 feed to 5% CO_2, which at first glance seems to be an ideal bulk-cut separation for a process involving membranes. TEA was found to be the most economical process, with the all-membrane process being the most expensive.

In the membrane/Benfield scheme both capital and operating costs were lower than those for the all-membrane system, but evaluated cost was still 10% higher than those were for TEA. Had the CO_2 content of the feed been slightly higher, the hybrid system would likely have been the lowest cost process [24].

Another study [25] involved a 4,400,000 m^3/day 90% CO_2 EOR stream at 1.86 MPa pressure. Membranes were used for bulk removal of CO_2 prior to a "polishing" step with conventional DEA unit. The hydrocarbon gas product and CO_2 enriched streams have pressure requirements of 4.39 and 3.0 MPa respectively.

Table 2.8 - Comparison among DEA Amine, Cryogenic, TEA/DEA Amine, membrane/DEA processes to treat a feed stream 4,400,000 m^3/day 90% CO_2 at 1.86 MPa pressure [25].

	PROCESS			
	DEA Amine	Cryogenic	TEA/DEA Amine	Membrane/DEA
Capital Investment (10^6 \$)	103.6	73.5	65	47
Operating Expenses (10^6 \$/year)				
▪ Ex. Utilities	10.8	7.8	6.9	8.8
▪ Utilities	24.9	18.0	15.9	10.7
▪ Product losses	0.1	2.6	0.1	0.9
Capital charge (10^6 \$/year)	28	19.9	17.6	12.7
Total Cost (10^6 \$/year)	68.2	48.2	40.4	33.1

A similar analysis was conducted by Monsanto on $3 \cdot 10^6$ m^3/day 80% CO_2 stream. Feed gas, available at 0.17 MPa, is compressed to 1.72 MPa for amine and membrane systems and to 3.44 MPa for the hot potassium carbonate process.

Table 2.9 - Process comparison for CO_2-enhanced oil recovery [25].

	PROCESS		
	Hot Potassium Carbonate	Cryogenic/DEA	Membrane/DEA
Capital Investment (10^6 \$)			
➢ CO_2 Removal unit	21.1	24.2	16.1
➢ DEA Treating	-	4.0	4.9
➢ Compression	19.9	17.3	18.1
➢ Other	4.0	-	4.01
➢ Total	45.0	46.4	43.1
Operating Expenses (10^6 \$/year)	11.8	8.5	6.5
Capital charge (10^6 \$/year)	12.2	12.5	11.6
Total Cost (10^6 \$/year)	24.0	21.0	18.1

In both the previous analyses the hybrid process, in which takes part a membrane unit, allows the lowest capital investment and operating costs.

Newer membranes and membrane processes should improve many of the economic comparison presented here.

Extensive tests were performed using a Prism separator within Monsanto plants working on petrochemical streams over a wide range of CO_2 compositions as reported in the following table.

The pilot units and the ARCO commercial unit have demonstrated that the system is capable to operate efficiently in presence of various contaminants (after two years the unit kept the 100% design performance).

Table 2.10 - Base-case operating conditions [25].

Feed pressure, psig	200-1000
CO_2 content	10%-90%
H_2S and other contaminants	As available in feed gas source
Operating temperature, °F	Up to 160
Water vapour	Saturated or 30°F-25°F above Dew Point

It only uses temperature control and liquid knock-out as feed pre-treatment. This level of pre-treatment is typical for refinery streams. The operating temperature of the separators is controlled to keep the process gas above its dew point during the separation step and to maximise the mass transport coefficient of H_2 and CO_2. As hydrogen or carbon dioxide is recovered and removed from the gas stream, the partial pressure of the remaining hydrocarbons increases. The hydrocarbons can condense on the membrane forming a resistance for gas permeation. The oil layer formed can only be partially removed. This oil layer will increase the resistance for mass transfer; the optimal temperature is between 130°F and 170°F. The feed stream is processed in a compression and gas/liquid unit followed by a preheat step before the separation in a Prism separator. The performance of a Prism separator system for a CO_2 separation application depends on:

-feed gas composition, operating pressure and temperature;
-nonpermeate CO_2 content requirement;
-permeate CO_2 purity requirement;
-hydrocarbon recovery requirement
-CO_2 recovery requirement.

For a feed gas containing 80% CO_2, a CO_2 purity of 95% can be achieved in a single stage separator, recommended as a topping or bulk CO_2 removal process. At pressures above 500 psi, a single stage permeator is capable of removing most of the CO_2 present in hydrocarbon stream to produce a pipeline quality gas. A two-stage scheme, with the permeate from the first stage fed after recompression to a second stage and the non permeate from the second stage recycled to a first stage, reduces CO_2 in the hydrocarbon in non permeate stream increasing the hydrocarbons recovery and recovers CO_2 at >95% purity. The economic attraction of membrane technology in EOR applications is particularly dependent on the price of oil and value of energy at any point in time. At price above $25/barrel, CO_2 recovery schemes become attractive. The economics of the application are also particularly sensitive to the selectivity and environmental stability of the membrane [26].

Natural Gas Treatment

Natural gas is approximately 80-99% methane, with the remainder being mostly other hydrocarbons (ethane, propane, butane, etc.) as well as some

nitrogen, oxygen, water, CO_2 and sulphur.
Sweetening by removal of CO_2 and H_2S in crude natural gas is done to:

1. Increase the heating value of the natural gas;
2. Decrease the volume of gas transported in pipelines;
3. Reduce corrosion during the transport and distribution of natural gas;
4. Prevent atmospheric pollution by SO_2, which is generated during combustion of natural gas containing H_2S;
5. Prevent CO_2 emission to the atmosphere to reduce the green house effect.

Cook and Losin have reported costs for upgrading natural gas in a membrane plant. Moreover, figures were also reported, which compared actual costs of the installed membrane plant with five amine/glycol plants, see Figure 2.24 (Table 2.11 shows plant design specifications). This comparison was based on information provided by amine plant manufacturers [27].

Table 2.11 - Treating plant design basis [27]

Pressure, psig	925
Temperature, °F	95
Composition, %vol	
- CO_2	11.0
- CH_4	86.4
- C_2H_6	2.0
- N_2	0.6
H_2O vapour, lb/MMSCF	<4
Flow rate, MMSCFD	30
Turn down volume, MMSCFD	5
Sales gas requirements	
- CO_2 content vol%	<2.0
- H_2O vapour content, lb/MMSCF	<4.0

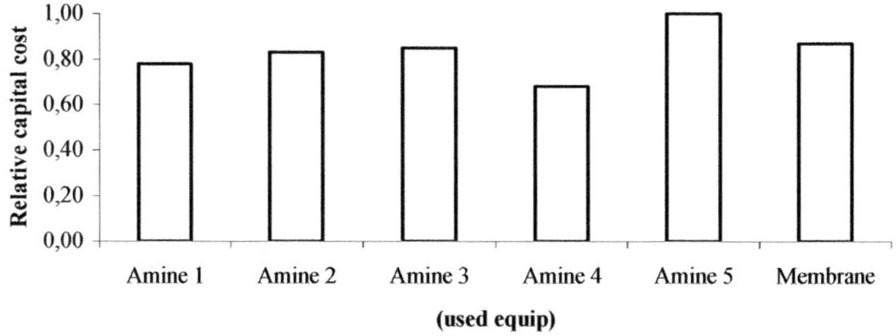

Figure 2.26 - Comparison of relative installed capital costs of membrane and amine systems [27]

Table 2.12 shows the estimated operating cost comparison for the membrane plant and an amine plant, and the actual operation costs based on 15 months of plant experience.

Table 2.12 - Estimated operating cost comparison (U.S.$) [27]

	Membrane plant (included compression)	Amine/glycol plant
Labour	50,000	135,000
Transportation/vehicles	15,300	25,000
Utilities	5,600	328,800
Fuel gas		
- Amine glycol reboiler		693,000
- Compressor/heaters	450,000	
Expendables		
Filter element/carbon/misc	35,000	65,000
Compressor maintenance	72,300	
Membrane elements	306,000	
Hydrocarbon loss	503,700	54,800
Amine/glycol		125,000
Total annual operation costs	1,437,900	1,426,600
Operating charge, $/MSCF	0.131	0.130
Actual operating charge after months of operation, $/MSCF	0.116	

The membrane plant was a two-stage membrane system, see Figure 2.27. The permeate of the first stage is recompressed and fed to the second membrane stage. In the second membrane stage, the CO_2 enriched gas is separated into a CO_2-rich permeate, which is vented to the atmosphere, and a hydrocarbon rich retentate which is recycled to the first membrane stage. The membrane plant included also pretreatment equipment, which provide key protection for the membrane elements.

The membrane plant (UOP's Separex) was installed in 1993 and extended in 1994 to 45 MMSCFD capacity. As seen in Table 3, the cost driving factors are those of membrane elements, compression and hydrocarbon loss. By using the permeate stream as fuel for the compression engine, costs could be reduced, but according to Cook and Losin it would require improved engines. After months of operation, the actual operation costs were found to be significantly lower than those projected in the planning phase, being only $0.116/MSCF. The main factors for the improvement were lower hydrocarbon loss in the permeate (average 2.06%), less membrane replacement and fuel usage than expected.

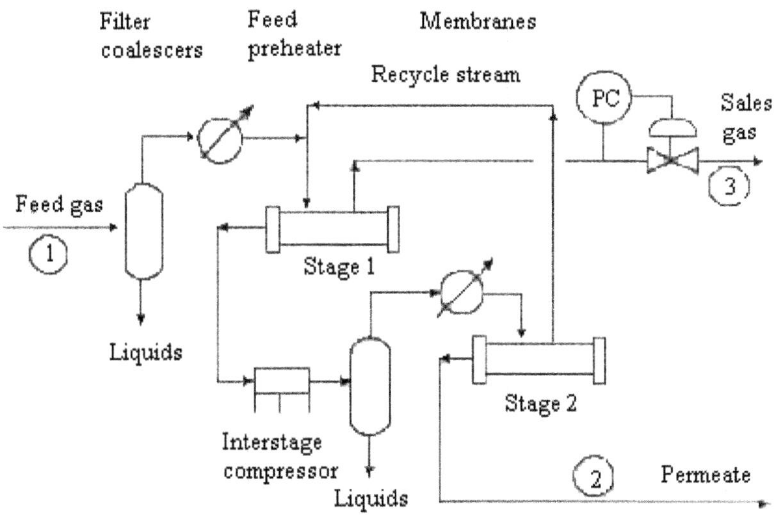

Stream number Stream name	1 Feed gas	2 Vent gas	3 Sales gas
Pressure, psig	925	5	910
Temperature, °F	86	94	95
Composition, mol%(dry)			
Carbon dioxide	11.0	81.1	1.9
Methane	86.3	18.7	95.2
Nitrogen	0.6	0.1	0.6
Ethane +	2.1	0.1	2.3
Flow, MMscfd	30.0	3.5	26.5

Figure 2.27 - Flow diagram of two-stage membrane unit [27].

Heinemann describes a membrane plant to reduce the CO_2 and H_2S contents from 6% and 10 ppm to pipeline specifications of 2.5 % and 3.2 ppm, respectively, in production of 15 TJ/d (dry gas flowrate from wells was equal to 31.5 MMSCFD) natural gas. Before building the membrane plant (at Beharra Springs, Northern Perth Basin, Australia), the costs were compared with an amine extraction plant, and it was found that both capital and operating costs were higher for the amine plant, see Table 2.13.

Table 2.13- Comparison of membrane separation and amine extraction cost [28].

	Membrane Separation	Amine extraction
Capital costs (relative)	1000	1543
Future Additional (including H_2S removal, additional flowline, refrigeration, field compression) relative capital costs	629	531
Operating costs (per year)	173	231
NPV Costs (15% discount)	2781	3638

Within a year from design activities started the skid mounted membrane plant was in operation. The economics of a single-stage plant with higher hydrocarbon loss and a two-stage plant with lower hydrocarbon loss, but high compression costs, were assessed. It was found that for the low production rate and the operation conditions (feed pressure 7.6 Mpa, T=40-50°C), a two-stage unit was not economically justified. The low pressure permeate from the single-stage process contained some 30% CO_2 and 100 ppm H_2S, while the retentate met the pipeline specifications. Figure 2.28 shows the flowsheet of the Beharra Springs gas processing facility.

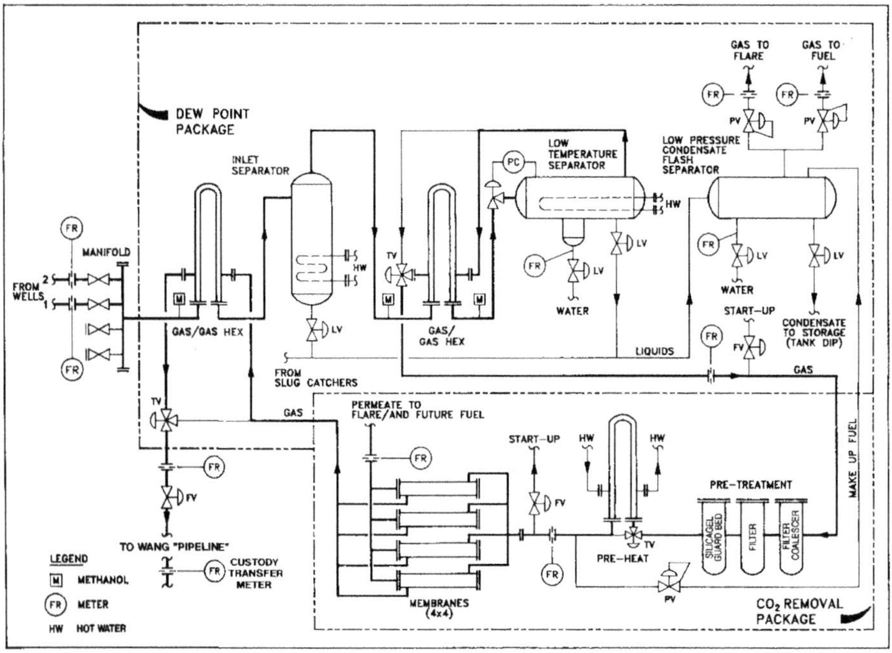

Figure 2.28 - Beharra Springs gas processing facility [28].

The world's largest stand alone membrane separation system which processes 180 MMSCFD gas at 1350 psig is in operation in Kadanwari in Pakistan. This plant permits simultaneous reduction of CO_2 and H_2S in natural gas stream from 12% to 3% and from 100 ppm to 3 ppm, respectively [10].

On the basis of this analysis membrane separation can be considered as an attractive technology in Natural Gas treatment in order to separate and to recover CO_2 for its further uses. Membranes are indicated for low to medium feed flow-rates (due to their modularity) with medium to high CO_2 feed concentration (driving force for the process) compared to amine systems.

Applications in which volume and composition of the gas change during the operation can be treated more easily by membranes than in conventional absorption plant. Another variable that influences the separation system performance is the cost of the feed; when it increases the region, where membrane processes are more convenient, decreases progressively.

A lot of membranes used for CO_2 removal have an inherent ability to dehydrate a gas stream, eliminating the need for a distinct dehydration unit as in amine systems. The presence of water in vapour phase does not constitute a problem for the membrane; however if liquid water is present the membrane performance could be damaged. Absorption methods are favourably applied for high feed flow rate (in which it is possible using large scale advantages) at low partial pressures of acid components and when are required highest purity levels. The increase of the feed cost tends to enlarge the operating area of this technology as consequence of reduced product loss. The combination of membrane and absorption units in a hybrid scheme can be the better solution in a more large range of operating conditions and typically expands the economical area of operability of membrane units. This combination takes the advantages of the two constituent units lowering the costs because each unit can work in its better conditions, but loses the characteristics of working simplicity of membrane unit alone. Moreover the presence of acid gases different of CO_2 at high concentration does not determine further problems in the separation. In fact, for natural gas containing H_2S (~1%) these systems have shown to offer cost savings compared to stand-alone membrane and amine units.

Cost evaluation apart, membrane units are preferred when needs of reduced weight and space or of localisation in remote areas become important.

2.4.6 Emerging technologies and future applications

Membrane contactors

A membrane contactor separates the CO_2-rich gas mixtures and the absorber liquids. When the liquid is not wetting the porous membrane surface the pores are left open for gaseous flow as shown schematically in Figure. 2.29.

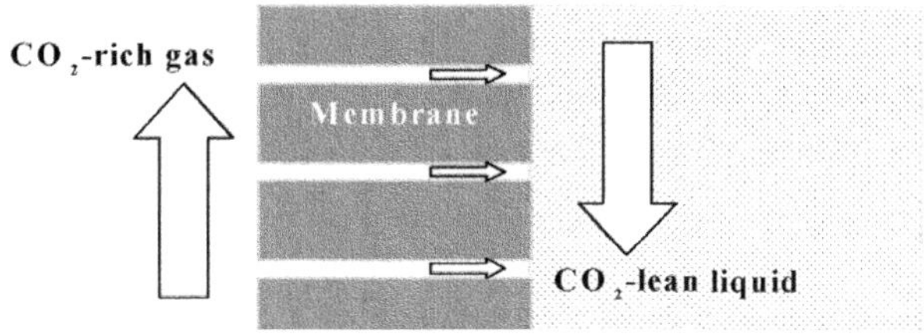

Figure. 2.29 -Schematic illustration of membrane contactor.

Membrane contactors can also be used in the reverse mode for CO_2 stripping. In this case water vapour providing heat can be used on the gas side for stripping off CO_2 from the liquid. The membrane contactors can be used to treat gas streams with very low CO_2 content, below 5-8% CO_2, a range where polymeric membranes are typically not used.

The Kværner membrane contactor concept is based on GORE-TEX, PTFE porous membranes, and amine solutions as absorbing media. The selectivity of the process depends on the absorber selectivity. The claimed advantages of the membrane contactors are [29]:

- High membrane packing density, 500-1500m^2/m^3
- Significant reduction in weight (~ 70%) and space (~ 65%)
- Reduced co-absorption of hydrocarbons
- No foaming, channelling or flooding
- High flexibility with respect to flow rates and solvent selection.
- Optimised boiler duty through optimised water/amine ratio and other chemicals
- The unit is insensitive to motion (important on ships and floating platforms)
- Flexibility with respect to orientation of the unit

A typical comparison between a membrane contactor unit and conventional column is given in Table 2.14.

Table 2.14 -Typical size of conventional absorber and membrane contactor (97200 tonnes CO_2/year)

Parameter	Column	Membrane Contactor
Size	O.D. = 4.5 m, Length = 22 m	4.0x4.5x5.0m
Weight (tonne)	70	24

Kværner has tested a large pilot unit at a gas terminal north of Aberdeen in Scotland. The membrane contactor is fed with a lean amine stream and sour natural gas from one of the plants existing gas treatments trains. The gas is processed in the membrane unit and returned to the plant. The main design parameters of this membrane unit are 88 barg, gas flow 5000 Nm^3/h and liquid flow 5 m^3/h. The CO_2 content is reduced from an average of approximately 6 % to 3.5% in the gas.

At Statoil gas terminal at Kårstø in Norway, another Kværner pilot unit is treating exhaust gas from a gas engine (520kW). The exhaust gas flow rate is 2610 kg/h and 85% (195 kg/h) of the CO_2 is separated from the exhaust gas. Exhaust treatment from gas turbines (CO_2 content from approximately 3 %

and above) is relevant for membrane contactors. Studies have been carried out to investigate the performance by a developed simulation programme and by real recovery testing at SINTEF. It has been verified that the simulation programme reproduces the performance of the contactor membrane units.

Applying membrane contactors for offshore and onshore gas turbine exhaust treatment are also considered. A combined cycle gas power plant of 57% efficiency would be reduced to approximately 50% efficiency with recovery of 85-90% of the CO_2 in the exhaust gas. This figure includes injection of CO_2 into the Norwegian subsea aquifer Utsira. The Kværner membrane contactor technology is currently under commercialisation.

Projected benefits of the membrane contactors, based on Kværner experience on different CO_2 sources, include reducing installed cost by 30-40% for offshore units, with similar operating cost reductions. Further cost and energy savings by removing constrains on the solvent, thereby allowing solvent optimisation [30].

TNO in The Netherlands has developed a novel membrane contactor which employs an aqueous absorption liquid, called CORAL, based on amino acids and alkaline salts. The claimed benefits of these liquids over the conventional amines are: i) the contactor can run stable with cheap membranes (for instance polyolefin membranes), ii) high oxygen stability and an improved corrosion resistance. iii) the vapour pressure is very low so there is insignificant loss of absorption liquids due to evaporation.

A pilot scale plant has been designed and built in at TNO, see Figure 2.30.

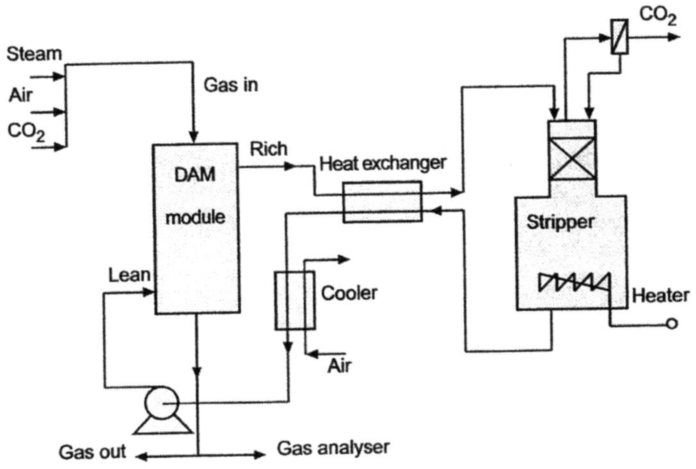

Figure 2.30 - Pilot plant flow sheet

The membrane contactor is equipped with a cross flow membrane module with 7.5 m^2 of polypropene hollow fibre membranes (pore size ~0.2 μm). The energy consumption of the process is mainly determined by the heat that is required for the regeneration of the absorption liquid.

Other studies on membrane contactors have been reported. Nii and. Takeuchi [31] studied the transport mechanism in hollow-fiber modules, with liquid stream into lumen and gas-liquid interfaces on the internal surface of the fiber. The authors compared the mass transport coefficients in membrane contactors and in packed beds with an advantage of the former in operations involving large flows, where flooding problems could be avoided. Nishikawa and coworkers [32] studied membrane fibre contactors in PTFE having a high degree of hydrophobicity with a stability more than 6600 hours. Using MEA flowing into the fibers and flue gas into the shell higher efficiency could be obtained compared to conventional systems.

Y.Lee and coworkers studied hollow fiber membrane contactors in which an aqueous potassium carbonate solution flows through the tube side and gas in the shell side in order to optimize the use of the whole length of the fiber. The permeation rate of CO_2 increaseed with gas pressure increases, due to the high CO_2 concentration at the liquid interface in equilibrium with gas. This effect was more marked in the case of pure water used as absorbent, which is a physical absorption. As the absorbent flow rate increased, the permeation rate of CO_2 increased, because the liquid boundary layer resistance decreases with the increase of the liquid velocity. Also, the transfer rate of CO_2 into the liquid increased as the concentration of K_2CO_3 increased, since the facilitated absorption ratio also increased. The saturation point moved towards to the direction of the outlet of the fiber as the liquid flow rate increased. However for relatively high flow rate such as 800 cm^3/min in fibers with radius size as large as 500 μm the liquid was not saturated at the module outlet [33].

Y.-I. Park and coworkers studied the separation of CO_2 by water-swollen hydrogel (WSH). The thickness of coating layer was between 0.8-1.2 mm, while that of the support was180-200 mm. The permeance of CO_2 was of the order of 10 GPU with selectivity CO_2/N_2=120 at room temperature. These values were verified in a small-scale separation equipment treating 0.05 Nm^3/min. The CO_2 permeation rate increased as the flow rate of the absorbent through hollow fiber membrane contactor increased, but for a gaseous mixture, this increase in CO_2 permeation rate is coupled to a decrease in the selectivity [34].

A comparison between membrane contactors over packing towers is summarised in Table 2.15.

Table 2.15 - Comparison between membrane contactors and packing towers [35].

	Liqui-Cel Contactors	Packed columns
Surface Area/Volume, ft^2/ft^3	> 1000	25-75
Space Efficiency	excellent	Fair
Capital Equipment Costs	good	Good
Installation Costs	good	Fair
Instrumentation Costs	good	Fair
Independent flow control for gas and liquid streams	excellent	poor
Vapor load per ft^2 contacting area for vacuum operation	low	high
Predictable scale-up	excellent	Fair
Low maintenance contactor design	good	good
Stripping gas/liquid feed ratio	low	high
Expandable for increased process capacity	excellent	poor
Expandable for increased efficiency	excellent	poor
Process turn-down capability	excellent	poor
No flooding and loading operating flow constraints	excellent	poor

Facilitated transport membranes

Significant research is on-going for improving the insufficient properties of CO_2 separation membranes for various application, particularly flue gas CO_2/N_2 separation. For instance, Chakma [36] has demonstrated separation factors as high as 244 and 140 in immobilized liquid DEA/PEG400 membranes for CO_2 and SO_2, respectively, with respect to nitrogen. The separation depends strongly on the partial pressure difference across the membrane. Lower pressure difference gave higher separation factors. Hiarayama et al. [34] have demonstrated for plasma polymerized composite membranes (skin layer of ethyleneoxide on support) a CO_2/N_2 separation of 70 with CO_2 permeability of 140 Barrer at 25°C, and CO_2/N_2 = 40 with permeability equal to 60 Barrer at 50°C.

Stable and highly selective immobilised liquid membranes (ILMs) have been developed for carbon dioxide separation by incorporating a variety of mobile non volatile carriers in glycerol immobilised in microporous hydrophilic flat films and hollow fibres. The performance of these liquid membranes was studied for separation of CO_2 from CO_2/N_2 and CO_2/O_2 mixtures. In presence of moisture in the feed gas, the non volatile carriers dissolved in glycerol (e.g., sodium carbonate, sodium glycinate etc.) drastically increase the flux of CO_2 with respect to N_2 or O_2 whose permeability is dramatically reduced by the highly viscous glycerol. The investigated carbonate concentration was in the range of 0-4.0 mol/dm^3. The feed gas RH range was 49-100%; the sweep gas was dry helium; CO_2 partial pressure range was 0.005-0.77 atm. Addition of Na_2CO_3 increased the CO_2 permeability drastically at lower carbonate concentrations; at higher Na_2CO_3

concentrations, this permeability increase is partly compromised by increased solution viscosity. N_2 permeability coefficient decreased with an increase in Na_2CO_3 concentration. At a feed RH = 50.7%, CO_2 partial pressure of 0.007 atm and Na_2CO_3 concentration of 1.0 mol/dm^3,the CO_2/N_2 separation factor observed was 3440. Prolonged runs lasting 14 days showed that the ILM permeation performances supported on poly(vinylidene fluoride) were quite stable [37]. When in the feed side the working conditions are the following: RH=60%, CO_2 partial pressure of 0.005 atm and glycine-Na concentration was 3.0 M, the CO_2/N_2 selectivity was over 5000 [38].

Both the carriers mentioned above require a solvent, either water or glycerol, to dissolve them. The use of novel liquid/oligomer carrier which do not require any additional solvents was explored for carbon dioxide separation. These carriers are non-volatile and have high densities of reactive functional groups. The use of such carriers eliminates the necessity of a solvent and the liquid membranes formed can be inherently stable. One such carrier in its pure form was found to provide very high selectivities (about 19,000) for carbon dioxide over nitrogen at low carbon dioxide partial pressure differences and high feed side relative humidities. These liquid membranes showed stable performance for more than one month [39].

Inorganic membranes in CO_2 mitigation

The three common process solutions discussed in CO_2 recovery from fossil fuel power plants, i) precombustion decarbonisation, ii) oxyfuel, iii) postcombustion CO_2 recovery, require different membrane solutions.

Inorganic membranes in precombustion decarbonisation (H_2, CO_2 separation) postcombustion CO_2 recovery (CO_2 separation)

In precombustion decarbonisation the gas separation may either be carried out at ambient or high temperature, the latter involving inorganic membranes. High temperature H_2 or CO_2 selective membranes operating in reforming and shift conversion would shift equilibrium or allow operation at lower temperatures in the equilibrium restricted conversions:

$$CH_4 + H_2O = CO + 3H_2 \qquad \text{reforming}$$

$$CO + H_2O = CO_2 + H_2 \qquad \text{water-gas shift}$$

In conventional steam reformers, high conversions of natural gas in the order of 85 - 90% or even higher, are obtained at reformer outlet temperatures around 850 -900 °C. The energy efficiency of steam reforming processes is relatively high, but the investments are substantial. Pure hydrogen can be produced at significantly lower temperatures by integrating in the reactor a membrane that selectively removes hydrogen during conversion. This is illustrated in Figure 2.31 for different membranes (F is permeance, α is separation factor) in production of 2.4 standard m^3 hydrogen per day at 600°C. A membrane area approximately equal to the typical heat exchanger area of 1000-2000 m^2 (2.4 mill. Sm^3H_2/day) is probably within an acceptable range for practical application. Potential savings in membrane reformer and downstream processing costs compared to conventional steam reforming must in many cases to be weighted against additional costs associated with recompression of the hydrogen permeate stream [40]. However, the concept facilitates CO_2 recovery from the high-pressure retentate CO_2 rich retentate stream.

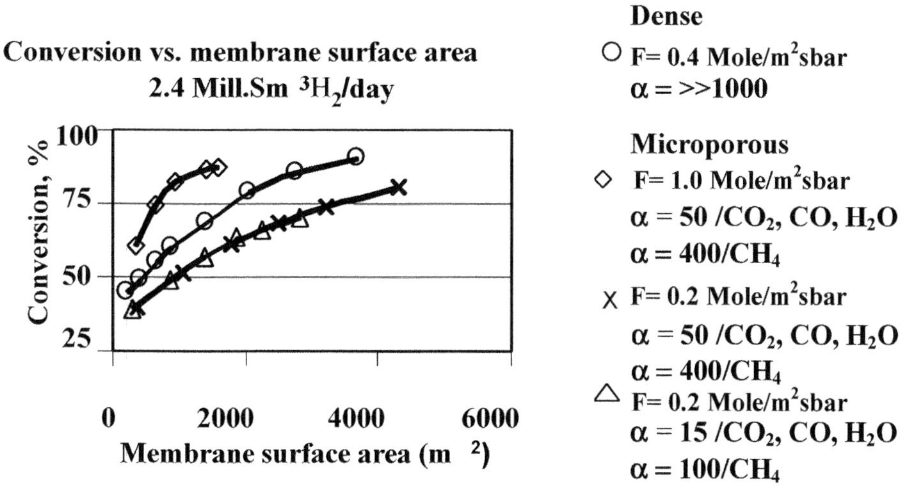

Figure 2.31 - Conversion of methane at 600 °C as a function of the membrane surface area for 4 different membranes with different hydrogen fluxes and separation factors.

The lower operation temperature results in lower natural gas consumption, and possible heat integration may further improve the economics and efficiency thereby giving further reductions in CO_2 and NO_X emissions. Further process integration is illustrated for a CO_2 selective membrane, Figure 2.32, which has the advantage that hydrogen is maintained on the high pressure side. For a hydrogen selective membrane hydrogen must be recompressed before combustion.

Another relevant example of a new power generation system with CO_2 recovery is based on a gas turbine combined cycle, named Hydrogen Decomposed Turbine Systems/HYDET. In this application the natural gas is reformed with simultaneously hydrogen removal using an high-temperature inorganic membrane. The residual gas, containing CH_4 and CO, is subject to combustion in an after burner using pure oxygen, with exhaust flue gas high concentrated in CO_2 directly for the recovery [42].

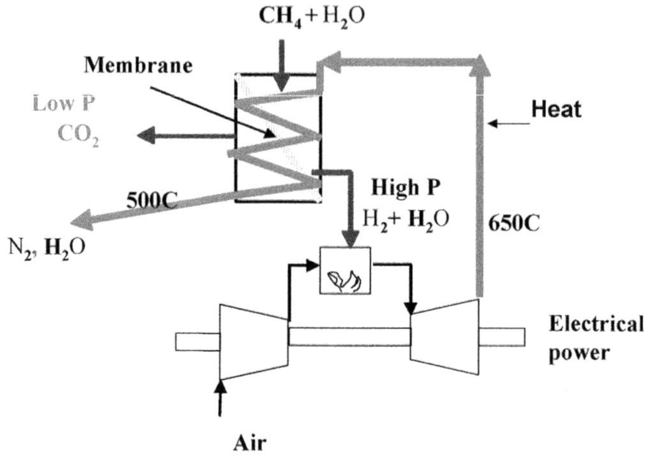

Figure 2.32- Integrated CO_2 membrane reactor and gas turbine. Low emission power generation [41]

The development of inorganic H_2 selective membranes has been ongoing for about four decades, while only the last 10-15 years CO_2 selective membranes has been given attention for high temperature applications. Small scale methanol reformers have been commercialised based on H_2 selective membranes. For large scale and higher temperatures significant cost reduction and improvements in performance are necessary before such processes could become a reality. Currently, the most focused H_2 selective membranes are metal based (mainly Pd systems), microporous (silica, silicon carbide and zeolites) and mixed conductors (proton+electron conducting perovskite, ...).

Inorganic membranes for CO_2 separation at high temperatures is an interesting option in end-of-pipe as well as for upstream decarbonisation process solutions. Since CO_2 is a heavier molecule than H_2, N_2 and CH_4, porous CO_2 selective membranes can not operate by Knudsen diffusion mechanism. Only if the membrane affinity for adsorption is higher towards CO_2 compared to these gases may the selectivity be switched. The polar nature of the CO_2 molecule promotes selective surface adsorption on many inorganic surfaces. In combination with affinity, the pore type has also great importance for the selectivity and permeance as demonstrated by Morooka et al. [43]. In Y-type zeolites, having high affinity for CO_2, the CO_2 molecules migrating along the pore walls are able to pass the few N_2 molecules migrating along the pore centre. This mechanism results in a high selectivity

(selectivity ≈ 80 at 30°C). Even at temperatures around 140°C, the selectivity was > 10, with a permeance of CO_2 of approx. 1.4 10^{-7} mol/m^2sPa. This is higher than for carbon membranes, having low CO_2 affinity but pores where CO_2 may pass the N_2 molecules, and silica membranes, having high affinity, but a pore shape giving limited possibility of CO_2 molecules passing N_2 molecules inside the pores. Higher selectivity may be obtained with molten carbonate liquid membranes suppported in a porous structure. These and other solutions need significantly further development in order to economically and technologically interesting.

As will be further described below, high temperature oxygen transport membranes may be integrated in power cycles. One possibility in precombustion decarbonisation processes is to employ an oxygen transport membrane to provide oxygen for partial oxidation of natural gas to produce syngas.

Inorganic membranes in oxyfuel solutions

In oxy-fuel and syngas process solutions, intergrated high temperature oxygen selective membranes may offer significant possibilities. Mixed conductors with high electronic and oxygen ion conductivity can be used for large scale oxygen production. In such membranes both oxygen ions and electronic defects are transported in the membrane material. Compared to the well known solid oxide fuel cell (SOFC) operation, mixed conduction simplifies considerably the configuration since electrodes and interconnect materials can be avoided (Figure 2.33). Oxygen permeation fluxes in the range of 10 cm^3/cm^2 min have been obtained in perovskites and perovskite related systems, e.g. (La.Sr)(Co,Fe)O_{3-d} and $SrFeCo_{0.5}O_x$, which is sufficiently high for cost reductions in syngas production [40]. Furthermore, methane to syngas conversion in lab-scale testing has shown one year stable performance with an average production rate of syngas of 50 cm^3/cm^2 min at 90% selectivity and 70 % conversion [44].

Integrated oxygen membranes in gas turbine electricity production is another promising application of mixed conductor membranes. Such a process is schematically shown in Figure 2.34 [45].

A recent patent developed the same concept using a mixture of O_2 and CO_2 as oxidant instead of air in stationary combustion processes [46]. The result is a CO_2 enriched flue gas stream that contains no nitrogen and requires much smaller process equipment to capture, concentrate, and then sequester. The flue composition is primarily CO_2 and water. A portion of the flue gas may also be recycled following hot gas clean-up operation for NOx, and SO_2, and particulate removal. In addition, O_2 -CO_2 combustion mode reduces furnace volume and therefore capital for new plants, reduces load to the flue gas desulfurisation and NOx reduction units, requires lower power for auxiliary load draws for primary and secondary combustion fans, lowers duty to the heat recovery and steam generator units. Oxygen derived by conventional cryogenic technology is not a viable option for oxygen-enriched combustion of coal-fired boilers.

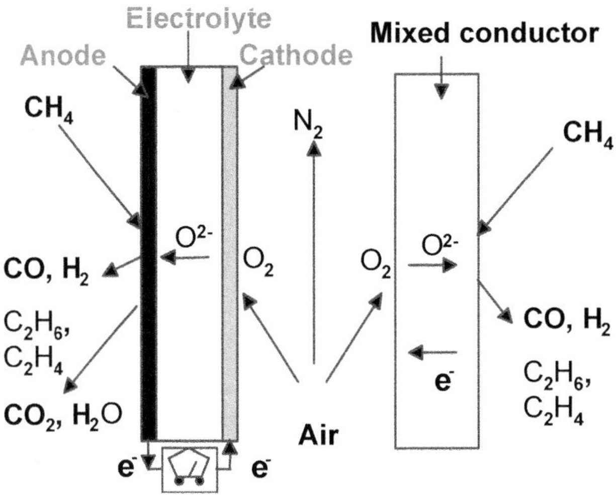

Figure 2.33- Left: Example of oxygen production with an oxygen ion conducting electrolyte membrane (SOFC configuration); right: With a mixed conducting membrane [46]

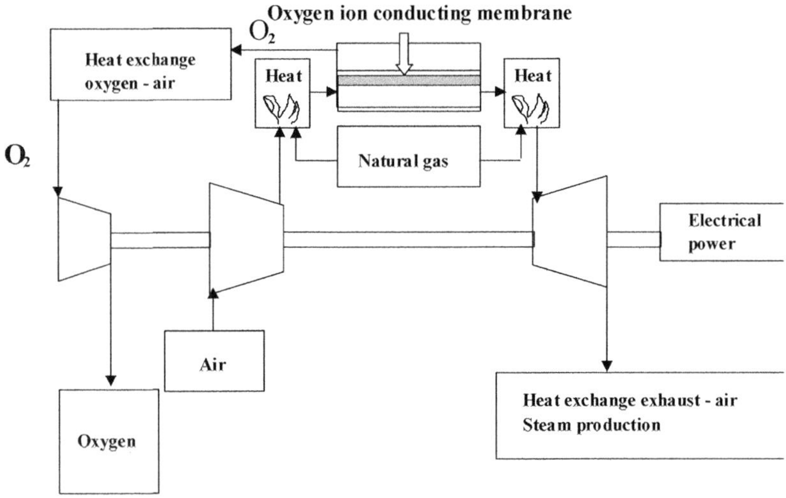

Figure 2.34- Integrated oxygen and power production [45]

This is because cryogenic technology requires extremely low temperature (below –290°F) where air becomes a liquid and oxygen is separated by distillation. A conventional power plant heats steam to over 1000°F for driving a turbine. With the availability of high temperature heat exchanger material, future power plants would be operated at much higher temperatures. This level of temperature spread, cryogenic on one side and very high temperature required for coal combustion on the other side, is not acceptable to electricity producing industries. A technology that would use the power plant's heat sources to thermally energise an air separation process and would produce low-cost oxygen, would enhance the acceptance of combustion modification by the electric utility industry sector, especially to comply with voluntary or involuntary carbon emission regulations. Availability of low-cost oxygen would also make oxygen-blown Integrated Gasification Combined Cycle an economic power producing choice.

With reference to a pulverised coal having the following composition:

Table 2.16 -Pulverised coal composition [46]

Component	% Mass
Carbon	61.5
Hydrogen	4.2
Oxygen	6.0
Ash	22.0
Sulphur	5.1
Nitrogen	1.2

Several options linked to different oxidation mixture compositions were analysed (see *Table 2.17*).

Table 2.17- Comparison of the feed composition effect on flue gas amount [46].

Case	Oxidation mixture composition, % vol.	Theoretical adiabatic flame temperature (T[a]), °F	Flue gas amount, lb/100lb ash-free coal
1.(air feed)	21% O_2, 79% N_2	3764	930
2.	30% O_2, 70% CO_2	3752	913
3.	40% O_2, 30% N_2, 30% CO_2	5481	615
4.	50% O_2, 50% CO_2	6414	554
5.50% O_2 + 50% flue gas from case 4	50% O_2, 40% CO_2, 10% H_2O	5303	608
6.40% O_2 + 60% flue gas from case 4	40% O_2, 48% CO_2, 12% H_2O	4463	724

In the first case, the conventional combustion with air, about three quarters of the flue gas are constituted by nitrogen and the stream separation cost results significant. In the second assumption analysed in this patent the flue gas has the same T^a without the additional cost of nitrogen separation: a easy CO_2 separation from water vapour must be performed. Another advantage is represented by no production of NO_x. In terms of flue gas flow rate normalised on ash-free coal no difference with respect the base case were obtained.

The significative increase of T^a means that less coal is required for a given power output. However in case 3 the flue gas contains N_2 that has to be separated out before recycling. The total quantity of flue gas is one third lower than the previous cases.

Furnace size reduction and consequent lower capital cost for new plant connected to a lower load to the flue gas desulfuration and NOx reduction units are the main results of the case 4. However the T^a is too high for the heat exchange materials used in conventional boilers. So a part of the flue gas must be recycled in order to obtain a reasonable T^a.

An oxidation mixture containing 50% O_2 and 50% flue gas recycled from the 50:50 O_2-CO_2 run (about 80% CO_2, the remainder H_2O) produces a higher amount of flue gas but with a T^a lower by over 1000°F compared to the 50:50 O_2-CO_2 assumption. The amount of flue gas is the same of that produced in the case 2.

The further case is characterised by a T^a still lower with respect to the previous case associated to a total quantity of flue gas increased of one third with respect to the case 4 in which a equimolar concentration of CO_2 and O_2 is used [46].

Fuel cells

Fuel cells based on polymeric electrolyte membranes transporting protons (Proton Exchange Membrane (PEM) Fuel Cell) are receiving much attention, particularly in the automotive industry. Hydrogen fuel gas is converted to protons and electrons on the anode side, and the electrons are transported in an external circuit to the cathode side where oxygen from air is reacting with the protons to water. The electrodes are covered with Pt catalysts to speed up the chemical conversions, and may easily become poisioned by CO on the anode side. High purity H_2 is therefore needed as fuel, which complicates the use of carbon containing chemicals for hydrogen storage. The PEM fuel cells is expected to replace traditional engines in over a period of 20 years. Technology for production and storage of hydrogen will be essential for utilising the possibilities of PEM technology. Today two main options may be foreseen, i) the fuel is pure hydrogen produced outside and stored in the vehicle, or ii) hydrogen is produced from chemical carriers inside the vehicle. In the latter case CO_2 emission may occur if the carrier contains carbon. Possible chemical carriers frequently mentioned are methanol, methane, LPG. Stationary large-scale production with distribution and storage of pure hydrogen in the vehicle would give possibilities of deep reductions in CO_2 emissions.

SOFC technology operating at ≈ 900-1000°C is much more robust with respect to the fuel compared to PEM, being able to run on different fuels as H_2, CO and CH_4. Much research is focused on lowering the operation temperature, either by generating thinner electrolytes or developing new types. Oxides with high proton conductivity may alternatively be used as electrolytes, however, these membrane materials (perovskites of cerates and zirconates) are still showing too low fluxes for commercial interest. Oxygen production with electrolyte membranes can be carried out in an opposite SOFC configuration (Figure 2.4.6.2.3). Instead of feeding the membrane sides by gasses with a large difference in oxygen partial pressure (e.g. air vs. H_2, or air vs CH_4), an electrical potential can be applied across the membrane to pump oxygen from one side to the other. This production method may give pure oxygen at high pressure from air at ambient pressure. Such devises are already commercialised for small-scale production.

Ideally in a SOFC operation H_2O and CO_2 are the products. However, it is not optimal to run the process to completion, thus the exhaust will contain a significant amount of burnable gases (H_2, CO, CH_4). To avoid dilution with N_2, the post burner process should be carried out in pure O_2. This is possible by combining at high temperature SOFC and MIEC (which produce the O_2) integrated with a (micro)turbine.

Membrane reactors

In the membrane field an innovative principle is represented by the combination of the separation and the reaction steps in a single device: a membrane reactor. In these units higher conversions and yields can be achieved with respect traditional fixed bed reactors avoiding side-reactions. Two main different configurations can be distinguish:

1. Inert membrane reactor, in which the membrane allows only the selective transport of specific species without having a direct role in the reaction phase. In this case the catalyst is typically contained into the lumen side of the membrane or entrapped inside the membrane depending on the application.
2. Catalytic membrane reactor in which the membrane is capable at the same time to act as catalyst and permit the selective passage of a reagent/product.

The development of inorganic materials allowed a rapid extension of the application field of the membrane reactors.

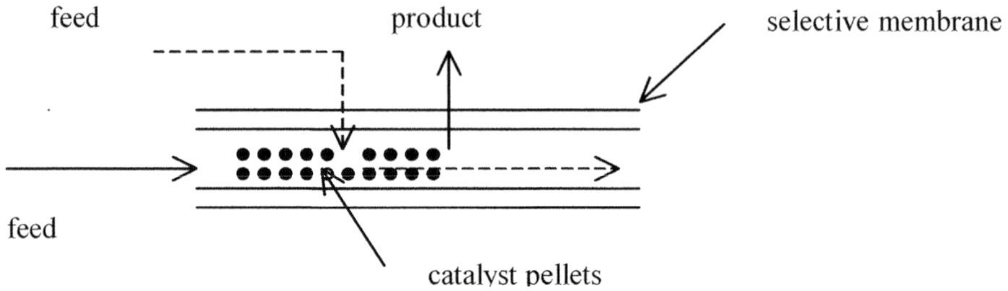

Figure 2.35- Scheme of inert membrane reactor with preferential removal of a product (solid line) or supplying of a reagent (dotted line).

The main advantages of catalytic membrane reactors are the following:

- Removal of at least one reaction product from the reaction environment by means of selective permeation through the membrane. This fact determines a shift toward the products for a reaction limited by thermodynamic equilibrium with consequent increase of its conversion according to Le Chatelier's law, the possibility to avoid undesired reactions of the products or their decomposition with catalyst poisoning.
- Controlled supply of one reactant to the reaction zone favoured by selective permeation through the membrane. In this case it is possible,, for example to supply a reactant in a particular reactive form apt to interact more actively on catalytic surface. Moreover this practice can be used to improve the reactant distribution on catalyst keeping low concentration level in order to avoid competitive reactions or deactivation phenomena for the catalyst.
- Two different reactions in the same unit. The membrane is selectively permeated by one product of the first reaction becoming a reagent for the second reaction, before its conversion to an undesirable specie. Also the heat can be transfer across the membrane e.g. from an exothermic to a endothermic reaction.

In the following some applications of major interest for the aim of this study are reported [47].

The conversion of the CO_2 thermal decomposition to give CO and oxygen, low at low temperatures in a traditional reactor, can be enhanced if oxygen is selectively removed from the reaction ambient. Moreover using a membrane reactor it is possible to obtain two streams respectively enriched in oxygen and in carbon monoxide.

A lot of applications of membrane reactors regarding equilibrium reactions in which hydrogen is also involved, because Pd and Pd-based alloys resulted extremely selective for this component. For example in syngas production all the principal alternative reactions (steam and dry reforming of the methane, catalytic partial oxidation of the methane) can be carried out in a membrane reactor with significant improving of the conversion levels. In the steam reforming of methane using palladium-membrane reactors it is possible to achieve conversion up to 100% at temperature as low as 500°C. The purity of recovered hydrogen is appropriate for its use in fuel cells.

Table 2.18 - Main reactions carried out in membrane reactor systems [47]

Reaction	Membrane type	Membrane material
Decomposition of CO_2	Dense	Zeolite Y stab. ZrO_2
Decomposition of H_2S	Porous	γ-Al_2O_3
	Dense	Pt
Decomposition of ammonia	Porous	ceramic
Decomposition of water	Dense	Pd
Dehydrogenation of ethane to ethylene	Porous	Al_2O_3
	Dense	Pd/Ag
Dehydrogenation of propane to propylene	Porous	Al_2O_3
Water-gas shift reaction	Dense	Pd
Dehydrogenation of methanol	Porous	γ-Al_2O_3
Dehydrogenation of n-butane	Porous	γ-Al_2O_3
Dehydration of isopropanol to propylene	Porous	Al_2O_3
Hydrogenation of butadiene	Dense	Pd
Hydrogenation of acetylene	Dense	Pd/Ag
Hydrogenation of nitrobenzene	Porous	γ-Al_2O_3
Hydrogenolysis of ethane to methane	Porous	Al_2O_3
Oxidative dehydrogenation of methanol	porous	Al_2O_3
Oxidative dehydrogenation of propane	porous	zeolite
Oxidative methane coupling	dense	Zr-Y-Ti oxides/perovskite-type oxides
	porous	Ceramic (with catalyst deposited on)
Oxidation of CO to CO_2	porous	Al_2O_3
Oxidation of ammonia to nitrogen	dense	Ag
Oxidation of methane to methanol	porous	ZrO-CaO- Al_2O_3
Fisher-Tropsch Synthesis	porous	carbon
Methane steam reforming	dense	Pd-alloy
Dry reforming of methane	dense	Pd-alloy
Partial oxidation of methane	dense	Perovskite, Pd-alloy
Methane conversion into higher hydrocarbons	dense	Pd-alloy

For dry reforming of methane in a reactor equipped with a thin palladium film supported on ceramic substrate, methane conversion of 99.1% can be obtained versus 26.3% achieved in a traditional reactor. Methane conversion

higher than 95% was obtained at 650°C for the same reaction using as membrane dense silica film supported on porous Vycor tubes.
Also in the partial oxidation of methane to syngas membrane reactors showed higher methane conversion with respect to traditional reactors (methane conversion of 90% was obtained at 500°C using a commercial tubular Pd-Ag/ceramic membrane produced by Johnson-Matthey with a specific feed gas composition).
In the table 2.18 are reported some reactions of industrial interest carried out in membrane reactors.

REFERENCES

1. Smith, I.M., Nilsson, C., Adams, D.M.H. "Greenhouse gases-perspectives on coal" IEAPER/12, IEA Coal Research, London, UK, 41 pp.
2. Yaws, C.L., Hopper, J.R., Wang, X., Rathinsamy, A.K., & Pike, R.W. "Calculating solubility & Henry's law constants for gases in water", Chemical Engineering, June 1999.
3. Kohl, A., Reisenfeld, F., "Gas purification", 4th edition, Gulf Publishing Company, Houston, TX, USA, 1985.
4. Ouidhuis, A.B.J. "Inventarisation of techniques for CO_2 removal from fuel gas or reformergas", Report No. ECN-C-92-043, 1992, from "Integrated research progrmme on carbon dioxideremoval and storage"
5. Allam, R.J., Spilsbury, C.G. "A study of the extraction of CO_2 from the flue gas of a 500MW pulverised coal fired boiler", Energy Conversion Mgmt. 33, 5-8, 373-378, 1992.
6. Brunauer, S., Deming, L.S., Deming, W. E., Teller, E. "A theory of the van der Waals adsorption of gases" J. Am. Chem. Soc. , 62, 1723-32, 1940.
7. Ishibashi, M., Otake, K., Kanamori S. Tokyo Electric Power Co. and A.Yasutake - Mitsubishi Heavy Industries Ltd - Proceedings of 4th International Conference on Greenhouse Gas Control Technologies.
8. Motohira, N., Shibuya, N., Kamiya, N., Ota, K., Kasai, H.and Hosaka, M.. "Electrochemical condensation of CO_2 by use of molten carbonate" Proc. 5th Int. Conf. Inorg. Memb., Nagoya, Japan. 108-111, June 22-26, 1998.
9. Bredesen R., "Key Points in the Development of Catalytic Membrane Reactors", Paper no. A7.0 in Proc. 13th Int. Congr. Chem Process Eng., Praha, Czech Republic, August 23-28 1998.
10. Puri, P.S. "Membrane gas separations: an opportunity for gas industry or just a niche market", Preprints of International conference on membrane science and technology (ICMST'98), Beijing, China, June 9-13 1998.
11. Wijmans H., "Emerging membrane gas and vapor separation processes", Proceeding of Engineering with membranes, I 14-19, Granada, Spain, June 3-6 2001.
12. Baker R., "Future directions of membrane gas separation technology", Membrane Technology, 138, 2000, 5-10, 2000.
13. Stern, S.A. "Polymers for gas separations: the next decade" J. Membrane Sci., 94, 1-65, 1994.

14. Park, J.Y., Paul, D.R. "Correlation and prediction of gas permeability in glassy polymer membrane materials via a modified free volume based group contribution method" J. Membrane Sci., 125, 23-39, 1997.
15. Golemme, G., Drioli, E., Lufrano, F., "Gas transport properties oh high performance polymers" Polymer Sci., 36 (11), 1647, 1994.
16. Herzog et al., Environmental Progr., 10(1), 64-74, 1991.
17. Rautenbach, R., Struck, A. "Assessment of different membrane gas separation processes for separations with low driving force", ICOM 1996, Yokohama, Japan.
18. Van Der Sluijs, J.P., Hendriks, C.A., Blok, K. "Feasibility of polymer membranes for carbon dioxide from flue gas", Energy Convers. Mgmt., 33(5-8), 429-436, 1992.
19. Karashima S. et al., Greehouse Gas Control Technol., Proc. Int. Conf. 4th, 1035-1037, 1999.
20. Hahm, M., Sohn, W.-I., Lee, Y., Kim, I., Kim, J.-H., Lee, S.-B. "Preparation of asymmetric polyethersulfone membrane and its gas separation performance" Memburein, 10(3), 130-138, 2000.
21. Hu, Q., Marand, E., Dhingra, S., Fritsch, D., Wen, J., Wilkes, G. "Poly(amide-imide)/TiO_2 nano-composite gas separation membranes: Fabrication and characterisation", J. Memb. Sci., 135, 65-79, 1997.
22. Rautenbach, R., Welsch, K. "Treatment of landfill gas by gas permeation-pilot plant results and comparison to alternatives", J. Membrane Sci., 87, 107-118, 1994.
23. Schell, W.J., Houston, C.D., Hopper W.L., "Membranes can efficiently separate carbon dioxide from mixtures" Oil Gas J., 81(33), 52-56, 1983.
24. Zolandz, R.R., Fleming, G.K. "Applications Membrane Handbook", W.S.W. Ho and K.K. Sirkar (Eds.), Van Nostrand Reinhold, New York, 78-94, 1992.
25. Spillman, R.W. "Economics of gas separation membranes", Chem. Eng. Progr., 51-62, 1989.
26. Henis, J.M.S. "Commercial and Practical Aspects of Gas Separation Membranes" in: D.R. Paul and Y.P. Yampol'skii (Eds.) CRC Press, Inc. Boca Raton Florida, 1994.
27. Cook, P.J., Losin, M.S. "Membranes provide cost-effective natural gas processing", Hydrocarbon Processing, April, 1995.
28. Heinemann, S.B., APEA Journal, 352-362, 1993.
29. Kværner Gas & Oil a.s. Information
30. Herzog, H., Falk-Pedersen, O. Fifth Int. Conf. on Greenhouse Gas Control Technologies, August 13-16, Cairns, Australia, 2000.
31. Nii, S., Takeuchi, H. "Removal of CO_2 and/or SO_2 from gas streams by a membrane absorption method" Gas Sep. Purif., 8 (2), 107-114, 1994.
32. Nishikawa, N. et al., "CO_2 removal by hollow-fiber gas-liquid contactor", Energy Convers. Mgmt., 36 (6-9), 415-418, 1995.
33. Lee, Y. et al., "Control of CO_2 removal rate by hollow fiber membrane contactor", in: Proc. of 2nd Korea-Italy workshop on "Membranes and membrane processes for the clean environment and clean energy", July 21-23, 1999 Muju Resort, Korea.
34. Park, Y.-I. et al. "Recovery of CO_2 by membrane separation", in: Proc. of 2nd Korea-Italy workshop on "Membranes and membrane processes for the clean environment and clean energy", July 21-23, 1999 Muju Resort, Korea.
35. Hoechst-Celanese brochure.
36. Chakma, A. "Separation of CO_2 and SO_2 from flue gas streams by liquid membranes", Energy Convers. Mgmt., 36 (6-9), 405-410, 1995.

37. Chen, H., Kovvali, A.S., Majurndar, S., Sirkar, K.K. "Selective CO_2 separation from CO_2-N_2 mixtures by immobilized carbonate-glycerol membranes" Ind. Eng. Chem. Res., 38(9), 3489-3498, 1999.
38. Chen, H., Obuskovic, G., Majurndar, S., Sirkar, K.K. "Immobilized glycerol-based liquid membranes in hollow fibers for selective CO_2 separation from CO_2-N_2 mixtures" J. Membrane Sci., 183(1), 75-88, 2001.
39. Kovvali, A.S., Chen, H., Obuskovic, G., Majumdar, S., Sirkar, K.K. "Immobilized liquid membranes for CO_2 separation" Preprints of Symposia - American Chemical Society, Division of Fuel Chemistry, 45(4), 665-667, 2000.
40. Bredesen, R., Sogge, J. "A technical and economic assessment of membrane reactors for hydrogen and syngas production", Proc. of the Seminar on the Ecological Applications of Innovative Membrane Technology in the Chemical Industry, Cetraro (Italy), 1-4 May 1996.
41. Bredesen, R. 3rd year Report Workpackage II, RUCADI Network, 2000.
42. Moritsuka, H., Greenhouse gas control technologies, 1999, 563-568.
43. Morooka, S., Kusakabe, K. in proc. of the SINTEF seminar on Novel Inorganic Membranes and Membrane Processes. February 4, 1999, Oslo, Norway
44. Bose, A.C., Stiegel, J.G., Sammells, A.F. Proc. 5th Int. Conf. Inorg. Memb., June 22-26, 1998, Nagoya, Japan, p. 6-9.
45. Stiegel, G.J. "Mixed conductiong ceramic membranes: a new paradigm for gas separation and reaction" Annual Membranes Technologies/Separation Planning Conf., USA, December 1998.
46. Bredesen, R., Mertin, F., Norby, T. "Measurements of surface exchange kinetics and chemical diffusion in dense oxygen selective membranes". Catal. Today 56, 315-324, 2000.
47. U.S. 6,173,663.
48. Drioli, E., Basile, A., Criscuoli, A. "High temperature membrane reactors for clean productions", Clean Products and Processes, 2, 179-186, 2000.

SECTION III

SUPERCRITICAL CARBON DIOXIDE

This Section summarizes the activity performed within WP2 of the BRRT-CT98-5089 RUCADI Project, "Recovery and Utilization of Carbon Dioxide" by the following Partners:

Alessandro Galia, *University of Palermo*, Italy
Giuseppe Filardo, *University of Palermo*, Italy
Andrew B. Holmes, *University of Cambridge*, UK
Tessa Early, *University of Cambridge*, UK
Louise Stamp, *University of Cambridge*, UK
Eugenio Quaranta, *Nowatech Ass.*, Italy
Herman Weyten, *VITO*, Belgium
Martyn Poliakoff, *University of Nottingham*, UK
Kulathu Seshan, *University of Twente*, The Netherlands
Johannes Penninger, *Sparqle International B.V.*, The Netherlands
Javid Hamid, *DERA*, UK
Olaf Walter, *FZK*, Germany
John T. McMullan, *University of Ulster*, UK

Chapter 3

PARTICLE FORMATION USING SUPERCRITICAL CARBON DIOXIDE

Luc Van Ginneken, Herman Weyten
Flemish Institute for Technological Research, Boeretang 200, B-2400 Mol, Belgium
Herman.Weyten@vito.be

1 PARTICLE FORMING PROCESSES

1.1 Conventional methods

Many of the products that are sold by the process industry – as bulk products, intermediates, fine chemicals, biochemicals, and food additives – as well as by the pharmaceutical industry are solids. The particle size and size distribution of these solids is frequently not desired for subsequent chemical reaction or use of these materials [1]. The particle size of these solids, therefore, has to be reduced. Conventional techniques for particle size redistribution are either mechanical (crushing, grinding and milling) or equilibrium controlled (crystallisation from solution) [1,2].

Many solids (such as certain chemicals, organics and pharmaceuticals) are difficult to process mechanically because of the compound's sensitivity to physical impaction, thermal degradation (as a consequence of the heat generated in the process), and chemical degradation (as a result of exposure to an oxidising atmosphere) [2,3]. In addition, conventional mechanical techniques may have a negative effect on the crystallinity of pharmaceutical compounds.

Conventional liquid-phase crystallisation, on the other hand, involves formation of a solid product from a liquid solution. In order for liquid-phase crystallisation to take place, the solution must be supersaturated; that is, the concentration of the solute in solution must exceed its solubility limit.

M. Aresta (ed.), Carbon Dioxide Recovery and Utilization, 123-136.

Conventional crystallisation methods that are applied in industry can be categorised according to the way supersaturation is generated. Cooling of the solution or evaporation of the solvent can be applied to create supersaturation [4]. What method is preferred depends on how the solubility varies with temperature. An alternative is to add an anti-solvent which is good miscible with the primary solvent to make a solvent mixture in which the solute has a reduced solubility, or to add a chemical reactant which produces a less-soluble product [4]. This causes a rapid crystallisation better known as precipitation. Several drawbacks, however, are associated with conventional crystallisation techniques, such as contamination with liquid remains, complex and costly post-processing treatments (for the complete elimination of liquid residues), considerable waste streams (due to the use of large volumes of liquid solvents), and high-energy requirements [3,5]. In addition, it is not always possible to obtain the desired particle size and particle size distribution (which are often critical quality attributes of pharmaceutical solids) with conventional crystallisation processes [2,5].

1.2 Supercritical crystallisation

A significant development in recent years has been the introduction of supercritical fluids (SCFs) in crystallisation processes. Although supercritical crystallisation is still a relatively new technology, it offers interesting possibilities of widening the range of products for which crystallisation could be used. Part of the appeal comes from the fact that supercritical crystallisation with low-critical-temperature fluids allows processing at lower temperatures than conventional particle reduction methods (such as milling and crystallisation), thus avoiding unwanted degradation of thermally unstable compounds. [1,2] This makes the process very favourable for precipitating pharmaceutical products. [6] In addition, (volatile) SCFs are easily separated from the solute after precipitation, leaving no traces in the product, contrary to liquid solvents that are used in conventional crystallisation techniques. [2,5] Furthermore, since the interesting properties and characteristics of SCFs can achieve very high supersaturation values and guarantee uniform conditions within the solution, the supercritical process offers the potential for producing uniform small particles with a narrow particle size distribution. [5,7,8] Finally, the supercritical crystallisation process produces reduced waste streams, as opposed to conventional liquid-phase crystallisation, since large volumes of liquid (anti-)solvents are not used and the volume of gas generated after pressure let down can easily be recovered. [2]

Carbon dioxide (CO_2) is the most commonly used SCF in supercritical crystallisation processes, although other SCFs have been considered and used. The use of CO_2 as precipitation medium has many advantages; besides its low critical temperature (T_c = 31.1 °C) and pressure (P_c = 73.8 bar) it is also non-flammable, non-toxic, environmentally acceptable, and inexpensive. [1,2] Probably the most important advantage is that it is a GRAS (Generally Recognised As Safe) solvent [6] and as such can be used in the food, pharmaceutical, and nutraceutical industries without restrictions. In addition, since CO_2 is inert, it allows the processing of chemical-sensitive compounds. [2]

2 CRYSTALLISATION METHODS USING SUPER-CRITICAL CO_2

Depending on the role of $scCO_2$ in the crystallisation process, three main approaches to form particles of a single compound have been developed. They are:

a) "Rapid Expansion of Supercritical Solutions (RESS)", in which the $scCO_2$ is used as a solvent;
b) "Supercritical Anti-Solvent" precipitation (generally referred to as SAS, although different acronyms have been used), in which the $scCO_2$ is used as an anti-solvent;
c) "Particles from Gas-Saturated Solutions (PGSS)", in which the $scCO_2$ is dissolved in the melt or a liquid suspension of the compound of interest.

To be able to choose an adequate $scCO_2$ process for particle formation, the concept and the state-of-the-art, as well as the advantages and drawbacks of the techniques considered, are presented below. In addition, some perspectives of application to various industrial fields are given.

2.1 Particle formation by Rapid Expansion of the Supercritical Solution (RESS)

2.1.1 Concept

In the RESS process, the compound of interest is dissolved in the $scCO_2$, after which the supercritical solution is rapidly expanded over a suitable nozzle or capillary into an expansion vessel. This rapid expansion leads to a dramatic change of the solute supersaturation ratio, causing subsequent precipitation of the solute.

2.1.2 Experimental set-up

The RESS process can be split up into two main parts; one is the extraction step were the $scCO_2$ is saturated with the solute, and the other is the precipitation step of the solute. Liquid CO_2 is charged into a high-pressure pump and pressurised to the desired extraction pressure. After exiting the pump, the pressurised CO_2 is preheated in a heat exchanger to ensure that it reaches the desired extraction temperature before it contacts the solute. The $scCO_2$ then passes through an extractor packed with the solute where it is saturated with that solute. Pressure and temperature in the extractor determine the equilibrium solubility of the solute in the $scCO_2$. After the saturated $scCO_2$ exits the extractor, it is expanded across a preheated nozzle (or a calibrated orifice) into the expansion vessel, which contains a suitable system for collection of the precipitated particles (such as a paper filter or a glass plate). Preheating of the nozzle up to an adequate pre-expansion temperature is necessary to prevent solvent condensation upon expansion. After depressurisation, the gaseous CO_2 is then either vented or recycled. In most cases, the pressure in the expansion vessel is atmospheric, but sometimes it is kept at intermediate pressures in order to reduce CO_2 recompression costs and to allow greater control over the average particle size and particle size distribution.

2.1.3 First time use

Although Smith and Wash of Batelle Institute [9] patented in 1983 a process according to the RESS concept, the first modern investigation of RESS, demonstrating the potential of this technique to drastically reduce the particle size of difficult-to-comminute materials (such as the heat-labile organics/pharmaceuticals ferrocene, dodecanolactam, navy-blue dye, β-estradiol, and lecithin), was carried out by Krukonis in 1984 [3] (at that time, however, Krukonis termed the process supercritical fluid nucleation instead of RESS). After this pioneering work more research groups started working on the RESS-phenomenon, and (as a consequence) many other studies rapidly emerged on the use of RESS for micronisation and particle formation purposes. Some key reviews on the subject [2,10] have been compiled.

2.1.4 Particle size and morphology

The very high supersaturation level obtained with the RESS process tends to decrease the particle size, and hence leads (in principle) to the production of small particles [7]. In addition, since the pressure changes travels at the speed of sound, the uniform conditions within the nucleating medium should result in a narrow particle size distribution [2,11]. These features clearly distinguish the RESS process from conventional crystallisation.

The size and morphology of the precipitated material can be tuned by changing the RESS process parameters, such as the equilibrium solubility of the solute in the $scCO_2$, the pre-expansion pressure and temperature, the nozzle geometry, and the pressure and temperature in the expansion vessel [7]. Many studies have been carried out to investigate the influence of these process parameters on the RESS precipitation of (bio-)polymers, inorganics, organics, and pharmaceutical compounds. Micronic particles (typically of 2-20 μm diameter) of different morphologies (such as spheres, fibres, and needles) have been obtained with the RESS process (for an extensive review of the literature, one can consult reference [10]). Some studies even report formation of nano-scale particles (typically of 0.1-0.5 μm) by using appropriate nozzle designs. [10]

2.1.5 *Industrial perspectives and applications*

RESS technology is more or less regarded as "open" literature, although a few patents [10] have been filed on the subject. Consequently, intellectual property infringements do not limit the use of RESS technology to a great extent. There are, however, several other reasons that limit the industrial applications of RESS technology. [6,8,10]

1. The RESS process is limited to compounds that have a relatively high solubility in $scCO_2$, i.e. low-molecular weight, non-polar compounds. Many interesting substances, however, are polar and therefore not very soluble in $scCO_2$, precluding industrial production at acceptable cost. Co-solvents may be used to enhance solubilities, but these added solvents affect the otherwise environmentally benign nature of the RESS process.
2. RESS is a semi-batch operation, as the $scCO_2$ is flowing continuously through the apparatus. The material processed by RESS must, therefore, be of high-added value, since a large amount of $scCO_2$ is needed to obtain even a small amount of product (due to the relatively low solubility of the product in the $scCO_2$). Typically, several hundred kilograms of CO_2 are required to obtain 1 kg of product. Although recycling can be used to reduce the loss of CO_2, the economics of the recycling process is unfavourable.
3. Successful scale-up of the RESS process to the production scale depends on a quantitative understanding of the underlying physical phenomena (such as nucleation and condensation) as well as insight in the relationship between the process parameters and product quality. In-depth theoretical information of many aspects of the process is, however, at the moment not at hand.
4. Scale-up of the nozzle configuration poses a significant challenge.

2.2 Particle formation by supercritical anti-solvent crystallisation

2.2.1 Concept

In the anti-solvent process, the compound of interest is dissolved in a conventional liquid solvent to form a solution, after which the solvent power of the liquid solution is decreased by saturating it with an anti-solvent (i.e., carbon dioxide in supercritical condition), causing precipitation of the dissolved compound.

In general, two conditions must be fulfilled for applying the anti-solvent approach; 1) the compound of interest is virtually insoluble in $scCO_2$, and 2) the conventional liquid solvent is completely miscible with $scCO_2$ at the crystallisation temperature and pressure. As such, the anti-solvent approach is useful for compounds that are insoluble in $scCO_2$ and, hence, cannot be processed with the RESS process.

2.2.2 Experimental set-up

Various ways of implementation (each designated with their own acronym) [8,10] are available, depending on how the contact between the conventional solvent and the anti-solvent ($scCO_2$) is brought about.

- In the first method, the precipitation vessel is partially loaded with the liquid solution of the compound of interest. The $scCO_2$ is then fed into the vessel (preferably from the bottom to achieve a better mixing of solvent and anti-solvent) to increase the vessel pressure up to the pressure required for precipitation. The $scCO_2$ dissolves into the liquid solvent, where it causes a volumetric expansion of this liquid solvent. As a result, the mixture solution becomes supersaturated since the CO_2-expanded solvent experiences a substantial decrease in its solvent strength, causing precipitation of the compound of interest. After a certain holding time, the precipitated particles are washed with $scCO_2$ by draining the expanded solution under isobaric conditions. The precipitation vessel is then slowly depressurised, and the precipitate is collected. This process has been termed "Gas Anti-Solvent[1] (GAS)" [12]. This mode of operation is also referred to as liquid batch operation, since the precipitation occurs in a liquid rich phase [5].
- In a second method, CO_2 is first fed into the precipitation vessel up to the desired pressure. CO_2 pumping is then either stopped, after which the

[1] Gaseous, liquid, as well as supercritical CO_2 (which are all, in general, designated as "dense" can expand the liquid solution. It is for this versatility of a dense gas to expand the liquid that the term "gas", rather than "supercritical fluid", was used in naming the process.

system is allowed to equilibrate for a certain period of time, or either continued at a constant flow rate. The liquid solution is then sprayed discontinuously (i.e., for a very short period of time) through a capillary nozzle (to obtain fine droplets) into the static or flowing $scCO_2$ (or sometimes liquid CO_2) to induce precipitation. The $scCO_2$ diffuses into the sprayed droplets and dissolves the solvent, causing precipitation of the dissolved compound. After compound precipitation, $scCO_2$ at the operating pressure is swept through the vessel to wash out solvent residues. The precipitation vessel is then slowly depressurised, and the precipitate is collected. This process, which originally has been used for polymer micronisation, has been designated "Precipitation with a Compressed fluid Anti-solvent (PCA)" [13,14]. This mode of operation can also be termed gas batch operation since precipitation occurs for a very short period of time in a supercritical fluid (or, in general, a dense gas) rich phase [5].

- The third method involves continuous feeding of the $scCO_2$ to the top of the precipitation vessel at a constant flow rate to pressurise the vessel up to the selected pressure. Once the system reaches steady state (pressure, temperature and flow), the liquid solution of interest is continuously sprayed (from another inlet point) at a constant flow rate through a nozzle into the flowing $scCO_2$. When in contact with the sprayed droplets, the $scCO_2$ very quickly dissolves into the solvent and *vice versa*, causing solute precipitation. The precipitated particles are collected on a stainless-steel frit filter at the bottom of the vessel. The fluid mixture ($scCO_2$ plus solvent) exits the vessel and flows to a second vessel, located after the precipitator that is operated at low pressure and used to recover the liquid solvent. After a desired amount of precipitate is obtained, the liquid solution feed is stopped but the flow of $scCO_2$ is maintained at the operating pressure to wash out solvent residues. After washing the particles, the precipitation chamber is gradually depressurised and the particles collected. This process has been termed "Supercritical Anti-Solvent (SAS)" process [15] or "Aerosol Solvent Extraction System (ASES)" [16,17].
- The fourth method, which is conceptually very similar to SAS and ASES, involves the co-introduction of a continuous flow of the liquid solution and the $scCO_2$ into the precipitation vessel via a specially designed coaxial nozzle, having two coaxial passages (one for the $scCO_2$ flow and one for the liquid solution flow) which terminate adjacent to one another at the outlet end. The liquid solution is dispersed (i.e., broken up into very small droplets) by the high velocity jet of $scCO_2$. Simultaneously with dispersion of the liquid solution, the solvent is extracted into the $scCO_2$. This leads to rapid supersaturation and, hence, particle nucleation and formation. In a more recent design, the nozzle

has three coaxial passages: an inner, an intermediate and an outer. This process is referred to as "Solution Enhanced Dispersion by Supercritical fluids (SEDS)" [18,19].

2.2.3 First time use

The first application of supercritical anti-solvent micronisation was proposed by Gallagher *et al.* [12] in 1989. They developed the GAS (liquid batch) process to produce fine crystals of the difficult-to-handle explosive nitroguanidine. It has since then been extended to the recrystallisation of other explosives, polymers, pharmaceuticals and fine chemicals (both organic and inorganic) using the different process arrangements and apparatuses described above. Excellent reviews on supercritical anti-solvent precipitation [5,6,10] have been published in recent years.

2.2.4 Particle size and morphology

The particle size, shape, and morphology resulting from the GAS recrystallisation process is – besides being a function of the initial concentration of solution, the type of solvent, the operating temperature, and the pressure profile –strongly influenced by the rate of addition of the $scCO_2$, since this determines the rate of supersaturation built-up in the solvent and the maximum level of supersaturation that is attained [20]. In general, the faster the $scCO_2$ is added, the smaller and more uniform the obtained particles are. The GAS recrystallisation process was initially used to tailor the particle dimensions of explosives, polymers, and inorganics, with obtained particle dimensions ranging from 0.5 μm to several hundreds of micrometers. In more recent years, however, also pharmaceuticals have been processed, with attained particle dimensions ranging from 0.01 μm to 20 μm. [10]
The PCA process has been predominantly used to control particle size and morphology of polymers by adjusting the anti-solvent temperature and density. Increasing temperature (at constant CO_2 density) appears to play an important role by decreasing flocculation and increasing agglomeration, and, hence, results in an increased particle size. Density (through a complex mechanism) affects both jet break-up hydrodynamics and mass transfer of CO_2 and the solvent. It is, however, suggested that mass transfer rather than jet break-up control particle size. [13] Since reduced densities result in reduced mass-transfer rates, larger and more agglomerated particles are produced. In addition, the morphology also depends on the concentration of the liquid solution. At a higher concentration, the increased viscosity of the solution stabilises the liquid jet and the rapid solvent mass transfer further raises viscosity. As a consequence, precipitation occurs before the jet break-

up, resulting in fibres in stead of small micro-spheres. [13] With the PCA process, mean particle diameters ranged from 0.1 to 3 μm for the micro-spheres, and from 20 to 150 μm for the fibres.

Various parameters (such as temperature, pressure, solvent type, concentration of the liquid solution, flow rates and nozzle type) can control the SAS (or ASES) precipitation process. Although much investigation has been accomplished on the SAS precipitation process, there is still limited knowledge on the quantitative aspects of the effect of the various operating parameters on the particle characteristics. [5] Until now, SAS has been applied to process many products, ranging from polymers, explosives, colouring matter, to pharmaceutical and inorganic powders. Many particle dimensions and morphologies have been observed [10].

The SEDS process couples the anti-solvent properties of $scCO_2$ with a physico-mechanical function by using a coaxial nozzle. This nozzle facilitates the simultaneous spraying of the liquid solution and the $scCO_2$, and achieves optimum dispersion of the liquid solution and mixing of the two streams. Control of the working conditions (such as pressure and temperature, and precise metering of the flow rates of the liquid solution and $scCO_2$ into the nozzle) allow controlling particle properties. In general, the precipitated products are nano-scale particles with a narrow size distribution and uniform morphology. [10] The SEDS technique, using a three-way coaxial nozzle, particularly facilitates the generation of bioactive particles (such as proteins) from aqueous solutions, since this type of nozzle improves dispersion and chemical interaction of the different streams.

2.2.5 Industrial perspectives and applications

Thus far, very interesting results have been obtained on a wide range of products. This triggered the interest of industry, since supercritical anti-solvent processing offers the possibility of obtaining new products with a quite efficient process. As a result of this increased interest, numerous patents have been filed, particularly in the pharmaceutical sector (for a comprehensive review see reference. [10] However, this resulted in a complex situation with regard to intellectual property, and this may limit future process applications.

The GAS and PCA processes, although relatively simple to perform, are in many cases not suitable for industrial production since they are operated in batch mode. The (semi-)continuous processes (SAS, ASES, and SEDS) on the other hand permit faster depletion of the solvent (and hence greater production rate of particles) relative to the batch processes. For these processes to be commercially viable, however, continuous production of particles with desired and reproducible product characteristics has to be demonstrated. [6] In particular, continuous harvesting of particles at high yield remains a challenge.

While some useful attempts have been made to interpret the effects of process variables on particle size and morphology, a rigorous mathematical model of the (semi-)continuous anti-solvent process is needed for a better mechanistic understanding. [5] Such an understanding is essential to perform the scale-up to pilot and industrial scale. Regarding mathematical modelling and model validation, supercritical anti-solvent techniques are, however, still at the very early stages of development. [5]

2.3 Particles from Gas-Saturated Solutions (PGSS)

2.3.1 Concept

The PGSS process consists in dissolving $scCO_2$ in the melt or a liquid suspension of the compound of interest, leading to a so-called gas-saturated solution, which is then expanded through a nozzle, leading to particle formation due to the sharp temperature decrease (below the solidification point of the treated compound) and volume increase caused by CO_2 expansion. The PGSS technique exploits the melting point depression of the compound of interest due to the pressurisation with dense CO_2.

2.3.2 Experimental set-up

The following basic set-up for the PGSS process is proposed. [21] The substance to be treated is first melted in a feed vessel. Subsequently, the molten compound is sucked into an autoclave. Next, $scCO_2$ is conveyed into the autoclave up to the required pressure using a high-pressure pump. The liquid phase is drawn off at the base of the autoclave and conveyed to the top of the autoclave using another high-pressure pump. This liquid-phase circulation intensifies material exchange between the liquid and $scCO_2$ phase. When the required CO_2 content in the liquid phase is reached, the gas-saturated solution is then expanded through a nozzle into the top of a spray tower. The combination of cooling and volume increase of the released CO_2 causes the substance to precipitate in a fine-particle form. The larger particles ($\approx$ 10 μm in diameter) are first separated from the resulting gas stream by sedimentation in the spray tower; then smaller particles ($>$ 1 μm) are removed in a cyclone separator; lastly, very small particles ($<$ 1 μm) are removed in an electrostatic precipitator or by using line filters. The residual gas is then passed out of the system, recompressed, and returned to the autoclave.

2.3.3 First time use

In 1994, Mandell *et al.* [22] first described the process concept of PGSS in a patent, which was assigned to Ferro Corporation. The Ferro patent incorporated a mechanical mixing step before pulverisation, but claimed only applications in paint coating materials. More or less at the same time, Weidner *et al.* [23] developed a very similar process for preparing particles and powders, which they called PGSS. Their patent [21,24] had a much wider claim (as compared to the Ferro patent) that encompassed almost any material but it did not incorporate use of a mechanical mixing step. Since then, many PGSS-type processes (using different implementations of the basic process concept) have been applied in various fields. For a recent review on PGSS and related processes, references [8,10,25] can be consulted.

2.3.4 Particle size and morphology

PGSS-type processes have been applied in various fields ranging from inorganic powders to pharmaceutical compounds. For a given system, the particle size of the final particles, which is correlated to the surface tension and the viscosity of the melt, can be adjusted by selecting the nozzle diameter (which is usually larger than 0.5 mm), pressure and temperature in the autoclave and temperature in the spray tower. [8] Although thus the particle size can be controlled, the production of nano-particles, however, does not seem to be feasible. [25]

2.3.5 Industrial perspectives and applications

PGSS-type processes are relatively new in the field of particle production, about which little information is available, because most is covered by patents. There is considerable overlap and similarity in the claims of these patents. Nevertheless, it is expected that many specific applications or process variations based on PGSS will be filed in the near future. [25]

The PGSS-type processes offer several advantages over the other high-pressure crystallisation techniques; PGSS can be operated at a considerably lower operating pressure in a fully continuous mode, the mass ratio between the $scCO_2$ and the compound to be treated is 2 to 3 orders of magnitude less than in the other high-pressure process techniques for producing solids, and it can process products which cannot be micronised by other processes. [21] Disadvantages might be that data on the solubility of $scCO_2$ in the compound of interest are not available for a whole range of products, and the inability of nano-particle production. In addition, process scale-up might also prove challenging. [8,25]

3 CONCLUDING REMARKS

The potential of using $scCO_2$ as medium to form particles has been outlined in the preceding part. It was shown that particle design of a compound by $scCO_2$ can be performed according to different techniques, which for convenience are summarised in Table 3.1.

Table 3.1.- Supercritical particle forming processes in a nutshell.

Process		Solvent	Anti-solvent	Principle
RESS		$scCO_2$	None	- Compound is dissolved in $scCO_2$ - $scCO_2$ solution is expanded over a nozzle - Δ P $\Rightarrow$ supersaturation $\Rightarrow$ precipitation
SAS	GAS	Conventional	$scCO_2$	- Compound is dissolved in conventional solvent - $scCO_2$ is fed into the solution - Expansion of solvent $\Rightarrow$ precipitation
SAS	PCA	Conventional	$scCO_2$	- Compound is dissolved in conventional solvent - Solution is discontinuously sprayed into the $scCO_2$ - Expansion of solvent $\Rightarrow$ precipitation
SAS	ASES	Conventional	$scCO_2$	- Compound is dissolved in conventional solvent - Solution is continuously sprayed into the $scCO_2$ - Expansion of solvent $\Rightarrow$ precipitation
SAS	SEDS	Conventional	$scCO_2$	- Compound is dissolved in conventional solvent - $scCO_2$ and solution are sprayed simultaneously through a coaxial nozzle - Extraction of the solvent into $scCO_2$ $\Rightarrow$ precipitation
PGSS		Compound	None	- $scCO_2$ is dissolved in the melt of the compound - Solution is expanded over a nozzle - Δ P $\Rightarrow$ supersaturation $\Rightarrow$ precipitation

Particle forming processes using supercritical CO_2 offer several attractive technological advantages over conventional techniques; the most important advantages are a reduced usage of conventional liquid solvents, the opportunity of processing thermally labile compounds, and the production of solvent-free products. In addition, supercritical-based particle forming processes offer a unique technology for producing micro- and nano-particles with a controlled particle size distribution and purity. The industrial implementation of this promising technology, however, faces some imperative challenges, such as continuous particle harvesting, demonstration of batch-consistency and reproducible product characteristics, process design, and processing costs. In addition, a quantitative understanding of the underlying physical phenomena as well as insight in the relationship between the process parameters and product quality are essential to perform the scale-up to pilot and industrial scale.

REFERENCES

1. McHugh, M.A., Krukonis V.J., "Supercritical Fluid Extraction: Principles and Practice" Newton, MA: Butterworth-Heinemann, 1994.
2. Tom, J.W., Debenedetti, P.G. "Particle formation with supercritical fluids – a review" J Aerosol Sci. 22, 555-584, 1991.
3. Krukonis, V. "Supercritical fluid nucleation of difficult-to-comminute solids" Paper 104 f, presented at the AIChE Annual Meeting, San Francisco, 1984.
4. Price, C.J. "Take some solid steps to improve crystallization" Chem Engin Progress, 9, 34-43, 1997.
5. Reverchon, E. "Supercritical anti-solvent precipitation of micro- and nano-particles" J. Supercrit. Fluids 15, 1-21, 1999.
6. Subramaniam, B., Rajewski, R.A., Snavely, K. "Pharmaceutical processing with supercritical carbon dioxide" J. Pharm. Sci. 86, 885-890, 1997.
7. Türk, M. "Formation of small organic particles by RESS: experimental and theoretical investigations" J. Supercrit. Fluids, 15, 79-89, 1999.
8. Bertucco, A. "Precipitation and Crystallisation Techniques." in Chemical Synthesis Using Supercritical Fluids, P.G. Jessop, W. Leitner, eds. Weinheim: Wiley-VCH, 1999.
9. Smith, R.D., Wash, R. "Supercritical fluid molecular spray film deposition and powder formation" Patent US 4582731, 1986.
10. Jung, J., Perrut, M. "Particle design using supercritical fluids: literature and patent survey" J. Supercrit. Fluids 20, 179-219, 2001.
11. Mohamed, R.S., Debenedetti, P.G., Prud'homme, R.K. "Effects of process conditions on crystals obtained from supercritical mixtures" AIChE J., 35, 325-328, 1989.
12. Gallagher, P.M., Coffey, M.P., Krukonis, V.J., Klasutis, N. "Gas Anti-solvent Recrystallisation: New Process to Recrystallise Compounds Insoluble in Supercritical Fluids." in Supercritical Fluid Science and Technology (ACS Symposium Series 406), K.P. Johnston, J.M.L. Penninger, eds. Washington, DC: American Chemical Society, 1989.
13. Dixon, D.J., Johnston, K.P., Bodmeier, R.A. "Polymeric materials formed by precipitation with a compressed fluid anti-solvent" AIChE J. 39, 127-139, 1993.

14. Mawson, S., Johnston, K.P., Betts, D.E., McClain, J.B., DeSimone, J.M. "Stabilised polymer micro-particles by precipitation with a compressed fluid anti-solvent. 1. Poly(fluoro acrylates)" Macromolecules 30, 71-77, 1997.
15. Yeo, S.-D., Lim, G.-B., Debenedetti, P.G., Bernstein, H. "Formation of micro-particulate protein powders using a supercritical fluid anti-solvent" Biotech Bioengng, 41, 341-346, 1993.
16. Bleich, J., Müller, B.W., Wassmus, W. "Aerosol solvent extraction system – a new micro-particle production technique" Int J. Pharm. 97, 111-117, 1993.
17. Fischer, W, Müller, B.W. "Verfahren zur Herstellung einer mindestens einen Wirkstoff und einen Träger umfassenden Zubereitung" Patent DE 3744329, 1989.
18. Hanna, M., York, P. "Method and apparatus for the formation of particles" Patent WO 95/01221, 1995.
19. Palakodaty, S., York, P., Pritchard, J. "Supercritical fluid processing of materials from aqueous solutions: the application of SEDS to lactose as a model substance" Pharm Res. 15, 1835-1843, 1998.
20. Gallagher, P.M., Coffey, M.P., Krukonis, V.J. "Gas anti-solvent recrystallisation of RDX: formation of ultrafine particles of a difficult-to-comminute explosive" J. Supercrit. Fluids, 5, 130-142, 1992.
21. Weidner, E., Knez, Z., Novak, Z. "Process for the production of particles or powders", Patent US 6056791, 2000.
22. Mandel, F.S., Green, C.D., Scheibelhoffer, A.S. "Method of preparing coating materials", Patent US 5399597, 1995.
23. Weidner, E., Knez, Z., Novak, Z. "PGSS (Particles from Gas-Saturated Solution) – a new process for powder generation" Proceedings of the Third International Symposium on Supercritical Fluids; 1994 October 17-19; Strasbourg, ISASF, 1994.
24. Weidner, E., Knez, Z., Novak, Z. "Process for preparing particles or powders", Patent WO 95/21688, 1995.
25. Tan, H.S., Borsadia, S. "Particle formation using supercritical fluids: pharmaceutical applications" Exp Opin Ther Patents, 11, 861-872, 2001.

Chapter 4

SUPERCRITICAL FLUID CHROMATHOGRAPHY (SFC)

Luc Van Ginneken, Herman Weyten
Flemish Institute for Technological Research, Boeretang 200, B-2400 Mol, Belgium
Herman.Weyten@vito.be

1. SFC THEORY AND FUNDAMENTALS

1.1 Chromatography as separation method

Chromatography is a separation technique whereby the components of a sample are resolved from each other by allowing them to be distributed between a moving fluid phase (called mobile phase) and a non-moving surface (called stationary phase) in varying proportions. Some of the components have a great affinity for the stationary phase and, hence, migrate slowly with the mobile phase; other components, on the other hand, interact weakly with the stationary surface and, hence, migrate more rapidly with the mobile phase. This difference in migration rate between the different components of the sample produces their separation at the end of the stationary phase, the degree of separation depending on the difference in the rates of migration. After separation, the components can be identified qualitatively, determined quantitatively, and collected separately. Besides the chemical nature of the stationary phase, the solvent power of the mobile phase determines the distribution behaviour of the components. [1,2]
Chromatography is classified according to the nature of the mobile phase. If this mobile phase is a gas then the technique is known as "Gas Chromatography (GC)"; if it is a liquid then the method is referred to as "Liquid Chromatography (LC)". Taking into account the nature of the

M. Aresta (ed.), Carbon Dioxide Recovery and Utilization, 137-148.

stationary phase, which is either solid or liquid, extends this classification. As such, four different chromatographic methods exist: gas-solid, gas-liquid, liquid-solid, and liquid-liquid. [1]

The stationary phase material can be spread evenly as a thin film on a flat surface (generally referred to as "thin-layer chromatography") or on a sheet of filter paper (called "paper chromatography"), or it can be inside a tube as a packing or a thin film (as in "column chromatography"). [1] As such, different chromatographic techniques do exist, some of which can be operated with different elution procedures (such as in elution chromatography, frontal chromatography, and displacement chromatography). [1,2] In addition, different modes of chromatographic separation (such as ion-exchange, adsorption, exclusion, affinity, and partition) are also available. [1]

Because of the wide selection of combinations of mobile and stationary phases, and the extensive modes of operation and separation that are available, chromatography has been applied to the separation of a wide variety of compounds. It has found widespread use as an analytical separation method, since its resolving power is unequalled among separation methods and many analytical tasks simply cannot be dealt with adequately by any other available method. In addition, since chromatographic techniques can be applied to an incredible concentration range and can be easily automated for routine analyses, these methods are a valuable and versatile separation tool in many scientific fields.

1.2 Use of supercritical fluids in chromatography

If the mobile phase in a chromatographic separation is a supercritical fluid (or a near critical gas or liquid), the method is – in accordance with GC and LC – referred to as "Supercritical Fluid Chromatography (SFC)". The appeal of using supercritical fluids (SCFs) as the mobile phase in chromatography comes from their unique properties, which are frequently described as being intermediate between those of a gas and a liquid. [3] Varying the fluid density can vary the solvent power of a SCF over an extremely wide range. [2,3] By manipulating the system pressure and temperature, the fluid density (and hence solvent power) can be controlled when operating in the critical region. [3,4] Solute solubility increases in general with increasing density of the fluid mobile phase. [2] In addition, temperature directly determines the vapour-pressure (and hence volatility) of the solutes. [2,4] Increasing temperature (at constant pressure) also increases continuously the diffusivity of solutes in the SCF to a value that is about one or two orders of magnitude higher than the corresponding value in liquid solvents, the most dramatic effect occurring at or slightly above the critical temperature. [3] As such, temperature and pressure can be employed in SFC as operational variables for tuning mobile phase characteristics and, hence, for influencing separation characteristics. [2]

SFC appears to be, on theoretical grounds, an alternative for GC, since the SCF mobile phase exerts its relatively high solvating capacity near the critical temperature of the mobile phase, which allows processing of thermo-labile or non-volatile compounds (for which GC is not suited). [2] In addition, SFC allows chromatographic separations to be made faster and at higher resolution than obtainable with conventional LC, since the higher diffusion rates of the solutes in the SCF mobile phase result in more frequent and rapid equilibrium of solute partitioning into and out of the stationary phase. [4,5] The use of SFC also reduces the use of organic solvents, making the technique more environmental friendly. By its very nature, SFC can cover an intermediate region between the more classical GC and LC [6].

1.3 Scale of SFC techniques

Chromatographic separations on the basis of SFC are of great interest to both academic and industrial researchers. Its use can roughly be broken down into two classifications, namely analytical SFC and preparative SFC. In *analytical* SFC, the goal is to separate a mixture of components, which allows identification and quantification of the separated components, usually in the picogram to milligram range. The separated components are not needed, and therefore, are not collected. In *preparative* SFC, the goal is to isolate and collect the components of a mixture in the mg to kg range. The preparative capabilities of SFC are applied at different scales of recovery, depending on the further use of the isolated components: small-scale or semi-preparative (milligrams up to grams); large-scale or pilot-scale (grams up to kilograms); production-scale or commercial-scale (kilograms up to tons).

2. ANALYTICAL-SCALE SFC

2.1 History

The first acknowledgement of the analytical potential for SCFs probably came from James Lovelock in 1958, who suggested that gases in their supercritical state might be used as mobile phases in chromatography [6]. This sharp observation, however, did not receive much attention for many years. In 1962, Klesper and his colleagues [7] were the first to demonstrate the feasibility of SFC by separating some porphyrins with dense dichlorodifluoromethane and monochlorodifluoromethane using a packed (LC-type) column. Although from then on a relatively large number of separations was made using packed-column SFC, it was not until the 1980s that SFC became popular as a result of the commercial introduction of SFC instrumentation [8]; in 1982 packed column instrumentation was introduced and in 1986 capillary (GC-like) instrumentation became available. As the limitations of capillary SFC became more apparent in subsequent years, interest in capillary SFC started to decline in the 1990s. By that time, packed column SFC was rapidly gaining ground when the technique was considered to be a suitable replacement for normal-phase (i.e., polar stationary phase and unpolar mobile phase) LC [8]. In recent years, however, there has been a decline in interest for SFC (although it is well established as an analytical technique), and it seems that SFC is now restricted to niche markets were it offers a real advantage over conventional analytical techniques. [6]

2.2 Stationary phases and columns

Two types of analytical columns are used in SFC, namely capillary (or open tubular) columns and packed columns. Capillary SFC columns resemble their GC counterparts except the inside diameter is typically smaller (due to the smaller diffusion rates in SCFs than in gases) and the stationary phases must be immobilized. [9] Extensive cross-linking of the stationary phase to the capillary column wall is necessary, since the solvent strength of the SCF mobile phase is sufficient to dissolve the stationary phase [6,10]. Capillary columns for analytical SFC are usually about 10-20 m in length and 50-100 μm in internal diameter. [2] The lower viscosities of SCFs, as compared to traditional liquids, results in a minimal pressure drop across the capillary column. As a consequence, longer columns may be used, leading to higher efficiencies. [2,9] Capillary columns are usually made of fused silica that is frequently coated with polysiloxanes and polyglycols [2], although other stationary phases have been used. [9]

Packed SFC columns, conventionally made of stainless steel, contain small deactivated support particles to which the stationary phases adhere. Packed-column SFC is most useful for the separation of mixtures usually separated by normal-phase high-performance LC (HPLC). [6] Since most of the packed SFC columns originate from HPLC, they are packed with chemically-bonded silica-based sorbents. [8,11] Since packed columns are relatively low in efficiency, a wide variety of stationary phases are used to optimise individual separations (for an extensive review on packed column stationary phases see reference [8]). Small-bore analytical columns of 0.5-2 mm internal diameter and a length of 25-50 cm are commonly used. In contrast with capillary SFC columns, packed SFC columns may suffer from a pressure drop across the column. [8,10] It is widely recognised that in a packed SFC column the density gradient, which is induced by the pressure drop, and the associated velocity gradient can result in a loss of column efficiency (with band broadening and peak deformation as a consequence). [8,12] To overcome this problem, packings with larger particle diameters than those used in LC were advised in order to reduce the pressure drop. [8]

2.3 Mobile phases

Although a wide range of SCFs (such as hexane, nitrous oxide, pentane, water, and xenon) have been studied as mobile phases in SFC, none of these fluids can compete with the advantages offered by CO_2, such as low critical conditions, readily availability, environmentally-benign nature, low cost, low health and safety risks. [6,10] As a consequence, CO_2 has been the mobile phase of choice for almost all commercial analytical-scale SFC systems. In addition, $scCO_2$ as mobile phase permits the use of a flame ionisation detector (FID), traditionally a GC detector, which provides universal detection for a wide range of organic compounds. [9]

Due to its non-polar nature (the polarity of $scCO_2$ is quite similar to hexane), $scCO_2$ is not able to elute very polar compounds. As a consequence, SFC using $scCO_2$ as the mobile phase is basically a normal-phase type of separation. [6] To overcome the non-polar nature of $scCO_2$, organic solvents (also called modifiers or entrainers) are added to enhance the solvating power of neat CO_2. These modifiers are, in general, organic liquids (such as methanol and acetonitrile) that are completely miscible with CO_2. Modifiers are especially used in packed column SFC, where their influence on elution strength is proportional to their concentration. When used in low concentrations, they bind to the most active sites of the stationary phase and as such prevent the analytes from being retained by these sites, enabling the elution of quite polar compounds [6]. If they are used in higher concentrations, they change the bulk polarity of the $scCO_2$ mobile phase. [6]

2.4 Instrumentation

The components of a supercritical fluid chromatograph generally include a CO_2 reservoir, a high-pressure CO_2 pump, a high-pressure modifier pump, a sample injector, a (packed or capillary) separation column, an oven to regulate column temperature, a detector for determining concentration of the eluted substances, a restrictor (which is placed either before or after the detector) for maintaining back-pressure, a sample collector, and a micro-processor for processing data and controlling the entire apparatus. [2] The column type determines the type of high-pressure pump used in SFC; for packed columns reciprocating pumps are used while for capillary columns syringe pumps are used. Ovens used in SFC are generally conventional GC- or LC-type ovens. Use of a restrictor permits the $scCO_2$ to flow through the column at the desired rate while its pressure is maintained at the desired level. Optical detectors, flame detectors, and spectroscopic detectors have been used.

In a review on SFC [6], Smith noted that many of the problems in implementing SFC as a routine technique in the 1980s came from the instrumentation. Particularly capillary SFC suffered from a number of drawbacks. To maintain a high efficiency, low flow-rates (μl/min) of $scCO_2$ must be used in capillary columns. It proved, however, difficult to maintain a reproducible constant flow-rate over a prolonged period of time, which is bad for chromatographic separations due to the loss in column efficiency. [12] In addition, clogging of the flow restrictors, used to maintain backpressure, was often a problem, which was due to the relatively large Joule-Thomson cooling during expansion of the CO_2. [6,12] Accurate injection of very small sample volumes also led to sophisticated and complex injection procedures [6,12]. These problems were largely overcome in second-generation instruments [6]. Packed column SFC, on the other hand, suffered less from instrumental problems, since it used equipment quite common to HPLC users.

2.5 Demonstrated applications

In Table 4.1 a reasonable indication of the major application areas of analytical SFC is given [5,6,9,10,14,15]. The entries in Table 1, which are not exhaustive, reflect the non-polar nature of carbon dioxide. A significant research effort has been put in soil analysis and the extraction of polycyclic aromatic hydrocarbons (PAHs), polychlorinated biphenyls (PCBs) and pesticides. Other major areas include food analysis, where the emphasis is on fats and oils, and chemical processing, with interest in petrochemical products and polymers. Although most pharmaceutical compounds are too polar for routine SFC, the pharmaceutical industry has adopted SFC because it is an easy and fast technique to carry out normal-phase separations. The enantioseparation of chiral analytes is probably the most successful application area of SFC (for a recent review see reference [13]). This is due to the fact that the normal-phase separation mode enhances the interaction between the analytes and stationary phase, and as such increases enantioselectivity [6].

3. PREPARATIVE-SCALE SFC

3.1 History

The preparative potential of SFC was already recognised as early as 1962, when Klesper and his co-workers [7] mentioned in their pioneering study on the analytical feasibility of SFC that "the porphyrins could be recovered at the outlet valve". Although small-scale preparative SFC has been investigated since then, instrumental problems initially limited the usage of preparative SFC on a large-scale. In 1982, with the coming of age of analytical-scale SFC, Perrut [16] first described (and patented) a preparative fractionation process by using elution SFC on a pilot-scale. This elution process was carried out with intermittent injections of the feed mixture, and hence operated discontinuously. In the next ten years, university laboratories undertook feasibility studies followed by development work and applications to valuable fractionation problems [17], which led to commercial development of preparative SFC in the early 1990s. However, at that time, preparative SFC was still a nascent market, with only one commercial supplier of large-scale preparative SFC equipment. From then on, research has been relayed by industry and applications and equipment improvements were developed. The advent in recent years of continuous (counter-current) preparative SFC processes, based on simulated moving bed (SMB) technology, might promote preparative SFC techniques, since it will permit faster and more-efficient production-scale purifications. [18]

3.2 Advantages in comparison to preparative GC and LC

Preparative SFC is a very useful tool, which has some distinct advantages in comparison to preparative GC and LC: in contrast with preparative GC, preparative SFC can be operated at low temperatures due to the high solvating capacity of the $scCO_2$ eluent and due to the fact that the mixture does not have to be vaporised; compared to preparative LC, fraction-eluent separation is much easier in preparative SFC, since the $scCO_2$ vaporises completely if the externally applied pressure is removed. [16,19] In addition, preparative SFC might also have beneficial economic implications.

Preparative SFC can be carried out either discontinuously (with intermittent injections of the feed mixture) or continuously (with continuous feeding of the mixture). Preparative batch SFC separations are carried out predominantly in elution mode, and the process is then referred to as preparative elution SFC. [2,16] By adopting SMB technology, the preparative batch SFC process can be converted into a continuous mode. [20]

3.3 Preparative elution SFC

Preparative elution SFC is characterised by 1) periodic injection of the feed mixture into a continuous flow of $scCO_2$, 2) chromatographic separation on the chromatographic column due to selective interactions of the feed components with both the $scCO_2$ and the stationary phase, 3) detection of the separated components at the column outlet, 4) fraction collection, 5) separation of the fractionated compounds from the $scCO_2$, and 6) purification and recycling of the $scCO_2$ when economical, which is particularly of interest in large-scale preparative SFC. [16,19]

Small-scale preparative elution SFC can be very useful for purifying small amounts of key products, which can be further used for determining physico-chemical properties or for identification purposes. [19] In general, it uses equipment derived from analytical-scale packed-column SFC. Most of the equipment used is quite similar, although different techniques of collection have been used, such as collection at atmospheric pressure, collection at high pressure, and adsorption on a solid followed by elution or dissolution in a liquid solvent. [17,19] Berger and Perrut [19] and Perrut [17] have reviewed the applications of small-scale preparative SFC.

Large-scale preparative elution SFC is very useful for final purification or fractionation of high-value materials (such as pharmaceuticals), since solvent-free fractions can be produced in large amounts. Its development, however, was initially limited by some technological difficulties, more particularly $scCO_2$ recycling, efficient periodic sample injection, $scCO_2$-product separation with high yields, and column packing technology were much more crucial than on a smaller scale. [19] With the design of high-performance cyclonic separators and the adoption of dynamic axial column-packing compression to large-scale preparative SFC, most of the problems could be overcome. [17,19] Illustrative examples of large-scale preparative SFC are the fractionation of polyunsaturated fatty acid esters on a dynamic axial compression column (60 mm in diameter and 230 mm in length) with neat $scCO_2$ flowing at 53 l/h [17], and the purification of an insecticide on a C-18 bonded silica column (100 mm in diameter and 250 mm in length) with neat $scCO_2$ as the eluent. [21]

Scale-up of preparative SFC can, in general, be achieved either by enlarging the dimensions of the chromatographic column, or by increasing the injected amount of sample. [2,22] However, since maintaining chromatographic separation conditions on large columns is not straightforward, it is generally preferred to increase the sample loading. However, this requires a good understanding of the band broadening mechanism under column overload conditions. [23] Therefore, in developing preparative separations, first conditions on an analytical scale are optimised and then under these conditions, the separation with sample overloading is studied. [2,23,24] For an excellent theoretical treatise of sample overloading in preparative SFC, the interested reader is referred to reference. [2]

3.4 Preparative simulated moving bed-SFC

In contrast to batch elution SFC, a SMB system consists of a number of individual chromatographic columns (each containing a solid stationary bed) which are connected together in series. This system, in which the sample mixture and $scCO_2$ are fed continuously and the analytes are collected continuously, is operated by shifting the sample mixture and $scCO_2$ injection points and the separated component collection points in a continuous manner so that the solid bed material appears to be moving with respect to the injection and collection points. Rather than moving the injection and collection points continuously, they are actually moved intermittently between the number of fixed columns. This is achieved by switching valves, which open or close different sections of the chromatographic columns. As such, a continuous counter-current moving bed is simulated. [18,20] The interested reader who would like to learn more about the general principle of SMBs should consult reference. [18] For an elaborate treatise of SMBs using a supercritical eluent, references [25] and [26] should be consulted.

As compared to classical SMB chromatography, the use of a SCF eluent in a SMB system offers the unique advantage that the elution strength of the SCF mobile phase can be tuned by density in the different zones of the system (i.e., pressure gradient mode) in order to optimise the separation performance. [25] As compared to preparative elution (batch) SFC, SMB-SFC offers considerable advantages, such as reduced solvent usage, 2) higher production yield (since it is operated in a continuous mode), 3) more easy downstream processing, 4) more cost-effective, and 5) more concentrated extract components can be obtained. [18,20] As a consequence, SMB-SFC can provide a better solution as a continuous method for producing large quantities of pure material at lower cost than a batch process. However, SMB-SFC requires more complex instrumentation and it is not as flexible as batch SFC. In addition, method development and optimisation are also more complex.

The key to successful operation of a SMB system is the proper selection of the flow rates and the valve-switching times of the feed and $scCO_2$ eluent streams. Development of a separation with a SMB system starts with the optimisation of the chromatographic separation on an analytical scale. [27]

In addition, the best operating pressure and temperature for the $scCO_2$ eluent has to be found. Next, the separation on the SMB system can be carried out, which basically involves the following steps [27]: 1) packing of the columns, (2) testing each of the columns by pulse injection, 3) pulse injection into the complete plant, and 4) separation of a dilute feed and productivity optimisation by increasing the feed concentration. This last step requires the knowledge of the component adsorption isotherms. Criteria for the choice of the operating conditions of SMB-SFC to achieve optimal separation performance are given in references [25] and [26].

Some very recent applications of SMB-SFC include the fractionation of polyunsaturated fatty acids [20], the separation of phyto isomers [27], and the enantioseparation of tetralol. [28]

Table 4.1 - Application areas of SFC in analytical chemistry [5,6,9,10,14,15]

CHEMICAL PROCESSING
Petroleum products (fuels and lubricants)
Organometallics and metal chelates
Oligomers, polymers, and polymer additives
Surfactants
Explosives
CLINICAL/PHARMACEUTICAL
Amino acids and peptides
Drugs
Enantiomers
Prostaglandins
Steroids
FOOD
Carbohydrates
Fats
Lipids
Tocopherols
Nitrosamines
Vitamins
Essential oils
NATURAL PRODUCTS
Alkaloids
Lipids
Pigments
Steroid hormones
Terpenoids
Waxes
ENVIRONMENTAL
Herbicides
Pesticides
Insecticides
Polychlorinated biphenyls (PCBs)
Polycyclic aromatic hydrocarbons (PAHs)

4. FUTURE PERSPECTIVES OF SFC

Although initially advertised as a technique that could overcome the limitations of both GC and LC, analytical SFC proved a partial disappointment in many laboratories. Smith, in his excellent review on the rise and fall of SFC [6], pointed out several reasons why the initial promise was not fulfilled; 1) SFC is primarily a normal-phase method that cannot compete with the widespread dominance of reversed-phase HPLC, 2) SFC is operationally complicated, and 3) little routine instrumentation is available. However, while SFC may never be as widely used as GC or HPLC, it still provides a useful complement to these chromatographies. Clearly there are a number of niche applications where SFC is superior to GC or HPLC. The considerable potential of SFC for enantioseparations of chiral mixtures probably holds the most promise for the future. [13]

Preparative SFC seems to be very promising for the final purification or fractionation of fine chemicals or natural products. Preparative batch SFC is more flexible and is easier to perform and optimise in the early stages of product development. Preparative SMB-SFC is essentially a production tool and it can provide a better solution as a continuous method for producing large quantities of pure material at lower cost than a batch process [18].

REFERENCES

1. Sewell, P.A., Clarke, B.P., "Chromatographic Separations" Analytical Chemistry by Open Learning Series, Chichester: John Wiley & Sons, 1987.
2. Brunner, G. "Gas Extraction" in An Introduction to Fundamentals of Supercritical Fluids and the Application to Separation Processes. Darmstadt: Steinkopff, 1994.
3. McHugh, M.A., Krukonis, V.J. "Supercritical Fluid Extraction: Principles and Practice" Newton, MA: Butterworth-Heinemann, 1994.
4. Payne, K.M., King, J.W. "Supercritical Fluid Extraction/Chromatography" in Separations Technology, W.P. Olson, ed. Buffalo Grove, IL: Interpharm Press, Inc., 1995
5. Chester, T.L., Pinkston, J.D., Raynie, D.E. "Supercritical fluid chromatography and extraction" Anal. Chem. 70, 301R-319R, 1998.
6. Smith, R.M. "Supercritical fluids in separation science – the dreams, the reality, the future" J. Chromatogr. A, 856, 83-115, 1999.
7. Klesper, E., Corwin, A.H., Turner, D.A. "High pressure gas chromatography above critical temperatures" J. Org. Chem. 27, 700-701, 1962.
8. Poole, C.F. "Progress in packed column supercritical fluid chromatography: materials and methods" J. Biochem. Biophys. Methods 43, 3-23,2000.
9. Chester, T.L., Pinkston, J.D., Raynie, D.E. "Supercritical fluid chromatography and extraction" Anal. Chem. 66, 106R-130R, 1994.
10. Park, G., Crull, A. "Supercritical Fluids: Their Time Has Come" Business Opportunity Report. Norwalk, CT: Business Communications Company, Inc., 1994.

11. Poole, C.F., Oudsema, J.W., Dean, T.A., Poole, S.K. "Stationary Phases for Packed Column Supercritical Fluid Chromatography." in Analysis with Supercritical Fluids: Extraction and Chromatography, B. Wenclawiak, ed. Berlin: Springer-Verlag, 1992.
12. Bouigeon, C., Thiébaut, D., Caude, M. "Long packed column supercritical fluid chromatography: influence of pressure drop on apparent efficiency" Anal. Chem. 68, 3622-3630, 1996.
13. Terfloth, G. "Enantioseparations in super- and subcritical fluid chromatography" J. Chromatogr. A 906, 301-307, 2001.
14. Chester, T.L., Pinkston, J.D. "Supercritical fluid and unified chromatography" Anal. Chem. 72, 129R-135R, 2000.
15. Lee, M.L., Markides, K.E. (eds.) Analytical Supercritical Fluid Chromatography and Extraction. Provo, UT: Chromatography Conferences, Inc., 1990.
16. Perrut, M. "Fractionation process for mixtures by elution chromatography with liquid in supercritical state and installation for its operation" Patent US 4478720, 1984.
17. Perrut, M. "Advances in supercritical fluid chromatographic processes" J. Chromatogr. A, 658, 293-313, 1994.
18. Nicoud, R.-M., Majors, R.E. "Simulated moving bed chromatography for preparative separations". LC-GC Europe, 13, 887-891, 2000.
19. Berger, C., Perrut, M. "Preparative supercritical fluid chromatography" J. Chromatogr., 505, 37-43, 1990.
20. Perrut, M., Nicoud, R.-M., Breivik, H. "Processes for chromatographic fractionation of fatty acids and their derivatives" Patent US 5719302, 1998.
21. Shaimi, M., Colopi, D., Jusforgues, P. "Preparative SFC: development, status and applications" Proceedings of the Fifth Meeting on Supercritical Fluids; March 23-25; Nice. Vandoeuvre: Institut National Polytechnique de Lorraine, 1998.
22. Yamauchi, Y., Kuwajima, M., Saito, M. "Sample introduction and elution method for preparative supercritical fluid chromatography" J. Chromatogr. 515, 285-293, 1990.
23. Cretier, G., Neffati, J., Rocca, J.L. "Experimental study of band broadening and solute interferences in preparative supercritical fluid chromatography" J. Chromatogr. A, 670, 173-179, 1994.
24. Bartle, K.D., Bevan, C.D., Clifford, A.A., Jafar, S.A., Malak, N., Verrall, M.S. "Preparative-scale supercritical fluid chromatography" J. Chromatogr. A, 697, 579-585, 1995.
25. Mazzotti, M., Storti, G., Morbidelli, M. "Supercritical fluid simulated moving bed chromatography" J. Chromatogr. A 786, 309-320, 1997.
26. Di Giovanni, O., Mazzotti, M., Morbidelli, M., Denet, F., Hauck, W., Nicoud, R.-M. "Supercritical fluid simulated moving bed chromatography II. Langmuir isotherm" J. Chromatogr. A, 919, 1-12, 2001.
27. Depta, A., Giese, T., Johannsen, M., Brunner, G. "Separation of stereoisomers in a simulated moving bed-supercritical fluid chromatography plant" J. Chromatogr. A, 865, 175-186, 1999.
28. Denet, F., Hauck, W., Nicoud, R.-M., Di Giovanni, O., Mazzotti, M., Jaubert, J.N., Morbidelli, M. "Enantioseparation through supercritical fluid simulated moving bed (SF-SMB) chromatography" Ind. Eng. Chem. Res., 40, 4603-4609, 2001.

Chapter 5

HOMOGENEOUS CATALYSIS IN SUPERCRITICAL CARBON DIOXIDE

Tessa R. Early[a], Andrew B. Holmes[a], Jin kyun Lee[a], Eugenio Quaranta[b] and Louise M. Stamp[a]
[a]*Melville Laboratory for Polymer Synthesis, University of Cambridge, Lensfield Road, Cambridge CB2 1EW, United Kingdom, abh1@cam.ac.uk;*
[b]*ICCOM-CNR, Dipartimento di Chimica, Campus Universitario, 70126 Bari, Italy, quaranta@chimica.uniba.it*

1. HOMOGENEOUS CATALYSIS

The utilization of SCFs as reaction media for metal-complex-catalyzed reactions has received growing attention in the last decade [1]. In fact, the peculiar properties of SCFs (high diffusivity, low viscosity, compressibility, tuneable solvent properties, etc) can make advantageous the use of SCFs as media for metal-complex catalyzed reactions in place of conventional solvents. Possible benefits are:

a) enhanced reaction rates and selectivities, as a result of the high solubility of gaseous reactants in SCFs, fast diffusion of solutes, weakening of solvation effects;
b) easier separation of reactants, catalyst, and products, after reaction;
c) elimination of solvent residues and wastes.

The solubility of metal complexes in $scCO_2$ is a factor of crucial importance for developing homogeneous catalysis in this medium. Often, the problem of catalyst solubility can be overcome successfully with a judicious choice of a CO_2-philic ligand environment around the metal [2]. CO_2-philic ligands should contain as little aromatic rings as possible and have siloxyl, perfluoroalkyl or branched alkyl groups as substituents. In general, in order to exhibit satisfactory solubility in $scCO_2$, metal complexes should be non-polar (or poorly polar), hydrophobic, quite volatile, uncharged and /or contain a CO_2-philic ligand framework. Nevertheless, even ionic complexes can exhibit sufficient solubility in $scCO_2$, if they have CO_2-philic counterions, such as triflate (trifluoromethanesulfonate) or BARF (tetrakis(3,5-bis(trifluoromethyl)-phenylborate) anion or lipophilic tetralkylammonium cations [3].

M. Aresta (ed.), Carbon Dioxide Recovery and Utilization, 149-168.

2. HYDROGENATION REACTIONS

Dihydrogen exhibits limited solubility in most organic solvents and, consequently, hydrogenation rates are often low, being controlled by H_2 diffusion in the liquid phase. Conversely, H_2 is totally miscible with $scCO_2$ and very high concentrations of H_2 can be attained in the SCF. This feature can be particularly advantageous in hydrogenation reactions in terms of enhanced rates and higher selectivities. However, it should be noted that very high concentrations of H_2 can have serious drawbacks on the solubility of catalysts and substrates in $scCO_2$. In fact, addition of H_2 in very high concentration can remarkably lower the density of $scCO_2$ and, as a result, the ability of the SCF medium to dissolve substrates and catalysts can be significantly reduced. Solving this problem requires the search for new catalysts and many efforts in this area are currently devoted to design and synthesize new CO_2-philic ligands able to improve the solubility of homogeneous catalysts in $scCO_2$, even in the presence of high mole fractions of H_2 [4].

2.1 Catalytic hydrogenation of supercritical carbon dioxide

Homogeneous catalytic hydrogenation of $scCO_2$ to formic acid (eq. 1), first accomplished by Noyori in 1994 [5,6], represents one of the first synthetic processes involving $scCO_2$ as both solvent and reagent. The process is promoted by Ru-catalysts, *cis*-$Ru(PMe_3)_4H_2$ (**1**) or *trans*-$Ru(PMe_3)_4Cl_2$ (**2**), that, because of their higher solubility in $scCO_2$, were shown to be more effective than *cis*-$Ru(PPh_3)_4H_2$, the most active catalyst in conventional solvents.

$$H_2 + CO_2 \xrightarrow{\text{Ru-cat.}} HC(O)OH \qquad (1)$$

The reaction was carried out at 50 °C under 85 bar of H_2 pressure and 120 bar of CO_2. Because of the unfavourable thermodynamics, the reaction was driven by the addition of a basic amine (triethylamine) to produce a stable ≅ 2:1 formic acid-amine adduct. Very interesting hydrogenation rates were obtained using co-catalysts or co-solvents such as water, MeOH or DMSO. Both catalytic efficiency (7200 TON for catalyst **2**) and reaction rates (TOF was 1400 h^{-1} for catalyst **1** over the first hour, and 1040 h^{-1} for catalyst **2**) were greater than any previously reported in conventional solvents.

$$H_2 + CO_2 + MeOH \xrightarrow{\text{Ru-cat. / scCO2 / 80 °C}} HC(O)OMe + H_2O \qquad (2)$$

$$H_2 + CO_2 + R_2NH \xrightarrow{\text{Ru-cat. / scCO2 / 80 °C}} HC(O)NR_2 + H_2O \qquad (3)$$

[R = Me, Et, *n*-Pr]

In the presence of alcohols or secondary aliphatic amines, this synthetic methodology can provide an interesting "one-pot" route to formate esters or formamides (eq. 2 and 3) [7,8]. In the latter cases, the hydrogenation of carbon dioxide to formic acid is coupled with subsequent reactions to give the formic acid derivatives.

Both reactions 2 and 3 are biphasic. The CO_2 hydrogenation step takes place in $scCO_2$, while the second, rate determining, step (esterification or aminolysis) occurs in a liquid phase. In the presence of **2** as catalyst, methylformate was obtained at 80 °C with selectivity close to 35 % and TONs up to 3500. The reaction rate was two orders of magnitude greater than in subcritical systems at the same temperature [7]. At 100 °C, using complex **2** as promoter, the hydrogenation of $scCO_2$, in the presence of dimethylamine, produced DMF with TON as high as 420000, in 99 % yield and 99 % selectivity [8]. Also, in this case, the catalytic efficiency was two orders of magnitude greater than any previously obtained in conventional solvents (3400 TON, 73 % yield).

The high TON values found for both methylformate and DMF synthesis are indicative of high catalyst activity and long catalyst lifetimes under the working conditions. Also TOF values are greater than any previously reported, for the same reactions, in liquid solvents. As already pointed out, the higher reaction rates in $scCO_2$ can be ascribed to the high concentration of H_2 in this medium, fast mass transfer between phases, weaker solvation of catalysts. Both processes 2 and 3 seem to be well suited to industrial application. The high reaction rates reported for these processes are ideal for larger scale continuous flow systems. If new technologies for H_2 production became readily available (H_2O electrolysis by using solar energy, for instance), this CO_2-based chemistry could replace the current processes based on toxic carbon monoxide.

The efficient formation of DMF has been reported also by Baiker using *trans*-$Ru(dppe)_2Cl_2$ (dppe = 1,2-bis(diphenylphosphino)ethane) or an heterogenized silica-grafted Ru complex [9,10].

The catalyst *cis*-$Ru(PMe_3)_4H_2$ also promotes the hydrosilylation of $scCO_2$ with $Me_2(Et)SiH$ to give the corresponding silylformate, although in very low yield (2% vs silane, after almost 2.5 days) [1g].

2.2 Hydrogenation of unsaturated organic substrates

Very few studies have been concerned with homogeneous hydrogenation of C=C double bonds in $scCO_2$; for the synthesis of achiral compounds. $HMn(CO)_5$ has been reported to catalyze the hydrogenation of a substituted cyclopropene in $scCO_2$ [11]. Moreover, the CO_2-philic catalyst $Rh(hfacac)(R_2PCH_2CH_2PR_2)$ [hfacac = hexafluoroacetylacetonate; R = m-$C_6H_4(CH_2)_2(CF_2)_6F$] promotes the hydrogenation of isoprene to give a mixture of isomeric isopentenes with very low formation of 2-methylbutane. Unfortunately, the reaction rate was lower than in organic solvents with analogous catalysts [12].

Most studies deal with the hydrogenation of prochiral systems, as enantiocontrol in these processes is sensitive to change in H_2 concentrations. The asymmetric hydrogenation, in $scCO_2$, of a prochiral alkene, such as (*E*)-2-methylbut-2-enoic acid (tiglic acid), has been first demonstrated by Noyori and Jessop [11] using the chiral catalyst $Ru(OAc)_2$[(*S*)-H_8BINAP] (**3**) (OAc = acetate).

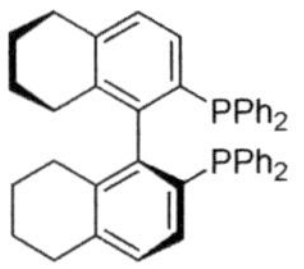

3 (*S*)-H_8BINAP

At 50 °C, the hydrogenation reaction took place smoothly in $scCO_2$ (P_{CO2} = 175-180 bar) to give (*S*)-2-methylbutanoic acid with enantioselectivity increasing with H_2 pressure, from 71% (P_{H2} = 5 bar) to 81% (P_{H2} = 30 bar). A quite different behaviour was observed in conventional solvents, such as hexane or methanol, where enantioselectivity was found to be practically constant (hexane) or decrease (methanol) with increasing H_2 pressure (in methanol: ee = 95% at P_{H2} = 5 bar, ee = 82% at P_{H2} = 30 bar; in hexane: ee = 74% at P_{H2} = 5 bar, ee = 73% at P_{H2} = 30 bar). At high H_2 pressure (30 bar), the results obtained in $scCO_2$ (81% ee, >99% yield) were comparable to the best results obtained in a conventional solvent (methanol, 82% ee, 100% yield), but, at low H_2 pressure (5 bar), the enantioselectivity in $scCO_2$ was lower than that reported in hexane or methanol.

The addition of fluorinated alcohol $CF_3(CF_2)_6CH_2OH$, soluble in $scCO_2$, enhanced the catalyst solubility and produced an increase of both conversion and enantioselectivity at low H_2 pressure (100% yield, 89% ee, at 5 bar of H_2). However, in the presence of $CF_3(CF_2)_6CH_2OH$, a different dependence of enantioselectivity from H_2 pressure was observed with respect to unmodified $scCO_2$, and, at 30 bar H_2 pressure, an ee as low as 76% was found. It is worth noting that no formation of the hydrogenated product has been observed when the reaction was carried out in liquid CO_2 at 20 °C.

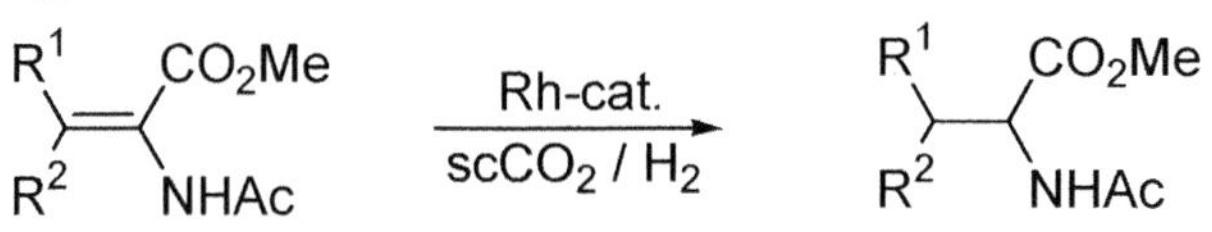

Scheme 1

A study by Tumas [3] *et al.* describes the enantioselective hydrogenation (40 °C, P_{H2} = 14 bar, total pressure = 330 bar) of prochiral α-enamide esters (Scheme 1), in $scCO_2$, in the presence of cationic Rh-catalysts [Rh(cod)((*R,R*)-Et-DuPHOS)]X [cod = 1,5- cyclooctadiene; X = triflate or BARF; (*R,R*)-Et-DuPHOS = (*R,R*)-1,2-bis(trans-2,5-diethylphospholano)benzene]. Very interesting enantioselectivities were observed in $scCO_2$. For β-monosubstituted α-enamides, the reaction proceeded cleanly and quantitatively to give the α-aminoacid derivatives (*R* isomer) with high enantioselectivities (90-99.7%) in both organic (hexane, methanol) solvents and in $scCO_2$. However, β,β-disubstituted α-enamide esters, which are very difficult to hydrogenate with high enantioselectivity, were reduced with significantly higher asymmetric induction in $scCO_2$ than in conventional solvents (hexane, methanol).

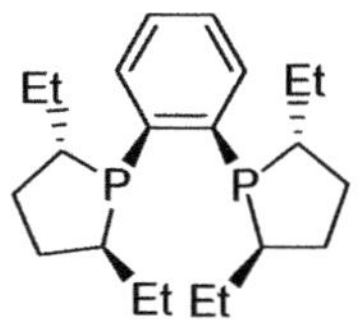

(*R*,*R*)-Et-DuPHOS

Quite recently, the use of $scCO_2$ as solvent for homogeneous hydrogenation of prochiral imines has been reported.

In conventional solvents, prochiral imines are reduced with high enantioselectivity by using chiral Ir catalysts and the process has also found practical application in industry [13]. Leitner and Pfaltz [14] have shown that such reaction can be performed in $scCO_2$ with comparable enatioselectivity and higher TON, by using cationic Ir complexes containing chiral phosphanodihydrooxazole ligands modified with perfluoroalkyl groups and having, as counteranion, lipophilic anions such as hexafluorophosphate, tetraphenylborate, or BARF.

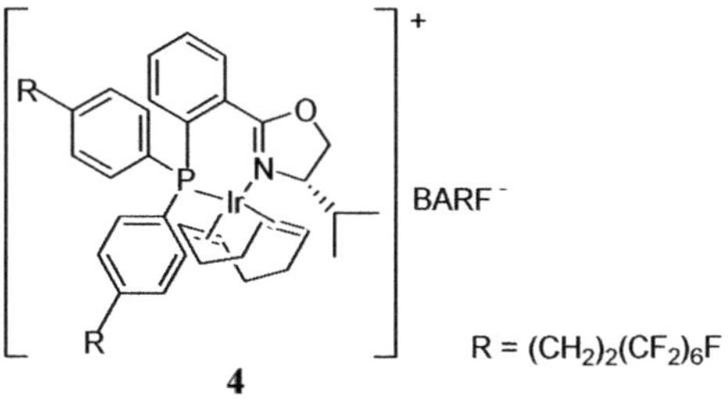

4

Interestingly, complex (**4**), containing the BARF anion, promotes (at 0.078% mol of catalyst) the hydrogenation of the industrially relevant imine phenyl-(1-phenyl-ethylidene)-amine to *N*-(1-phenylethylidene)aniline with high yield (98%) and asymmetric induction (81%). The enantioselectivity in $scCO_2$ well compares with the reaction in dichloromethane and the related industrial process. However, catalyst efficiency was remarkably higher in $scCO_2$ (6800 TON after 6 h) than in CH_2Cl_2 (1400 TON after 22 h), even using a lower load of catalyst. The catalyst, after separation from the product, could be reused several times with only a very modest decrease in the enantioselectivity.

All these studies emphasize the potential of $scCO_2$ as novel, non-toxic and eco-compatible reaction medium for asymmetric catalysis and, more specifically, for asymmetric hydrogenation reactions. Such processes are of great relevance for the synthesis of biologically active compounds, such as pharmaceutical, cosmetics and food additives, for which purity requirements are very stringent. Within this context, the use of $scCO_2$ as reaction medium could be very profitable, even if the enantioselectivities were not superior to those attained in conventional solvents.

3. OXIDATION REACTIONS

The use of $scCO_2$ as reaction medium for oxidation reactions is very attractive. Because of its inertness towards oxidation, $scCO_2$ represents a safe solvent and allows the avoidance of problems related with the formation of side-products from solvent oxidation. An early report describes the use of a bi-phasic water-CO_2 medium for the catalytic oxidation of cyclohexene to adipic acid, using $NaIO_4$ as oxidant and RuO_2 as catalyst. However, a very low catalytic activity was observed (TON=5) because of catalyst deactivation in the aqueous phase [2b]. A few other studies deal with the use of with *tert*-butylhydroperoxide (TBHP) as the oxidant. $Mo(CO)_6$ a very soluble species in $scCO_2$, effectively promotes the oxidation of olefinic substrates with TBHP in $scCO_2$ [15-17]. *trans*-1,2-Diols were formed using aqueous TBHP solutions (70 %), while, with anhydrous hydrocarbon solutions of TBHP, the epoxide was obtained [15a]. The highest yields and fastest rates of diol and epoxide formation were observed with *cis*-alkenes, while *trans*-alkenes were considerably less reactive. Cyclooctene was quantitatively converted into the epoxide in 3 h, at 86 °C and 220 bar of CO_2 pressure. Phenyl-substituted olefins showed a tendency to cleave to their corresponding aldehyde. *trans*-2-Heptene and *trans*-stilbene were totally unreactive.

Allylic and homoallylic alcohols have been easily oxidized with TBPH in liquid CO_2, at 25 °C, using $VO(O\text{-}^iPr_3)_3$ as catalyst. [16] Substrate conversion was quantitative and high selectivities (95-100%) for epoxide formation were found. Rates were higher than in hexane solutions.

Diastereoselective epoxidation of allylic alcohols with aqueous TBHP (70 %) has been achieved in $scCO_2$ using a vanadyl-salen system as catalyst [15b]. Satisfactory yields (50-100 %) were obtained under mild conditions (40 °C) after 24 h. Secondary allylic alcohols were preferentially converted, with interesting diastereoselectivity, into the *erythro*-product.

Recent studies have demonstrated that compressed CO_2 could be a suitable solvent also for asymmetric epoxidation of allylic alcohols [16]. However, rates and enantioselectivities are still too low, when compared with those observed in conventional solvents.

In oxidation catalysis, great attention is devoted to the use of molecular oxygen as selective "green" oxidant for synthetic purposes. Because of its complete miscibility with O_2, its inertness to oxidation, and its safety, $scCO_2$ can be a suitable solvent also for catalytic oxidation reactions using molecular oxygen as selective oxidant. However, studies in this area are still rare [18-24].

Koda and coworkers reported the aerobic oxidation, in both liquid and $scCO_2$, of cyclohexane in the presence of acetaldehyde, using an iron-porphyrin species, containing pentafluorophenyl groups, as catalyst [19]. Cyclohexanone and cyclohexanol were obtained as main oxidation products, but in very poor yield. Iron-porphyrin systems have been studied by Tumas as catalysts for oxidation of cyclohexene with O_2 in $scCO_2$ [20]. Selectivity of epoxidation reaction was modest (up to 34%) because of side-formation of allylic oxidation products. Nevertheless, in $scCO_2$, the epoxide formation was more selective than in CH_2Cl_2 (up to 21%), under otherwise similar conditions.

Recently, Jiang has described the O_2 oxidation of acrylic acid esters to acetals in $scCO_2$ [21]. The process is promoted by $PdCl_2$ and CuCl or $CuCl_2$, as co-catalyst. At 40 °C, in the presence of methanol, methylacrylate was converted into $(MeO)_2CHCH_2C(O)OMe$, as the main product, with high conversion and selectivity (up to 97%, at 10 bar O_2 pressure).

The oxidation of a variety of aromatic substrates in $scCO_2$ has been studied [22,23]. Recent reports show that a few Co(II) salts and complexes are active catalysts in $scCO_2$ for the aerobic oxidation of a number of aromatic hydrocarbons and substituted phenols.

A major drawback of using supercritical carbon dioxide for homogeneous catalytic oxidations is the limited number of homogeneous catalysts that are sufficiently soluble in unmodified $scCO_2$. Recently, to overcome this and related drawbacks, a new process has been developed in which the conventional solvent is almost totally replaced by dense CO_2 [24]. This novel reaction medium has been termed CO_2-expanded solvent medium. The conventional solvent is volumetrically expanded by dense carbon dioxide, while retaining in solution the substrate, solvent and catalyst. With respect to unmodified-$scCO_2$ or liquid-solvent processes, the new process shows several advantages, such as increased O_2 concentration in the CO_2-expanded solvent, no restriction in the choice of catalyst, high oxidation rates and low process pressure.

The examples discussed above well illustrate the considerable potential of $scCO_2$ as solvent for homogeneous catalytic oxidations of several organic substrates with O_2 or other oxidants. As a whole, the results so far obtained are quite encouraging and suggest that, in the future, this field could be a very promising research area.

4. POLYMERIZATION REACTIONS

Relatively few examples of utilization of $scCO_2$ as medium for metal complex promoted polymerization reactions have been reported to date.

A few groups have studied the ring opening metathesis polymerization (ROMP) of norbornene (relevant to the synthesis of the polymer Norsorex) and other cyclic alkenes, both in liquid and supercritical CO_2, using, as promoters $[Ru(H_2O)_6](Tos)_2$ [25], or the conventional metathesis catalysts (**5**) and (**6**) [26].

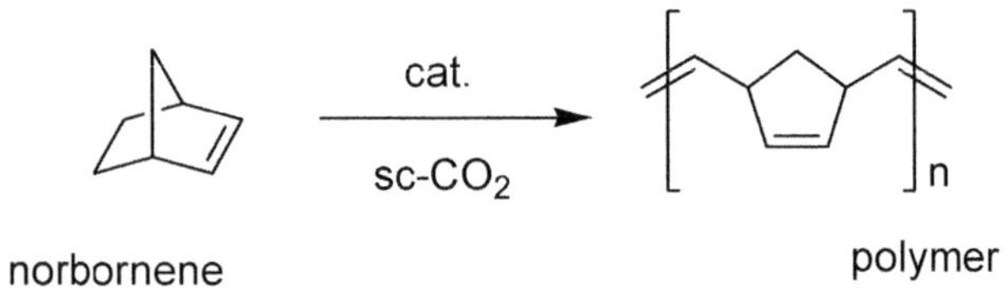

Scheme 2

Catalysts **5** and **6** promote the formation of norbornene polymer in excellent yield (93 and 97%, respectively). Polymer yield and *cis:trans* ratio were found to be similar in both liquid and $scCO_2$. Molecular weight distribution was narrower in $scCO_2$ (2.0) than in CH_2Cl_2 (2.3) or liquid CO_2 (3.2). The *cis:trans* ratio was dependent on the catalysts used, with **6** favouring the *trans*-configuration (78:22) and **5** the *cis*-geometry (66:34). However, the *cis:trans* ratios were very similar to those observed, with the same catalysts, in CH_2Cl_2. Interestingly, catalyst **6** promotes the complete conversion of the monomer in very short times (1 h), even with a substrate to catalyst ratio of 5300:1.

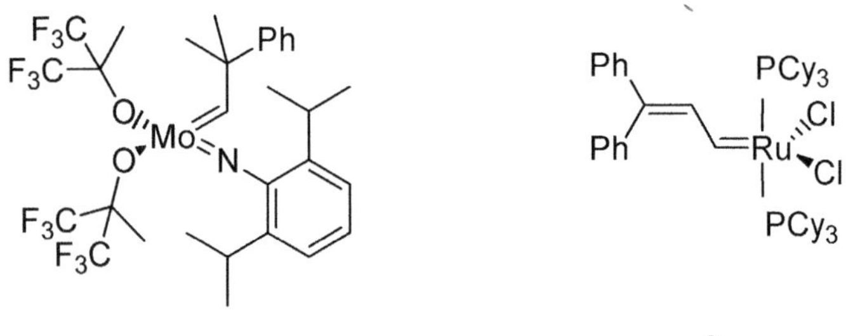

5 **6**

The polymerization of phenylacetylene, in liquid- or $scCO_2$, has been recently investigated by Leitner using Rh-based systems as catalysts [27]. Using (2,5-nbd)Rh(acac), as catalyst precursor, and a basic amine as co-catalyst, a substantially insoluble catalytic system was obtained, able, however, to promote, in compressed CO_2, the polymerization reaction with good selectivity for the *cis*-cisoidal isomer. It is worth noting that, in polar solvents (THF, for example) Rh-based catalysts usually promote, very selectively, the formation of the *cis*-transoid product. Preferential formation of *cis*-cisoidal polyphenylacetylene (PPA) was also obtained in *n*-hexane, but with lower polymerization rates and stereoselectivities than in compressed CO_2. A very soluble catalytic system was obtained from (nbd)Rh(acac) and the fluorinated phosphine $P[p\text{-}C_6H_4\text{-}(CH_2)_2(CF_2)_6F]_3$ in the presence of quinuclidine. The addition of

the ligand remarkably affected the *cis*-cisoid/*cis*-transoid ratio, increasing the selectivity for the formation of the *cis*-transoidal isomer.
An attractive example of metal-complex promoted polymerization in $scCO_2$ deals with the formation of polycarbonates by co-polymerization of epoxides with CO_2. Interestingly, this reaction can be promoted by both homogeneous [28] and heterogeneous [29] zinc or chromium-catalysts where $scCO_2$ serves not only as a solvent, but also as reactant.

The systems described above are fairly promising and demonstrate that $scCO_2$ is a useful, alternative solvent not only for radical and ionic polymerization reactions [30], but also for polymerisation processes promoted by metal complexes. This methodology can prove to be quite useful, especially if conventional techniques are not required, in the subsequent purification steps, to remove the metal from polymer.

5. HYDROFORMYLATION REACTIONS

Hydroformylation reactions are similar to hydrogenation reactions in that they require the use of gaseous reagents (H_2 and CO) in addition to solid metal catalysts. The benefits of using $scCO_2$ which is miscible with the reagent gases, allowing high concentrations of the reagent gases in the SCF reaction mixture are as applicable to hydroformylation as to hydrogenation, discussed in Section 5.1.

Rathke reported the first example of the utilization of $scCO_2$ in hydroformylation reactions in 1991. The hydroformylation of propene was studied using a cobalt carbonyl catalyst [31]. A higher selectivity (88%) for the linear product, *n*-butanal, was achieved in $scCO_2$, using $Co_2(CO)_8$ (10 mol %) and H_2 and CO (42 bar). This compared favourably to the reaction in benzene where higher pressures (80 bar H_2 and CO) and catalyst loading (15 mol %) were used to give an 83% yield of *n*-butanal. Guo *et al.* found a slight influence on the linear to branched ratio of products depending on the pressure and temperature. The temperature was held at 88 °C and the linear selectivity was found to increase from 73 to 81 % as the pressure doubled. A kinetic study of the reactions showed the activation energy of the reaction in $scCO_2$ to be 23 ± 1.4 kcal mol^{-1}. This is comparable to or lower than those measured in organic solvents, 27-35 kcal mol^{-1} [32]. Drawbacks of this reaction include the high catalyst loading required as well as elevated temperatures.

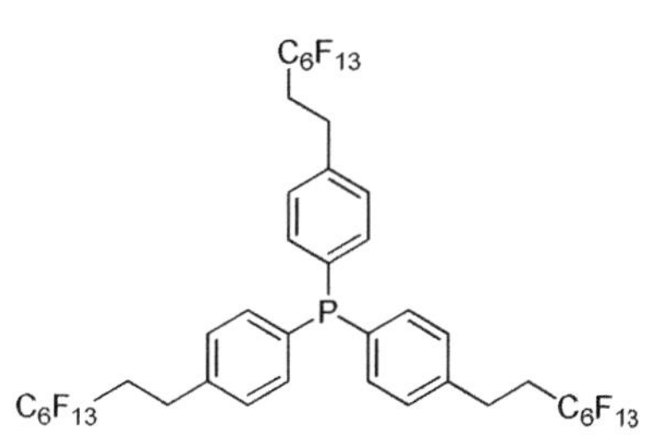

7

Other studies mainly deal with rhodium catalysts, which are commercially preferred to cobalt, due to their greater stability and higher reactivity allowing milder reaction conditions. In order to enhance the CO_2 solubility and therefore the catalytic activity, Leitner and Erkey independently developed fluorinated Rh-catalysts for hydroformylation reactions in $scCO_2$ [4, 33].

Leitner [34] described the use of Rh-phosphine [35] or Rh-phosphite modified catalysts obtained from Rh(hfacac)(cod) and a perfluoroalkylsubstituted triarylphosphine or triarylphosphite such as phosphines **7** [$P(p\text{-}C_6H_4\text{-}(CH_2)_2(CF_2)_6F)_3$] and **8** [$P(m\text{-}C_6H_4\text{-}(CH_2)_2(CF_2)_6F)_3$] and the perfluorinated phosphite **9** [$P(O\text{-}p\text{-}C_6H_4\text{-}(CH_2)_2(CF_2)_6F)_3$]. Comparative studies on the hydroformylation of 1-octene using the systems [Rh(hfacac)(cod)]/phosphine **8** in $scCO_2$ and [Rh(hfacac)(cod)]/PPh_3 in toluene under similar conditions indicated that the $scCO_2$ system showed a slightly higher activity and a remarkably higher selectivity towards the linear aldehyde. A high *normal:iso* ratio was observed also in the reactions with the phosphite **9** in $scCO_2$. Moreover, in the latter case, no evidence was found for double bond isomerization, a side reaction usually observed with Rh-phosphite modified catalysts in liquid solvents. In addition the use of [Rh(hfacac)(cod)] with any of the perfluorinated phosphines **7-9** allowed extraction of the product aldehydes from the reaction mixture by changing the temperature and pressure after complete conversion. The catalyst could then be re-used in a series of five catalytic runs without noticeable change in selectivity and activity. Interestingly the use of unmodified [Rh(hfacac)(cod)] in the hydroformylation of a range of olefins in $scCO_2$ also gave remarkably higher rates compared with control experiments in toluene under otherwise identical conditions. No conversion was observed in liquid CO_2. Higher rates were also observed in the hydroformylation of alkyl acrylates in $scCO_2$, with the use of a [Rh(hfacac)(cod)]/phosphine **7** catalytic system [36]. Changing the reaction medium from toluene to $scCO_2$ resulted in a 10-20 fold increase in TOF. It was postulated that this significant enhancement is due to CO_2 acting as an electron acceptor, stabilising key intermediates and hence promoting the overall reaction.

Erkey [37, 38] described the use of phosphine ligands bearing small trifluoromethyl groups to achieve large increases in soublility in $scCO_2$ without adversely affecting the catalyst activity. The catalysts *trans*-$RhCl(CO)P(p\text{-}CF_3C_6H_4)_3)_2$ and the superior $RhH(CO)(P(p\text{-}CF_3C_6H_4)_3)_3$ were used in the hydroformylation of 1-octene. Near complete conversion was seen in 3.5 h with a *normal:iso* ratio of 3.0. This is comparable to the selectivity of the conventional catalyst $RhH(CO)PPh_3$ in benzene [39]. Further studies on the effects of ligand modification indicated catalytic activity increased with decreasing basicity of the phosphine according to the series: $P(3,5\text{-}(CF_3)_2C_6H_3)_3)_3$ > $P(4\text{-}CF_3C_6H_4)_3)_3$ > $P(4\text{-}OCF_3C_6H_4)_3)_3$ [40].

The use of simple, $scCO_2$-soluble trialkylphosphines such as PEt_3 was reported by Cole-Hamilton as an alternative to the expensive fluorinated derivatives in the hydroformylation of hex-1-ene. The catalyst was prepared *in situ* from [$Rh_2(OAc)_4$] and PEt_3 and proceeded in a similar rate to that in toluene but with a slightly improved *normal:iso* ratio [41]. The use of insoluble phosphites $P(OPh)_3$ and $P(OC_6H_4$-*p*-$C_9H_{19})_3$ with the same Rh-source showed high activity and much higher selectivity for the straight chain product compared with comparable conditions in conventional organic solvents [42]. However, the ligand showed degradation over extensive usage. Carbonylated phosphine **10**, synthesized by Xiao, was shown to have enhanced solubility to PPh_3 in $scCO_2$. However, hydrogenation rates using this phosphine were disappointingly low in $scCO_2$ compared with toluene [43].

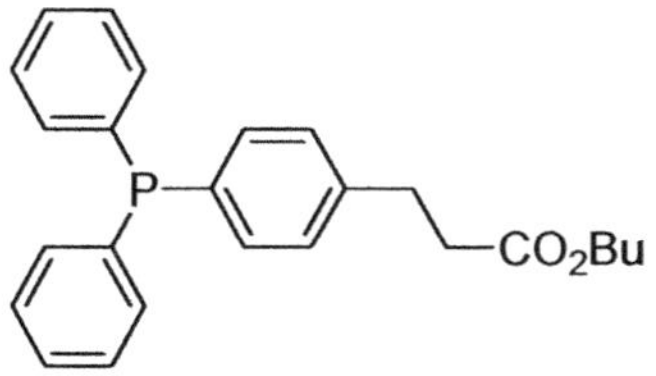

10

The asymmetric hydroformylation of prochiral olefins also has been investigated. The attempted use of catalysts bearing the chiral ligand (R,S)-BINAPHOS, **11** were not effective in homogeneous hydroformylation reactions in $scCO_2$ due to its low solubility. A total loss of enantioselectivity was observed under conditions with a homogeneous $scCO_2$ phase [44].

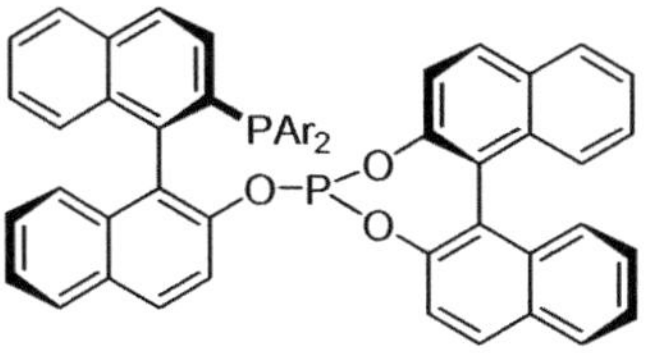

11 : Ar = Ph,
12 : Ar = *m*-C_6H_4-$(CH_2)_2(CF_2)_5CF_3$

The use of perfluoroalkyl-substituted BINAPHOS derivative **12** was found to promote the hydroformylation of prochiral olefines with very high level of enantiocontrol in $scCO_2$ [45, 46]. The catalytic activity and enantiocontrol obtained in $scCO_2$ were comparable to those obtained in conventional solvents with BINAPHOS **11**. In addition, unprecedented high regioselectivity for the branched aldehydes was obtained. This regioselectivity appears to be derived from the CO_2-philic substituent, as the mproved regioselectivity was also observed using the fluorinated derivative **12** in benzene and under neat conditions.

6. CARBON-CARBON BOND FORMING REACTIONS

6.1 Palladium-catalysed coupling reactions

A number of examples of Pd-promoted C-C coupling reactions of aryl halides in $scCO_2$ have been studied. Conventional palladium / phosphine catalysts are very insoluble in $scCO_2$ and early studies used fluorinated ligands to enhance the solubility of the catalytic species.

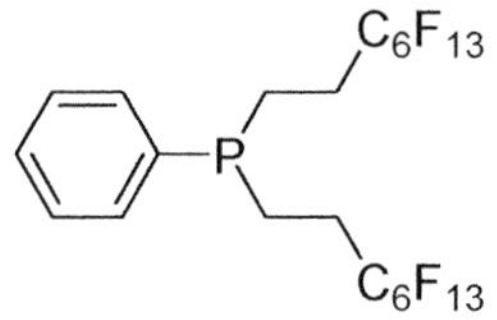

13

The perfluorinated phosphine **13** was developed by Holmes [47] and was able to solubilize its palladium complex, made *in situ* with $Pd(OAc)_2$ in $scCO_2$, a significant achievement considering the insolubility of Pd catalysts in $scCO_2$. The Heck coupling of electron poor alkenes gave higher yields, but a lower TON than the comparable reaction in acetronitrile at the same temperature. For example, the reaction between iodobenzene and methyl acrylate was achieved in 91 % yield (5 mol %, 18 TON) compared with a 50 % yield in acetonitrile (1 mol %, 50 TON) [48]. Suzuki and Sonogashira reactions were also achieved, although yields were lower than in conventional solvents [49, 50].

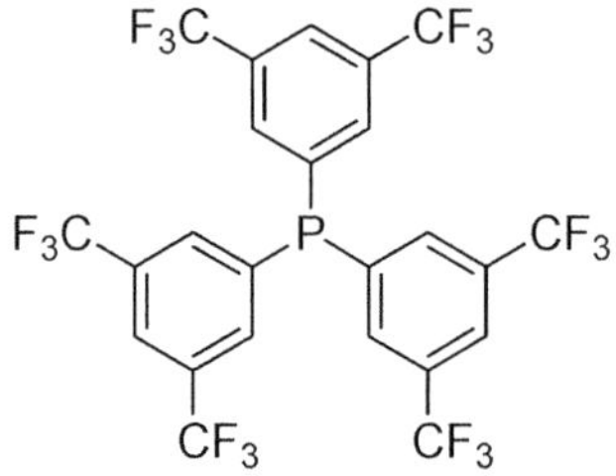

14

Around the same time, Tumas also reported the Heck reaction of iodobenzene in $scCO_2$ [51]. Fluorinated derivatives of triphenylphosphine, **14** $P(3,5\text{-}(CF_3)_2\text{-}C_6H_3)_3$ and **15** $P(C_6F_5)_3$ were used to enhance the solubility of these catalysts in $scCO_2$. However, even with these ligands the reaction phases were described as dark and opaque, preventing visual determination of whether the reaction was homogeneous. The yield of the reaction was comparable to that in toluene, and it has been speculated that the reaction may have been heterogeneous, occurring in a liquid phase [52]. The

Stille reaction was also investigated, and found to proceed in higher yields with the fluorinated phosphine **14**, than conventional, non-fluorinated phosphines. The improvement in yield along the series PPh_3 < P(2-furyl)$_3$ < phosphine **14** was suggested by Tumas to be due to the increasing solubility of the phosphine, although the rate of reaction also increased when the fluorinated phosphine was used in toluene.

The use of commercially available phosphines is considered an important breakthrough if the use of $scCO_2$ in palladium catalysis is to be widely adopted. Rayner reported the use of palladium acetate with triphenylphosphine and tris(2-furyl)phosphine for the Heck reaction with iodobenzene and methyl acrylate [53]. However, high catalyst loadings (14 mol %) and long reaction times (40 h) were required for good conversions. The use of fluorinated palladium sources such as $Pd(OCOCF_3)_2$ gave more active catalytic systems, allowing the reduction of the catalyst loadings to 2 % with good conversions. An intramolecular Heck cyclisation was also reported by Rayner, using the same catalytic system, where both higher yields and suppression of isomerisation of the exocylic double bond formed in the reaction was observed in $scCO_2$ [54].

Recently Holmes has reported the use of tri(*t*-butylphosphine) as a highly active ligand in $scCO_2$ [55]. This ligand has been used by Fu over the last few years in many palladium-catalysed reactions in conventional solvents. Its compatability with $scCO_2$ may be in part due to its aliphatic nature, which may confer some solubility in $scCO_2$. A combination of palladium acetate with P(*t*-Bu)$_3$ was shown to catalyze Heck and Suzuki reactions with low catalyst loadings and good yields, compared with previous studies in $scCO_2$, although inferior to those obtained by Fu [56, 57]. The first coupling of bromides was also reported, which represents another significant step towards a viable industrial process.

While it has been demonstrated that palladium-catalysed reactions can proceed in $scCO_2$, in general there are few examples of faster reaction or higher yields than are achievable in conventional solvents. The benefit of using $scCO_2$ is therefore currently limited to the environmental advantages provided by the use of $scCO_2$, and potential opportunities for processing.

6.2 Ring closing metathesis

Ring closing metathesis (RCM) is an important key step for the preparation of molecules of biological interest. This reaction is promoted by Mo- or Ru carbene complexes in chlorinated or aromatic solvents and requires high dilution for the preparation of medium and large rings, that can find practical application in pharmaceutical industry. Very recently, the RCM of a variety of substrates has been successfully accomplished in $scCO_2$ by Leitner, Fürstner and coworkers [26], by using the same catalytic systems employed in conventional solvents. This approach has been followed for the synthesis of the natural product epilachnen. The metathesis reaction is strongly affected by pressure. The RCM product is formed in high yield at high pressure, while oligomers are mainly obtained at low pressure, via acyclic diene metathesis. Another application of CO_2 was as a protective medium for amines capable of forming a carbamic acid. Evidence for the reversible formation of the corresponding carbamic acid was shown by ^1H NMR experiments, obtained in compressed CO_2.

6.3 Diels-Alder reactions

The Diels-Alder reaction has been studied extensively in CO_2 [58]. An early example was reported in 1987 by Paulaitis and Alexander on the Diels-Alder reaction of maleic anhydride and isoprene [59]. It was found that overall the rate increased slightly with pressure but at 35 °C, close to the critical pressure a dramatic variation in rate was observed. This was said to be a result of a change in activation volume close to the critical point.

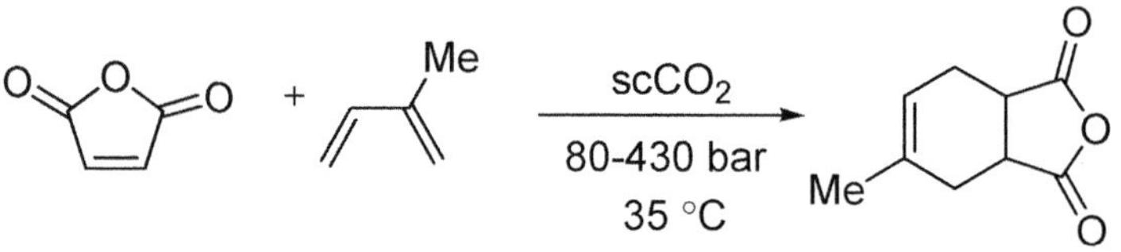

Scheme 3

Ikushima also reported an enhancement in the reaction rate of methyl acrylate and isoprene when carried out in $scCO_2$ [60]. The pressure also had an effect on the product distribution. In conventional solvents the 1,4-product predominated over the 1,3-product and the same was found when the reaction was carried out in CO_2 except when the conditions reached near critical and reverse selectivity was observed.

The calculations from data reported by Ikushima [59] and Paulaitis and Alexander [58] were repeated by Danheiser and coworkers and gave good agreement for rate enhancement [61]. The regiochemical selectivity was also re-investigated for the methyl acrylate-isoprene system reported by Ikushima [59]. However, it was found that selectivity was similar to that observed in conventional solvents with no dramatic reversal in selectivity close to the critical point.

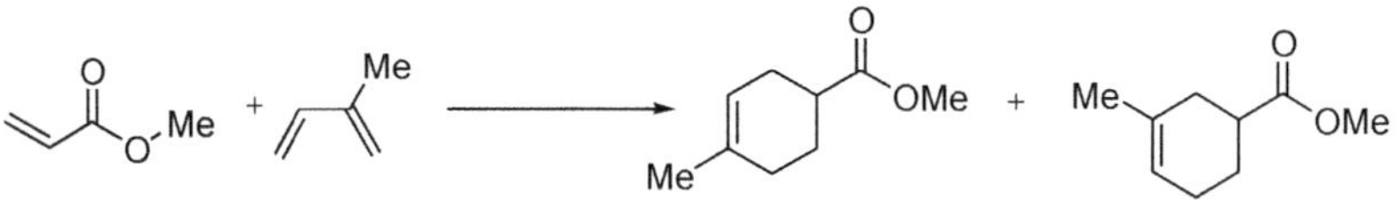

Scheme 4

Danheiser and coworkers have also reported a silica-catalysed Diels-Alder reaction between a number of dienes and dienophiles in $scCO_2$ [62]. The silica-catalysed reaction was found to enhance product yield, rate and selectivity, although the selectivity of the reaction was found to be independent of pressure. However, the product yield was found to decrease when the pressure was increased owing to the higher solubility of the reactants at higher pressures. This meant that fewer reactants were adsorbed on the surface of the catalyst, therefore causing a shift away from the catalytic process.

Rayner and coworkers developed a Lewis acid-catalysed Diels-Alder reaction involving scandium trifluoromethane sulfonate [63]. The catalytic system was optimized for the reaction between *n*-butyl acrylate and cyclopentadiene. It was found that quantitative conversion could be achieved at 50 °C, in 15 hours with 6.5 mol % catalyst loading. Increasing the temperature to 80 °C allowed complete conversion with a reduced catalyst loading of 1 mol %. Dramatic control of the *endo* to *exo* ratio was observed when the pressure was varied. The ratio reached a maximum of 24:1 before falling when the pressure was increased further. The use of $scCO_2$ as a solvent for this reaction allowed optimization of the product selectivity, which was significantly higher than conventional solvents such as toluene (10:1) and chloroform (11:1).

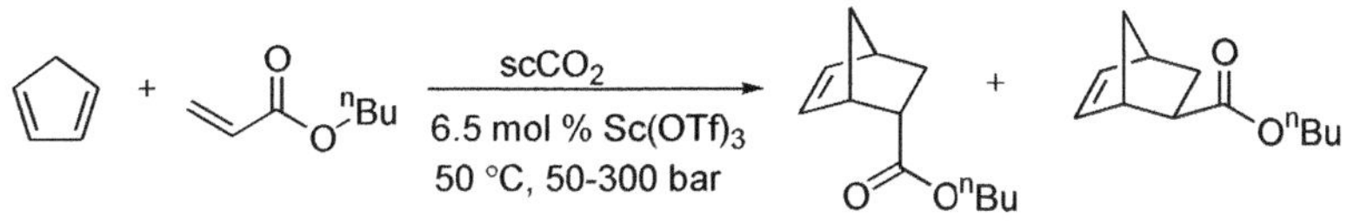

Scheme 5

Kobayashi also reported Diels-Alder reactions in $scCO_2$ using scandium perfluoroalkanesulfonates [64]. The catalyst activity was improved by increasing the length of the perfluoroalkyl chain and therefore enhancing solubility.

The Diels-Alder reaction can be successfully carried out in CO_2 using a number of dienes and dienophiles. However, the effects of fluid pressure and density on rates and selectivities have shown conflicting results in several groups [1a].

6.4 Other cyclisation reactions

The Pauson-Khand reaction involving the co-cyclisation of an alkyne with an alkene and CO to obtain cyclopentenone rings was found by Jeong to proceed smoothly in $scCO_2$ using $Co_2(CO)_8$ as catalyst [65]. Rh- and Ru-carbonyl complexes have been also used as catalysts for the Pauson-Khand reaction, but their activity in $scCO_2$ was lower than in organic solvents.

The nickel-catalysed coupling of two alkyne molecules with CO_2 to give 2-pyrones was achieved in $scCO_2$ by Reetz, who coupled 3-hexyne with CO_2 to give tetraethyl-2-pyrone [66]. The use of an *in situ* generated catalyst from $Ni(cod)_2$ and the chelating phosphine, **16** $Ph_2P(CH_2)_4PPh_2$ gave a modest TON and selectivity for pyrone with a lower conversion than in benzene. Replacing phosphine **16** with the simple alkylphosphine $P(CH_3)_3$ gave a more active catalytic system that allowed the synthesis of pyrone at lower temperatures and with improved selectivity, though still with a lower rate than that observed for the same system in conventional solvents. In addition, a shorter catalyst lifetime was observed in $scCO_2$. This was attributed to the formation of nickel carbonyl complexes formed by nickel catalysed reduction of CO_2 by $P(CH_3)_3$, which causes serious concern for processes utilizing trialkylphosphines in $scCO_2$ [67].

Avilés has recently achieved a cyclotrimerization of alkynes in $scCO_2$ using a colbalt complex as activator [68]. $CpCo(CO)_2$ displayed a high solubility in $scCO_2$. In this case the possibility of CO_2 incorporation to form pyrones was not observed, and a mixture of cyclotrimers were obtained. However, yields in $scCO_2$ were lower than those obtained in toluene or neat alkyne.

REFERENCES

1. a) The use of supercritical fluids in synthetic organic chemistry. R. S. Oakes, A. A. Clifford, C. M. Rayner, J. Chem. Soc., Perkins Trans. 1, 2001, 917 - 941; b) Chemische Synthese in überkritischem Kohlendioxid – die bessere Lösung. W. Leitner, Erdöl, Erdgas, Kohle, 2000, 116, 137. c) P. J. Jessop, W. Leitner, in 'Chemical synthesis using supercritical fluids', P. J. Jessop, W. Leitner (eds), Wiley-VCH, Weinheim, 1999, p. 351, ISBN 3527296050. d) R. A. Roger in 'Current topics in flavours and fragrances. Towards a new millennium of discovery', K. A. D. Swift (ed.), Kluwer, Boston, 1999, p. 59, ISBN 075140490x. e) Homogeneous catalysts for application in supercritical carbon dioxide as a 'green' solvent, W. Leitner, C.R. Acad. Sci., Ser.IIc: Chim, 2000, 3, 595 - 600. f) Carbon dioxide as an environmentally benign reaction medium for chemical synthesis, W. Leitner, Appl. Organomet. Chem., 2000, 14, 809-814. g) Homogeneously-catalyzed syntheses in supercritical fluids, P. G. Jessop, Top. Catal., 1998, 5, 95-103. h) Supercritical carbon dioxide as a substitute solvent for chemical synthesis and catalysis, D. A. Morgenstern, R. M. LeLacheur, D. K. Morita, S. L. Borkowsky, S. Feng, G. H. Brown, L. Luan, M. F. Gross, M. J. Burk, W. Tumas in 'Green Chemistry : designing chemistry for the environment', ACS Symp. Ser. 626, P. T. Anastas, T. C. Williamson (eds), ACS, Washington DC, 1996, pp. 132–151, ISBN 0841233993. i) E. Dinjus, R. Fornika, M. Scholz in 'Chemistry under extreme or non-classical conditions', R. van Eldik, C.D. Hubbard (eds), Wiley, New York, 1996, p. 219, ISBN 0471165611.
2. Design, synthesis, and evaluation of novel, highly CO_2-soluble chelating agents for removal of metals, A. V. Yadzi, E. J. Beckmann, Ind. Eng. Chem. Res., 1996, 35, 3644-3652.
3. Asymmetric catalytic hydrogenation reactions in supercritical carbon dioxide, M. J. Burck, S. Feng, M. F. Gross, W. Tumas, J. Am. Chem. Soc., 1995, 117, 8277-8278.
4. a) Verwendung perfluoralkylsubstituierter Phosphorverbindungen als Liganden für die homogene Katalyse in überkritischem Kohlendioxid, W. Leitner, S. Kainz, D. Koch, K. Wittmann, Ger. Offen. DE 19702025 A1 1998 (Chem. Abstr., 1998, 129, 175353b). b) Synthesis and characterization of {[(COD)Rh(bis-(2R,3R)-2,5-diethylphospholanobenzene)](+)BARF(-)} for use in homogeneous catalysis in supercritical carbon dioxide, B. Guzel, A. M. Omary, J. P. Fackler and A. Akgerman, Inorg. Chem. Acta, 2001, 325, 45-50. c) Highly efficient enantioselective catalysis in supercritical carbon dioxide using the perfluoroalkyl-substituted ligand (R,S)-3-H^2F^6-BINAPHOS, G. Francio, K. Wittmann, W. Leitner, J. Organomet. Chem., 2001, 621, 130-142. d) Mechanistic aspects of

dihydrogen activation and transfer during asymmetric hydrogenation in supercritical carbon dioxide, S. Lange, A. Brinkmann, P. Trautner, K. Woelk, J. Borgou, W. Leitner, Chirality, 2000, 12, 450-457. e) Novel rhodium(I) catalysts for ecologically benign homogeneous catalysis in supercritical carbon dioxide, M. A. Omary, Z. K. Lopez, B. Lin, J. P. Fackler, A. Akgerman, Abstr. Pap. Am. Chem. Soc. 2000, 220: 176-INOR Part 1.

5. a) Production of formic acid. P. G. Jessop, T. Ikariya, R. Noyori, JP7173098, 1995. b) Homogeneous catalytic hydrogenation of supercritical carbon dioxide, P. G. Jessop, T. Ikariya, R. Noyori, Nature, 1994, 368, 231-233. c) Homogeneous hydrogenation of carbon dioxide, P. G. Jessop, T. Ikariya, R. Noyori, Chem. Rev., 1995, 95, 259-272.
6. Homogeneous catalysis in supercritical fluids: Hydrogenation of supercritical carbon dioxide to formic acid, alkyl formates, and formamides, P.G. Jessop, Y. Hsiao, T. Ikariya, R. Noyori, J. Am. Chem. Soc., 1996, 118, 344-355.
7. Methyl formate synthesis by hydrogenation of supercritical carbon dioxide in the presence of methanol, P.G. Jessop, Y. Hsiao, T. Ikariya, R. Noyori, J. Chem. Soc., Chem. Commun., 1995, 707-708.
8. Catalytic production of dimethylformamide from supercritical carbon dioxide, P. G. Jessop, Y. Hsiao, T. Ikariya, R. Noyori, J. Am. Chem. Soc., 1994, 116, 8851-8852.
9. Highly active ruthenium complexes with bidentate phosphine ligands for the solvent-free catalytic synthesis of N,N-dimethylformamide and methyl formate, O. Krocher, R. A. Koppel, A. Baiker, J. Chem. Soc., Chem. Commun., 1997, 453-454.
10. Silica hybrid gel catalysts containing Group(VIII) transition metal complexes: Preparation, structural, and catalytic properties in the synthesis of N,N-dimethylformamide and methyl formate from supercritical carbon dioxide, O. Krocher, R. A. Koppel, M. Froba, A. Baiker, J. Catal., 1998, 178, 284-298.
11. Asymmetric hydrogenation of α,β-unsaturated carboxylic acids in supercritical carbon dioxide, J. Xiao, S. C. A. Nefkens, P.G. Jessop, T. Ikariya, R. Noyori, Tetrahedron Lett., 1996, 37, 2813-2816.
12. S. Kainz, D. Koch, W. Leitner, in 'Selective reactions of metal activated molecules', H. Werner, W. Schreier (eds), Viewey, Wiesbaden, 1998, p. 151, ISBN 352806918x.
13. Enantioselective catalysis for agrochemicals. The case histories of (S)-metolachlor, (R)-metalaxyl and clozylacon, H .U. Blaser, F. Spindler, Top. Catal., 1997, 4, 275-282.
14. Iridium-catalyzed enantioselective hydrogenation of imines in supercritical carbon dioxide, S. Kainz, A. Brinkmann, W. Leitner, A. Pfaltz, J. Am. Chem. Soc., 1999, 121, 6421-6429.
15. a) Oxidation of alkenes in supercritical carbon dioxide catalyzed by molybdenum hexacarbonyl, G. R. Haas, W. Kolis, Organometallics, 1998, 17, 4454-4460. b) The diastereoselective epoxidation of olefins in supercritical carbon dioxide, G. R. Haas, J. W. Kolis, Tetrahedron Lett., 1998, 39, 5923-5926.
16. Selective epoxidation in dense phase carbon dioxide, D. K. Morita, D. R. Pesiri, W. H. Glaze, W. Tumas, Chem. Commun., 1998, 1015-1016.
17. Organometallics of transition metals in supercritical carbon dioxide: Solubilities, reactions, catalysis, U. Kreher, S. Schebesta, D. Walther, Z. Anorg. Allg. Chem., 1998, 624, 602-612.
18. L. Locker, D. Koch, W. Leitner, in 'Selective oxidation in petrochemistry', G. Emig, C. Kohlpaintner, B. Lucke (eds), DGMK, Hamburg, 1998, p. 209, ISBN 3931850447.
19. Aerobic oxidation of cyclohexane catalyzed by Fe(III)(5,10,15,20-tetrakis(pentafluorophenyl)porphyrin)Cl in sub- and super-critical CO_2, X.-W. Wu, Y. Oshima, S. Koda, Chem. Lett., 1997, 1045-1046.
20. Metalloporphyrin-catalyzed homogeneous oxidation in supercritical carbon dioxide, E. R. Birnbaum, R. M. Le Lacheur, A. C. Horton, W. Tumas, J. Mol. Catal. A: Chem. 1999, 139, 11-24.
21. Palladium(II)-catalyzed oxidation of acrylate esters to acetals in supercritical carbon dioxide, L. Jia, H. Jiang, J. Li, Chem. Commun., 1999, 985-986.

22. Catalytic oxidation reactions in supercritical carbon dioxide, T. Aida, Chorinkai Saishin Gijutsu, 2000, 4, 72-76 (Chem. Abstr., 2000, 133, 334954z).
23. Autoxidation of substituted phenols catalyzed by cobalt schiff base complexes in supercritical carbon dioxide, G. T. Musie, M. Wei, B. Subramaniam, D. H. Busch, Inorg. Chem., 2001, 40, 3336-3341.
24. a) Catalytic oxidation of organic substrates by transition metal complexes in organic solvent media expanded by supercritical or subcritical carbon dioxide, B. Subramaniam, D. H. Busch, M. Wei, G. T. Musie, PCT Int. Appl. WO 01/77052, 2001, (Chem. Abstr., 2001, 135, 305486j). b) Homogeneous catalytic oxidation of organic substrates in CO_2-expanded solvent media, M. Wei, G. T. Musie, D. H. Busch, B. Subramaniam, Abstr. Pap. Am. Chem. Soc. 2001, 222: 294-INOR Part 1.
25. Stereochemistry of ring-opened metathesis polymers prepared in liquid CO_2 at high pressure using $Ru(H_2O)_6(Tos)_2$ as catalyst, J. G. Hamilton, J. J. Rooney, J. M. DeSimone, C. Mistele, Macromolecules, 1998, 31, 4387-4389.
26. a) Olefin metathesis in compressed carbon dioxide, A. Fürstner, D. Koch, K. Langemann, W. Leitner, Angew. Chem. Int. Ed. Engl., 1997, 36, 2466-2469 b) Olefin metathesis in supercritical carbon dioxide, A. Fürstner, L. Ackermann, K. Beck, H. Hori, D. Koch, K. Langemann, M. Liebl, C. Six, W. Leitner, J. Am. Chem. Soc., 2001, 37, 9000-9006.
27. a) Rhodium-catalyzed phenylacetylene polymerization in compressed carbon dioxide, H. Hori, C. Six, W. Leitner, Macromolecules, 1999, 32, 3178-3182. b) Microstructure and morphology of polyphenylacetylene prepared in compressed carbon dioxide, H. Hori, C. Six, W. Leitner, Appl. Organomet. Chem., 2001, 15, 145-147.
28. a) Copolymerization of 1,2-epoxycyclohexane and carbon dioxide using carbon dioxide as both reactant and solvent, M. Super, E. Berluche, G. Costello, E. Beckman, Macromolecules, 1997, 30, 368-372. b) Copolymerization of CO_2 and cyclohexene oxide, M. Super, E. Beckman, Macromol. Symp., 1998, 127, 89-108.
29. a) Supercritical carbon dioxide as solvent for the copolymerization of carbon dioxide and propylene-oxide using a heterogeneous zinc carboxylate catalyst, D. J. Darensbourg, N. W. Stafford, T. Katsurao, J. Mol. Catal., 1995, 104, L1-L4. b) Copolymerization of CO_2 and 1,2-cyclohexene oxide using a CO_2-soluble chromium porphyrin catalyst, S. Mang, A. I. Cooper, M. E. Colclough, N. Chauhan, A. B. Holmes, Macromolecules, 2000, 33, 303-308.
30. a) Polymerizations in supercritical carbon dioxide, L. Kendall, D. A. Canelas, J. L.Young, J. M. DeSimone, Chem. Rev., 1999, 99, 543-563. b) Polymer supported chromium porphyrin as catalyst for polycarbonate formation in supercritical carbon dioxide, L. M. Stamp, S. A. Mang, A. B. Holmes, K. A. Knights, Y. R. de Miguel, I. F. McConvey, Chem. Commun., 2001, 2502-2503.
31. Propylene hydroformylation in supercritical carbon dioxide, J. W. Rathke, R. J. Klingler, T. R. Krause, Organometallics, 1991, 10, 1350-1355.
32. Hydroformylation of propylene in supercritical carbon dioxide, Y. Guo, A. Akgerman, Ind. Eng. Chem. Res., 1997, 36, 4581-4585.
33. Perfluoroalkyl-substituted arylphosphanes as ligands for homogeneous catalysis in supercritical carbon dioxide, S. Kainz, D. Koch, W. Baumann, W. Leitner, Angew. Chem. Int. Ed. Engl., 1997, 36 1628-1630.
34. Rhodium-catalyzed hydroformylation in supercritical carbon dioxide, D. Koch, W. Leitner, J. Am. Chem. Soc., 1998, 120, 13398-13404.
35. Hydroformylierung mit unmodifizierten Rhodiumkatalysatoren in überkritischem Kohlendioxid, W. Leitner, D. Koch, Ger. Offen. DE 19730783 A1. 1999 (Chem. Abstr., 1999, 130, 140776k).
36. Rapid hydroformylation of alkyl acrylates in supercritical CO_2, Y. Hu, W. Chan, A. M. Banet Osuna, A. M. Stuart, E. G. Hope, J. Xiao, Chem. Commun., 2001, 725-726.
37. Homogeneous catalytic hydroformylation of 1-octene in supercritical carbon dioxide using a novel rhodium catalyst with fluorinated arylphosphine ligands, D. R. Palo, C. Erkey, Ind. Eng. Chem. Res., 1998, 37, 4203-4206.

38. Homogeneous hydroformylation of 1-octene in supercritical carbon dioxide with $[RhH(CO)(P(p\text{-}CF_3C_6H_4)_3)_3]$, D. R. Palo, C. Erkey, Ind. Eng. Chem. Res., 1999, 38, 2163-2165.
39. Homogeneous hydroformylation of alkenes with hydridocarbonyltris(triphenylphosphine)rhodium(I) as catalyst, K. C. Brown, G. Wilkinson, J. Chem. Soc. A., 1970, 2753-2764.
40. Effect of ligand modification on rhodium-catalyzed homogeneous hydroformylation in supercritical carbon dioxide, D. R. Palo, C. Erkey, Organometallics, 2000, 19, 81-86.
41. Hydroformylation of hex-1-ene in supercritical carbon dioxide catalysed by rhodium trialkylphosphine complexes, I. Bach, D. J. Cole-Hamilton, Chem. Commun., 1998, 1463-1464.
42. Hydroformylation reactions in supercritical carbon dioxide using insoluble metal complexes, M. F. Sellin, D. J. Cole-Hamilton, J. Chem. Soc., Dalton Trans., 2000, 11, 1681-1683.
43. Carbonylated phosphines as ligands for catalysis in supercritical CO_2, Y. Hu, W. Chen, L. Xu, J. Xiao, Organometallics, 2001, 20, 3206-3208.
44. Catalytic asymmetric hydroformylation in the presence of compressed carbon dioxide, S. Kainz, W. Leitner, Catal. Lett., 1998, 55, 223-225.
45. Highly regio- and enantio-selective rhodium-catalysed asymmetric hydroformylation without organic solvents, G. Francio, W. Leitner, Chem. Commun., 1999, 1663-1664.
46. Highly efficient enantioselective catalysis in supercritical carbon dioxide using the perfluoroalkyl-substituted ligand (R,S)-3-H^2F^6-BINAPHOS, G. Francio, W. Leitner, J. Organomet. Chem., 2001, 621, 130-142.
47. Palladium-catalysed carbon-carbon bond formation in supercritical carbon dioxide, M. A. Carroll, A. B. Holmes, Chem. Commun., 1998, 1395-1396.
48. A polymer-bound bidentate-phosphine palladium complex as a catalyst in the Heck arylation, P. W. Wang, M. A. Fox, J. Org. Chem., 1994, 59, 5358-5364.
49. The palladium-catalyzed cross-coupling reaction of phenylboronic acid with haloarenes in the presence of bases, N. Miyaura, T. Yanagi, A. Suzuki, Synth. Commun., 1981, 11, 513-519.
50. A convenient synthesis of acetylenes : catalytic substitutions of acetylenic hydrogen with bromoalkenes and bromopyridines, K. Sonogashira, Y. Tohda, N. Hagihara, Tetrahedron Lett., 1975, 4467-4470.
51. Palladium-catalyzed cross-coupling reactions in supercritical carbon dioxide, D. K. Morita, D. R. Pesiri, S. A. David, W. H. Glaze, W. Tumas, Chem. Commun., 1998, 1397-1398.
52. Homogeneous catalysis in supercritical fluids, P. G. Jessop, T. Ikariya, R. Noyori, Chem. Rev., 1999, 99, 475-493.
53. Use of fluorinated palladium sources for efficient Pd-catalysed coupling reactions in supercritical carbon dioxide, N. Shezad, R. S. Oakes, A. A. Clifford, C. M. Rayner, Tetrahedron Lett., 1999, 40, 2221-2224.
54. Suppression of double bond isomerisation in intramolecular Heck reactions using supercritical carbon dioxide, N. Shezad, A. A. Clifford, C. M. Rayner, Tetrahedron Lett., 2001, 42, 323-325.
55. Palladium-catalysed cross-coupling reactions in supercritical carbon dioxide, T. R. Early, M. A. Carroll, R. S. Gordon, A. B. Holmes, I. F. McConvey, R. E. Shute, Chem. Commun., 2001, 1966-1967.
56. Heck reactions in the presence of $P(t\text{-}Bu)_3$: Expanded scope and milder reaction conditions for the coupling of aryl chlorides, F. Littke, G. C. Fu, J. Org. Chem., 1999, 64, 10-11.
57. Versatile catalysts for the Suzuki cross-coupling of arylboronic acids with aryl and vinyl halides and triflates under mild conditions, F. Littke, C. Dai, G. C. Fu, J. Am. Chem. Soc., 2000, 122, 4020-4028.
58. Reactions at supercritical conditions - applications and fundamentals, P. E. Savage, S. Gopalan, T. I. Mizan, C. J. Martino, E. E. Brock, AIChE J., 1995, 41, 1723-1778.
59. Reactions in supercritical fluids - A case-study of the thermodynamic solvent effects on a Diels-Alder reaction in supercritical carbon dioxide, M. E. Paulaitis, G. C. Alexander, Pure Appl. Chem., 1987, 59, 61-68.

60. A Diels-Alder reaction in supercritical carbon dioxide medium, Y. Ikushima, S. Ito, T. Asano, T. Yokoyama, N. Saito, K. Hatakeda, T. Goto, J. Chem. Eng. Jpn., 1990, 23, 96-98.
61. Kinetic correlation of Diels-Alder reactions in supercritical carbon dioxide, R. D. Weinstein, A. R. Renslo, R. L. Danheiser, J. G. Harris, J. W. Tester, J. Phys. Chem., 1996, 100, 12337-12341.
62. Silica-promoted Diels-Alder reactions in carbon dioxide from gaseous to supercritical conditions, R. D. Weinstein, A. R. Renslo, R. L. Danheiser, J. W. Tester, J. Phys. Chem. B, 1999, 103, 2878-2887.
63. Use of scandium tris(trifluoromethanesulfonate) as a Lewis acid catalyst in supercritical carbon dioxide: efficient Diels-Alder reactions and pressure dependent enhancement of endo : exo stereoselectivity, R. S. Oakes, T. J. Heppenstall, N. Shezad, A. A. Clifford, C. M. Rayner, Chem. Commun., 1999, 1459-1460.
64. Lewis acid catalysis in supercritical carbon dioxide. Use of scandium tris(heptadecafluorooctanesulfonate) as a Lewis acid catalyst in Diels-Alder and aza Diels-Alder reactions, J. Matsuo, T. Tsuchiya, K. Odashima, S. Kobayashi, Chem. Lett., 2000, 178-179.
65. Catalytic Pauson-Khand reaction in supercritical fluids, N. Jeong, S. H. Hwang, Y. W. Lee, J. S. Lim, J. Am. Chem. Soc., 1997, 119, 10549-10550.
66. Supercritical carbon dioxide as a reaction medium and reaction partner, M. T. Reetz, W. Konen, J. Strack, Chimia, 1993, 47, 493.
67. Organometallics of transition metals in supercritical carbon dioxide: Solubilities, reactions, catalysis, U. Kreher, S. Schebesta, D. Walther, Z. Anorg. Allg. Chem., 1998, 624, 602-612.
68. CPCO$(CO)_2$-catalysed cyclotrimerisation of alkynes in supercritical carbon dioxide, F. Montilla, T. Avilés, T. Casimiro, A. A. Ricardo, M. N da Ponte, J. Organomet. Chem., 2001, 632, 113-118.

Chapter 6

HETEROGENEOUS REACTIONS IN SUPERCRITICAL CARBON DIOXIDE

R. Amandi, J. Hyde, M. Poliakoff
School of Chemistry, Nottingham University, University Park, Nottingham, NG7 2RD, UK
martyn.poliakoff@nottingham.ac.uk.

1. INTRODUCTION

Supercritical Carbon Dioxide ($scCO_2$) is being increasingly seen as a commercially viable and environmentally acceptable solvent for chemical transformations and processing. Resent technologies approaching commercialization have included dry cleaning, polymer synthesis [1] and catalytic reactions. [2] Of these, heterogeneous catalysis is the technology which promises both the widest and the most immediate implementation in chemical manufacture. The high level of research activity in this field has led to the publication of extensive reviews over the past 3 years; these reviews include those by Baiker [3], Meehan *et al.*[4] and Hyde, Leitner and Poliakoff [5] and Hyde *et al.*[6] The aim of this chapter is not to repeat the contents of these reviews but rather to give a convenient source of references, highlighting some of the key developments over the past few years. The chapter is arranged into sections based on the different types of types of reaction involved in heterogeneously catalyzed processes. It should be pointed out that much information is now available from internet websites, but the inherently transient nature of these sites makes their inclusion in a reference work such as this inappropriate.

M. Aresta (ed.), Carbon Dioxide Recovery and Utilization, 169-179.

2. GENERAL CONSIDERATIONS

Any chemical process can be run either continuously or as a batch. Traditionally the majority of commercial reactions in the fine chemicals industry have been run as batch processes because of both convenience and quality control. Reactions in $scCO_2$ involve high pressures and hence more elaborate (and therefore more costly) equipment than reactions in conventional solvents. Thus, although it is quite feasible to run heterogeneous catalytic experiments in $scCO_2$ as a batch process [6] on the laboratory scale, most larger scale reactions have been or will be run as continuous processes. [7-9]

The key question in any application of $scCO_2$ is "what are the advantages of running this process supercritically compared to more conventional processes?" The references given at the end of this chapter begin to answer this question from the scientific point of view. $scCO_2$ is an effective solvent for a range of catalytic reactions, which do not involve conventional solvents with their associated problems of pollution and waste disposal. Scientifically, reactions in $scCO_2$ have been shown to go faster, at least on some occasions than those in conventional solvents. In the case of hydrogenation, reactions can be optimized for a specific product more effectively than in many traditional gas phase or liquid phase reactions. [7] There are indications that acid catalyzed reactions in $scCO_2$ also have advantages over their more conventional counterparts. [10] There is still argument whether $scCO_2$ can decoke the surface of many heterogeneous catalysts, prolonging their lifetime. [11-15] Answering the question in economic terms is more difficult. It is not yet clear how the costs of supercritical processing compare with those of traditional chemical manufacture, but there are signs that $scCO_2$ may well generate cost savings. These savings may not be in the cost of the immediate process (*e.g.* hydrogenation) but in the fact that the product of the supercritical process may be generated in a form that requires no further purification. It is the elimination of purification stages which will result in cost savings.

Theoretical considerations and calculations can never be regarded as experimental proof. The same is true both in scientific research and in the economics of manufacture. The supercritical fluid community is, therefore, very fortunate that one company (Thomas Swan and Co. Ltd.) has undertaken to build an $scCO_2$ demonstration plant for chemical manufacture at their Consett site at County Durham, UK. This plant is based on Swan-SCF® technology developed jointly between our laboratory and those at Thomas Swan. The object of the plant is to explore the feasibility of large-scale (1000 tones pa) chemical manufacture in $scCO_2$. The plant was commissioned in March 2002, and within 2 months had demonstrated the first aspect of our thesis namely that purification stages can be rendered unnecessary. In the hydrogenation of isophorone, to generate 3,3,5-trimethylcyclohexanone, [7] the product was generated with saleable purity without additional work up. This large-scale experiment in Green Chemistry is the first of its type and will need to be repeated for each of the emerging clean technologies before the value of these new processes can be firmly established.

3. HYDROGENATION [16-18,3,19-35,7,36-49,8,50-56]

Catalytic hydrogenation is $scCO_2$ is one of the most active aspects of heterogeneous catalysis in $scCO_2$. The field was pioneered by K-H Pickel and colleagues at Hoffman Roche, Switzerland. The work was then extended in a number of laboratories including our own at Nottingham. [35,7,44-46] The key advantage to hydrogenation in supercritical fluids over traditional solvents, is the mass transport of hydrogen to the catalytic surface, which in conventional solvents, has been identified as the rate limiting step. [41] In $scCO_2$ however, this is overcome by employing the complete miscibility of hydrogen in $scCO_2$. The lack of phase boundary increases rate of mass transport, but on the other hand, reduces the effective concentration at the catalysts surface to that of the bulk. More recently, it has become clear that mass transport limitations can be overcome even where two phases are present.[22] Nevertheless, hydrogenation in $scCO_2$ can be performed on a wide variety of functionalities and has been proven to be as, if not more, effective than conventional solvents.[7] A wide spectrum of reactions have been studied under supercritical conditions ranging from the simple reduction of alkenes to asymmetric synthesis of high value pharmaceutical intermediates. Much research has focused on the hydrogenation of fats and oils,[16,36,39,40,51,52] as these processes are ideally suited to continuous flow reactors. The limitation of $scCO_2$ hydrogenation reactions, is found in the reduction of non-aromatic primary amines, which largely undergo addition of CO_2 to form solid carbamates; a problem which is easily rectified by the use of other supercritical fluids, such as scPropane.

4. ACID CATALYZED REACTIONS [56,57,10,58,6,59-62,51,63,64]

The range of reactions that can be performed under acid catalyzed conditions is extensive, ranging from Friedel-Crafts alkylation to etherification. [10,64] Traditionally, the acid catalyst in question is used in near stoichiometric quantities, often in excess, which involves considerable costs in clean-up and disposal after use. A relatively new consideration is to use solid acid catalysts in $scCO_2$ continuous flow equipment in an attempt to minimize the volume of catalyst and the associated problems of isolation / disposal.[10] One of the key advantages to this approach, besides the localization of the catalyst, is that under supercritical conditions many of the commercially available catalysts, such as Nafion®[64]can change their properties as a function of temperature and pressure, often yielding different products. Etherification reactions are a particularly promising avenue for investigation, as ether synthesis is commercially expensive and produces environmentally damaging waste products, which must be disposed of. $scCO_2$ has been shown to provide a useful medium for this class of reactions under continuous conditions. Many symmetric ethers can be produced and by simply altering the conditions may produce mono- or bis- ethers from alcohol starting materials. Formation of cyclic ethers, such as tetrahydrofuran (THF) and acetal moieties can be formed from alcohols.[10] Transesterification reactions have mostly been studied in the field of fuels, as oils such as those obtained from rapeseed, can be transformed in to viable fuels by an acid catalyzed reaction with methanol.[60]

The references given here deal mainly with the latter two subjects, as Freidel-Crafts chemistry is covered separately in a later section.

5. HECK COUPLING AND CARBON-CARBON BOND FORMATION [65-71,37,72-75]

Carbon-Carbon bond formation is widely used in chemistry to increase the carbon chain length in the synthesis of many compounds. The principal technique is Heck-Coupling, which is a palladium(0) catalyzed reaction.[72] This reaction is also possible in $scCO_2$, although it is performed mainly in batch processes due to the by-products of the reaction. Continuous flow Heck-Coupling is possible in $scCO_2$, however, the use of high flow rates is required to overcome the problem of blockages due to the halide salt precipitation. Suzuki-Coupling is another class of C-C bond forming reactions which may also be performed in $scCO_2$.[70,71] Many of the catalysts used for these classes of reactions are supported homogeneous catalysts, adapted from analogues, which have been found to perform particularly well in batch homogeneous processes.

6. FRIEDEL-CRAFTS [11,76,58,77]

This section is closely linked to carbon-carbon bond formation and acid catalyzed reactions, in the sense that Friedel-Crafts acylation and alkylation are among the most fundamental and useful reactions of carbon-carbon bond formation in aromatic rings. Unfortunately, traditional Friedel-Crafts reactions are also some of the most wasteful in terms of by-product formation, catalyst usage and atom efficiency.[2] Current industrial techniques utilize traditional organic solvents in the presence of excess amounts of either strongly Lewis-acidic materials such as aluminum trichloride ($AlCl_3$) or alternatively, strong Brønsted acids like concentrated sulfuric acid and hydrofluoric acid. The application of both these reagent types and solvent systems leads to serious problems both environmental and economic upon work up. This is an area where $scCO_2$ may provide cleaner, more sustainable routes to the chemicals which previously caused high levels of eco-toxicity.

Besides the advantages that heterogeneous catalysts offer in terms of recovery, Friedel-Crafts chemistry is particularly advantageous in $scCO_2$, as it has been shown that the selectivity for mono-alkylated products can be tuned as a function of both temperature and pressure.[58] Many commercially available catalysts can be used for these reactions, such as Nafion®[76] and Amberlyst™, which makes the reaction in $scCO_2$ even more appealing especially under continuous flow conditions.[11] Although the reaction can be performed under other supercritical fluids, such as propane, which offers both an SCF and alkylating agent in a single component, $scCO_2$ is more effective at reaching desirable levels of selectivity, especially towards the mono- products.[58]

7. HYDROFORMYLATION [78-82,22,83,25,84-89,72,90-94]

Hydroformylation reactions contribute a major role in the fine chemicals and petroleum industries. The vast scale of the catalytic cracking and reforming reactions of the petroleum industry make this an extremely lucrative chemical transformation. Traditionally, the reaction performed homogeneously, requires the reclamation of highly expensive catalysts, which are usually rhodium species, and therefore highly toxic and environmentally damaging. The use of heterogeneous catalysts offers the immediate advantage of the catalysts recovery. Many catalysts have been developed in laboratories world-wide, but relatively few are used industrially as they are inherently costly. The homogeneously catalysed reaction of hydroformylation has been extensively reviewed in $scCO_2$, [95-98] but there are few examples of genuine heterogeneously catalysed hydroformylation.[79,84,90] Supported homogeneous catalysts however, make up the majority of available heterogeneous catalysts.[22,85] Compared to traditional catalysts, these offer equal, if not better TOFs and selective *n:iso* ratios when the reaction is performed in $scCO_2$.[89]

8. CONCLUSIONS

The field of catalytic chemistry in $scCO_2$ has continued to expand over the past five years. This growth has had the twin drivers of intellectual curiosity and of the quest for more environmentally acceptable chemical processes. Particularly encouraging have been the appearance of several groups new to the field and the major advances in homogeneous catalysis and the commercialization of the first plants for heterogeneous reactions. We believe that the field will continue to expand dramatically. Important new developments can be expected soon in both heterogeneous and homogeneous catalysis in SCFs.

This bibliography contains a selection of the more important papers involving inorganic chemistry in $scCO_2$, together with a number of other papers of relevance to reaction chemistry in $scCO_2$. It does not include any Patents and only a few papers, published in Conference Proceedings. The papers cover the period January 1998 to May 2002.

REFERENCES

1. Giles, M. R., Hay, J. N. & Howdle, S. M. "The copolymerisation of methyl and ethyl methacrylate in supercritical carbon dioxide" Macro. Rapid Commu. 21, 1019-1023, 2000.
2. Hyde, J. R., Licence, P., Carter, D. & Poliakoff, M. "Continuous catalytic reactions in supercritical fluids" Appl. Catal. A-Gen. 222, 119-131, 2001.
3. Baiker, A. "Supercritical fluids in heterogeneous catalysis" Chem. Rev. 99, 453-473, 1999.
4. Bertucce, A., Vetter, G. "High Pressure Process Technology: Fundamentals and Applications" Industrial Chemistry Library, Vol. 9, 2001.
5. Hyde, J. R., Poliakoff, M. & Leitner, N. K. V. in High Pressure Chemistry (Ed. van Eldik, R.) 371-397 (VCH Weiley, York), 2001.
6. Kerler, B., Martin, A., Jans, A. & Baerns, M. "Partial oxidation of propane in compressed carbon dioxide using CoO_x/SiO_2 catalysts" Appl. Catal. A-Gen. 220, 243-252, 2001.
7. Hitzler, M. G., Smail, F. R., Ross, S. K. & Poliakoff, M. "Selective catalytic hydrogenation of organic compounds in supercritical fluids as a continuous process" Org. Process Res. Develop. 2, 137-146, 1998.

8. Tschan, R., Schubert, M. M., Baiker, A., Bonrath, W. & Lansink-Rotgerink, H. "Continuous semihydrogenation of a propargylic alcohol over amorphous $Pd_{81}Si_{19}$ in dense carbon dioxide: effect of modifiers" Catal. Lett. 75, 31-36, 2001.
9. Wandeler, R., Kunzle, N., Schneider, M. S., Mallat, T. & Baiker, A. "Continuous platinum-catalyzed enantioselective hydrogenation in 'supercritical' solvents" Chem. Commun. 673-674, 2001.
10. Gray, W. K., Smail, F. R., Hitzler, M. G., Ross, S. K. & Poliakoff, M. "The continuous acid-catalyzed dehydration of alcohols in supercritical fluids: a new approach to the cleaner synthesis of acetals, ketals, and ethers with high selectivity" J. Am. Chem. Soc. 121, 10711-10718, 1999.
11. Clark, M. C. & Subramaniam, B. "Extended alkylate production activity during fixed-bed supercritical 1-butene/isobutane alkylation on solid acid catalysts using carbon dioxide as a diluent" Ind. Eng. Chem. Res. 37, 1243-1250, 1998.
12. Giles, M. R., Hay, J. N., Howdle, S. M. & Winder, R. J. "Macromonomer surfactants for the polymerisation of methyl methacrylate in supercritical CO_2" Polymer 41, 6715-6721, 2000.
13. Niu, F. H. & Hofmann, H. "Investigation of coke extraction from zeolite-HY under supercritical and near-critical conditions" Can. J. Chem. Eng. 75, 346-352, 1997.
14. Subramaniam, B., Arunajatesan, V. & Lyon, C. J. "Catalalysis Deactivation" 63-77 (Elsevier Science Publ B V, Amsterdam) 1999.
15. Trabelsi, F. S., Abaroudi, K., Larrayoz, M.A., Recasens, F., Sueiras, J.E., "Coking and ex-situ catalysts reactivation using supercritical CO_2: a preliminary study" Ind. Eng. Chem. Res. 39, 3666-3670, 2000.
16. Andersson, M.B.O., King, J.W. & Blomberg, L.G. "Synthesis of fatty alcohol mixtures from oleochemicals in supercritical fluids" Green Chem. 2, 230-234, 2000.
17. Arunajatesan, V., Subramaniam, B., Hutchenson, K. W. & Herkes, F. E. "Fixed-bed hydrogenation of organic compounds in supercritical carbon dioxide" Chem. Eng. Sc. 56, 1363-1369, 2001.
18. Bach, I. & Cole-Hamilton, D. J. "Hydroformylation of hex-1-ene in supercritical carbon dioxide catalysed by rhodium trialkylphosphine complexes Chem. Commun. 1463-1464, 1998.
19. Baiker, A., Grunwaldt, J. D., Muller, C. A. & Schmid, L. "Catalytic materials by design" Chimia 52, 517-524, 1998.
20. Baker, R. T. et al. « Rhodium(I) complexes with phosphorus ligands containing partially fluorinated "pigtails" as catalysts for hydrogenation and hydroboration of alkenes in supercritical carbon dioxide" Abstr. Pap. Am. Chem. Soc. 218, 568-INOR, 1999.
21. Bertucco, A., Canu, P., Devetta, L. & Zwahlen, A. G. "Catalytic Hydrogenation in Supercritical CO_2: Kinetic Measurements in a Gradientless Internal-Recycle Reactor" Ind. Eng. Chem. Res. 36, 2627, 1997.
22. Bhanage, B.M., Ikushima, Y., Shirai, M. & Arai, M. "Multiphase catalysis using water-soluble metal complexes in supercritical carbon dioxide" Chem. Commun. 1277-1278, 1999.
23. Bonilla, R.J., James, B.R. & Jessop, P.G. "Colloid-catalysed arene hydrogenation in aqueous/supercritical fluid biphasic media" Chem. Commun. 941-942, 2000.
24. Brown, R. A. et al. "Asymmetric hydrogenation and catalyst recycling using ionic liquid and supercritical carbon dioxide" J. Am. Chem. Soc. 123, 1254-1255, 2001.
25. Bruneau, C. "Green chemistry and catalysis" Actual. Chimique, 127-129, 1999.

26. Campestrini, S. & Tonellato, U. "Manganeseporphyrin-catalysed heterogeneous and unusually chemioselective oxidations of sulfides by monopersulfate in supercritical CO_2" J. Mol. Catal. A-Chem. 164, 263-272, 2000.
27. Carroll, M. A. & Holmes, A. B. "Palladium-catalysed carbon-carbon bond formation in supercritical carbon dioxide" Chem. Commun. 1395-1396, 1998.
28. Cavazzini, M., Pozzi, G., Quici, S., Maillard, D. & Sinou, D. "Palladium-catalysed asymmetric allylic alkylation in the presence of a chiral 'light fluorous' phosphine ligand" Chem. Commun. 1220-1221, 2001.
29. De Jong, A. et al. "Solvent screening for the supercritical hydrogenation of polyunsaturated hydrocarbons using VLE calculations" Ind. Eng. Chem. Res. 40, 3225-3229, 2001.
30. DeSimone, J., Selva, M. & Tundo, P. "Nucleophilic displacements in supercritical carbon dioxide using silica-supported phase-transfer agents" J. Org. Chem. 66, 4047-4049, 2001.
31. Devetta, L., Canu, P., Bertucco, A. & Steiner, K. "Modelling of a trickle-bed reactor for a catalytic hydrogenation in supercritical CO_2" Chem. Eng. Sci. 52, 4163, 1997.
32. Dinjus, E. "Advances in Chemical Conversions for Mitigating Carbon Dioxide" 127-140 (Elsevier Science Publ B V, Amsterdam), 1998.
33. Farkas, G. et al. "Effect of carbon support properties on enantioselective hydrogenation of isophorone over palladium catalysts modified with (-)-dihydroapovincaminic acid ethyl ester" J. Mol. Catal. A-Chem. 153, 215-219, 2000.
33. Haughton, L., Williams, J.M.J. "Catalytic applications of transition metals in organic synthesis" J. Chem. Soc.-Perkin Trans. 1, 3335-3349, 2000.
35. Hitzler, M.G., Poliakoff, M. "Continuous hydrogenation of organic compounds in supercritical fluids" Chem. Commun. 1667-1668, 1997.
36. King, J.W., Holliday, R.L., List, G.R., Snyder, J.M. "Hydrogenation of vegetable oils using mixtures of supercritical carbon dioxide and hydrogen" J. Am. Oil Chem. Soc. 78, 107-113, 2001.
37. Li, J.H., Jia, L.Q., Jiang, H.F. "Transition metal-catalyzed reactions in supercritical carbon dioxide" Chin. J. Org. Chem. 20, 293-298, 2000.
38. Li, Y.W., He, D.H., Yuan, Y.B., Cheng, Z.X., Zhu, Q.M. "Synthesis of isobutene from CO hydrogenation over nanosized zirconium dioxide-based catalysts" Chin. J. Cat. 23, 185-190, 2002.
39. Macher, M.B., Härröd M., van den Hark, S., Møller, P. "Hydrogenation at supercritical single-phase conditions. Abstr. Pap. Am. Chem. Soc. 220, 72-IEC, 2000.
40. Macher, M.B., Hogberg, J., Møller, P., Härröd, M. "Partial hydrogenation of fatty acid methyl esters at supercritical conditions" Fett-Lipid 101, 301-305, 1999.
41. Martins, E.P., Aranda, D.A.G., Pessoa, F.L.P., Zotin, J.L. "Hydrogenation of diesel aromatic compounds in supercritical solvent environment" Braz. J. Chem. Eng. 17, 361-365, 2000.
42. Naota, T., Takaya, H., Murahashi, S.I. "Ruthenium-catalyzed reactions for organic synthesis" Chem. Rev. 98, 2599-2660, 1998.
43. Pillai, U.R., Sahle-Demessie, E. "Selective hydrogenation of maleic anhydride to gamma-butyrolactone over Pd/Al_2O_3 catalyst using supercritical CO_2 as solvent" Chem. Commun. 422-423, 2002.
44. Poliakoff, M., George, M.W. "Intermediates in organometallic and organic chemistry: spectroscopy, polymers, hydrogenation and supercritical fluids" J. Phys. Org. Chem. 11, 589-596, 1998.
45. Poliakoff, M. et al. "Supercritical fluids: Clean solvents for green chemistry" Chin. J. Chem. 17, 212-222, 1999.

46. Poliakoff, M., Meehan, N.J., Ross, S.K. "A supercritical success story" Chem. Ind. 750-752, 1999.
47. Qiao, Z., Wang, M. "Progress in catalytic asymmetric hydrogenation" Chin. J. Org. Chem. 21, 325-340, 2001.
48. Shirai, M., Tanaka, T., Arai, M. "Selective hydrogenation of alpha-, beta-unsaturated aldehyde to unsaturated alcohol with supported platinum catalysts at high pressures of hydrogen" J. Mol. Catal. A-Chem. 168, 99-103, 2001.
49. Suh, D.J., Park, T.J., Lee, S.H., Kim, K.L. "Nickel-alumina composite aerogels as liquid-phase hydrogenation catalysts" J. Non-Cryst. Solids 285, 309-316, 2001.
50. Tschan, R. et al. "Semihydrogenation of a propargylic alcohol over highly active amorphous $Pd_{81}Si_{19}$ in "supercritical" carbon dioxide" Appl. Catal. A-Gen. 223, 173-185, 2002.
51. van den Hark, S., Härröd, M. "Hydrogenation of oleochemicals at supercritical single-phase conditions: influence of hydrogen and substrate concentrations on the process" Appl. Catal. A-Gen. 210, 207-215, 2001.
52. van den Hark, S., Härröd, M., Møller, P. "Hydrogenation of fatty acid methyl esters to fatty alcohols at supercritical conditions" J. Am. Oil Chem. Soc. 76, 1363-1370, 1999.
53. van den Hark, S., Härröd, M., Møller, P. "Hydrogenation of fatty acid methyl esters to fatty alcohols at supercritical conditions" J. Am. Oil Chem. Soc, 76, 1363, 1999.
54. Wai, C.M., Hunt, F., Ji, M., Chen, X.Y. "Chemical reactions in supercritical carbon dioxide" J. Chem. Educ. 75, 1641-1645, 1998.
55. Yu, J.Q., Spencer, J.B. "Heterogeneous transfer hydrogenation involves pairwise hydrogen transfer from the same position of two molecules of formic acid" Chem. Commun. 1935-1936, 1998.
56. Zhao, F.Y., Ikushima, Y., Shirai, M., Ebina, T., Arai, M. "Influence of electronic state and dispersion of platinum particles on the conversion and selectivity of hydrogenation of an alpha,beta-unsaturated aldehyde in supercritical carbon dioxide" J. Mol. Catal. A-Chem. 180, 259-265, 2002.
57. Ghaziaskar, S.H., Ikushima, Y. "Continuous synthesis of oleyl oleate in supercritical carbon dioxide using solid p-toluenesulfonic acid as catalyst" Iran J. Chem. Chem. Eng.-Inter. English Ed. 19, 79-83, 2000.
58. Hitzler, M.G., Smail, F.R., Ross, S.K., Poliakoff, M. "Friedel-Crafts alkylation in supercritical fluids: continuous, selective and clean" Chem. Commun. 359-360, 1998.
59. Kim, Y.-C.L., Hyeon Sup. "Selective synthesis of furfural from xylose with supercritical carbon dioxide and solid acid catalysts" J. Ind. and Eng. Chem. (Seoul, Republic of Korea) **7**, 424-429, 2001.
60. Ma, F.R., Hanna, M.A. "Biodiesel production: a review" Bio. Tech. 70, 1-15, 1999.
61. Poliakoff, M.H., Martin, G., Swan, T.M., Ross, S., Fiona, R., Tacke, T., Wieland S., Environ. Progress, 1998.
62. Subramanian, B. C., Michael C. "Extended alkylate production activity during fixed-bed supercritical 1-Butene/isobutene alkylation on solid acid catalysts using carbon dioxide as a diluent" Ind. Eng. Chem. Res. 37, 1243-1250, 1998.
63. Van den Hark, S., Härröd, M., Møller, P. Studies in surface science and catalysis. 130A International Congress on catalysis, 2000, Pt. A 539-544, 2000.
64. Wei, H.C., DeYoung, J.P., Kipp, B.E., DeSimone, J.M. "A quantitative kinetic study of alpha-methylstyrene dimerization using nafion (R) solid acid catalyst in supercritical carbon dioxide" Abstr. Pap. Am. Chem. Soc. 216, 441-POLY, 1998.

65. Beletskaya, I.P., Cheprakov, A.V. "The Heck reaction as a sharpening stone of palladium catalysis" Chem. Rev. 100, 3009-3066, 2000.
66. Bhanage, B.M., Ikushima, Y., Shirai, M., Arai, M. "Heck reactions using water-soluble metal complexes in supercritical carbon dioxide" Tetrahedron Lett. 40, 6427-6430, 1999.
67. Bhanage, B.M., Ikushima, Y., Shirai, M., Arai, M. "Heck reactions in supercritical carbon dioxide using organometallic complex catalysts" High Press. Res. 20, 131-142, 2001.
68. Cacchi, S., Fabrizi, G., Gasparrini, F., Villani, C. "Carbon-carbon bond forming reactions in supercritical carbon dioxide in the presence of a supported palladium catalyst" Synlett, 345-347, 1999.
69. Cao, C.T., Fadeev, A.Y., McCarthy, T.J. "Reactions of organosilanes with silica surfaces in carbon dioxide" Langmuir 17, 757-761, 2001.
70. Early, T.R. et al. "Palladium-catalysed cross-coupling reactions in supercritical carbon dioxide" Chem. Commun. 1966-1967, 2001.
71. Gordon, R.S., Holmes, A.B. "Palladium-mediated cross-coupling reactions with supported reagents in supercritical carbon dioxide" Chem. Commun. 640-641, 2002.
72. Morita, D.K., Pesiri, D.R., David, S.A., Glaze, W.H., Tumas, W. "Palladium-catalyzed cross-coupling reactions in supercritical carbon dioxide" Chem. Commun. 1397-1398, 1998.
73. Shezad, N., Clifford, A.A., Rayner, C.M. "Suppression of double bond isomerisation in intramolecular Heck reactions using supercritical carbon dioxide" Tetrahedron Lett. 42, 323-325, 2001.
74. Tumas, W.L.F., Brown, G. in *221st INOR-039* (ed. Abstr. Pap. Am. Chem. Soc.), 2001.
75. Yeung, L.K., Lee, C.T., Johnston, K.P., Crooks, R.M. "Catalysis in supercritical CO_2 using dendrimer-encapsulated palladium nanoparticles" Chem. Commun. 2290-2291, 2001.
76. Harmer, M.A., Hutchenson, K.W. "Super acid catalysis in supercritical fluid reaction media for the formation of linear alkyl benzenes" Chem. Commun. 18-19, 2002
77. Santana, G.M., Akgerman, A. "Alkylation of isobutane with 1-butene on a solid acid catalyst in supercritical reaction media" Ind. Eng. Chem. Res. 40, 3879-3882, 2001.
78. Abraham, M., Snyder, G. "Conversion of propylene to 1-butanal through heterogeneous hydroformylation in supercritical carbon dioxide" Abstr. Pap. Am. Chem. Soc. 218, 80-IEC, 1999.
79. Abraham, M.A., Snyder, G. "Development of a heterogeneous hydroformylation catalyst for use with supercritical CO_2" Abstr. Pap. Am. Chem. Soc. 219, 18-CATL, 2000.
80. Abraham, M.A., Snyder, G. in 5th International Symp. on Supercritical Fluids 7, Atlanta, 2000.
81. Abraham, M.A., Snyder, G., Tadd, A. "Development of heterogeneous catalysts for hydroformylation in supercritical CO_2" Abstr. Pap. Am. Chem. Soc. 220, 166-IEC, 2000.
82. Bach, I., ColeHamilton, D.J. "Hydroformylation of hex-1-ene in supercritical carbon dioxide catalysed by rhodium trialkylphosphine complexes" Chem. Commun. 1463-1464, 1998.
83. Bischoff, S., Kant, M. "Efficient catalysts for the two-phase hydroformylation of long-chain alpha-olefins" Ind. Eng. Chem. Res. 39, 4908-4913, 2000.
84. ColeHamilton, D.J., Sellin, M.F. "Heterogeneous catalytic hydroformylation of hex-1-ene in supercritical carbon dioxide" Abstr. Pap. Am. Chem. Soc. 218, 344-INOR, 1999.
85. Dharmidhikari, S., Abraham, M.A. "Rhodium supported on activated carbon as a heterogeneous catalyst for hydroformylation of propylene in supercritical carbon dioxide" J. Supercrit. Fluids 18, 1-10, 2000.

86. Hu, Y.L., Chen, W.P., Osuna, A.M.B., Iggo, J.A., Xiao, J.L. "Fast and unprecedented chemoselective hydroformylation of acrylates with a fluoropolymer ligand in supercritical CO_2" Chem. Commun. 788-789, 2002.
87. Hu, Y.L. et al. "Rapid hydroformylation of alkyl acrylates in supercritical CO_2" Chem. Commun. 725-726, 2001.
88. Ke, J., Han, B.X., George, M.W., Yan, H.K., Poliakoff, M. "Acoustic measurements of critical points for four-component mixtures in hydroformylation reactions in carbon dioxide" Fluid Phase Equil. 185, 327-337, 2001.
89. Meehan, N.J. et al. « Continuous, Selective Hydroformylation in Supercritical Carbon Dioxide using an Immobilised Homogeneous Catalyst" Chem. Commun. 1497, 2000.
90. Sellin, M.F., Cole-Hamilton, D.J. "Hydroformylation reactions in supercritical carbon dioxide using insoluble metal complexes" J. Chem. Soc.-Dalton Trans. 11, 1681-1683, 2000.
91. Sinou, D., Maillard, D., Pozzi, G. "Synthesis of a family of triarylphosphanes with fluorous phase affinity" Euro. J. Org. Chem. 269-275, 2002.
92. Snyder, G., Tadd, A., Abraham, M.A. "Evaluation of catalyst support effects during rhodium-catalyzed hydroformylation in supercritical CO_2" Ind. & Eng. Chem. Res. 40, 5317-5325, 2001.
93. White, S., Bennett, B.L., Roddick, D.M. "Organometallics in acidic media: Catalytic dimerization of ethylene by (perfluoroalkyl)phosphine complexes of platinum and palladium in trifluoroacetic acid" Organomet. 18, 2536-2542, 1999.
94. Xiao, J.L., Banet, A., Chen, W.P. "Ligand effects in hydroformylation in supercritical carbon dioxide" Abstr. Pap. Am. Chem. Soc. 220, 168-IEC, 2000.
95. Jessop, P.G. "Homogeneously-catalyzed syntheses in supercritical fluids" Topics in Catal. 5, 95-103, 1998.
96. Jessop, P.G., Ikariya, T., Noyori, R. "Homogeneous catalysis in supercritical fluids" Chem. Rev. 99, 475-493, 1999.
97. Leitner, W. "Reactions in supercritical carbon dioxide ($scCO_2$)", Modern Solvents in Organic Synthesis 107-132, Springer-Verlag Berlin, Berlin, 1999.
98. Leitner, W. "Homogeneous catalysts for application in supercritical carbon dioxide as a 'green' solvent" Comptes Rendus De L Academie Des Sciences Serie Ii Fascicule C-Chimie 3, 595-600, 2000.

Chapter 7

POLYMER SYNTHESIS IN SUPERCRITICAL CARBON DIOXIDE

Giuseppe Filardo, Alessandro Galia, Alberto Giaconia
Dipartimento di Ingegneria Chimica dei Processi e dei Materiali, Università di Palermo, Italy; filardo@dicpm.unipa.it

1. INTRODUCTION

It is current opinion of most of the people involved in supercritical fluids (SCFs) technology that as the end of the twentieth century has been devoted to enlarge exploitation of compressed gases in separation processes the beginning of the twenty-first will be mainly aimed to improve their utilisation as solvents for reactions.

In this context the last decade has been characterised by a rapidly increasing interest on the possibility of performing polymerisation reactions in dense carbon dioxide as a dispersant/solvent.

In this chapter a brief overview on the main features of these new polymerisation techniques is reported. Particular emphasis is given to the synthesis of fluoropolymers which have recently reached the level of industrial exploitation as a brilliant result of an academic–industrial research collaboration between DuPont and University of North Carolina chemists resulted in the commercialisation of a new Teflon® fluoropolymer process at DuPont's Fayetteville (North Carolina) plant, in a new $40 million facility using supercritical carbon dioxide instead of water as the solvent.

For more detailed informations on these processes the authors suggest to refer to the numerous useful reviews in the recent open literature [1,2,3,4].

M. Aresta (ed.), Carbon Dioxide Recovery and Utilization, 181-195.

2. GENERAL CONSIDERATIONS

When CO_2 is considered as a polymerisation medium its solvent strength towards the reagents and the products is of major concern. Carbon dioxide is a low dielectric constant solvent which behaves with good approximation like a hydrocarbon solvent for what concern the solubility of volatile non-polar molecules of low molar mass. On the other hand it is a weak Lewis acid and it has a significant quadrupole moment which gives a relevant contribution to its solubility parameter thus allowing it to dissolve some polar molecules such as methanol. As a result CO_2 is a good solvent for most vinyl monomers but is an exceedingly poor solvent (akin to fluorocarbons) for most high molar mass polymers [1]. The only classes of polymers which have exhibited good solubility in supercritical CO_2 under relatively mild conditions ($T<100°C$, $P< 35$ MPa) are amorphous fluoropolymers, silicones and, more recently investigated, polycarbonate-polyether [5,6,7]. Owing to this solubility consideration carbon dioxide has been used as a free-radical homogeneous polymerisation media only for highly fluorinated amorphous polymers [8], while heterogeneous techniques have been adopted in the case of other vinyl monomers as reported in recent exhaustive reviews [3,4].

In addition to solubility consideration a further relevant feature of dense CO_2 of major concern for polymer synthesis is its capability of highly plasticize polymers thus lowering their glass transition temperature (T_g). The plasticization effect is particularly important in heterogeneous processes where polymer rich phases, with relatively high local viscosity, are formed during the process. The plasticization effect leads to an increase in the free volume of the polymer chains which in turn increases the value of the diffusion coefficient of the species inside the coagulated particle thus allowing the progression of the polymerisation process up to high value of monomer conversion. Also translational and segmental diffusivity of the growing polymer chains are enhanced by this effect and this leads to a shift in the onset of the gel effect at higher value of the conversion or to a practical suppression of it with a consequent advantage for the thermal control of the process.

An additional advantage of the CO_2 swelling effect is the possibility of extracting from the synthesised polymer highly toxic initiator residues, unreacted monomers, decomposition products or any other added species (e.g.: a surfactant) with high extraction efficiencies achieved using an "easy-to-handle" plant facility like carbon dioxide. Moreover, unique new materials and morphologies may be accessible as a result of polymer plasticization by carbon dioxide and/or by the addition of appropriately designed amphiphatic surfactants [9].

Last but not least, from a kinetically point of view, dense CO_2 is inert towards chain transfer reactions in free radical polymerisations thus making easier the control of the quality of the polymer product.

3. SYNTHESIS OF FLUOROPOLYMERS

Chain free radical polymerisation of fluorinated vinyl monomers is a key synthetic route for the preparation of high performance specialty polymers exhibiting the unique combination of excellent chemical resistance, high thermal stability, low dielectric constant and dissipation factor, unusual surface properties, low water absorptivities, excellent weatherability and low flammabilities. All this bonanza arises from the effect of fluorine substituents in the chain which on the other hand leads to a more difficult control of the polymerisation process with respect to what observed in the case of hydrocarbon homologues. Owing to the high electrophilicity of fluorinated macroradical species propagating the chain a strong kinetic competition exists between addition of fluorinated free radicals to π-bonds of unreacted monomer and chain transfer reactions involving H-atom donor species [10].

Fluoropolymers are usually obtained through heterogeneous (emulsion or suspension) polymerisation techniques in aqueous systems. As a consequence of the adopted aqueous initiators unstable carboxylic acid and acid fluoride end groups are generated which, owing to their thermal instability, must be removed before processing the polymer melt.

A possible way to reduce the amount of thermally unstable end-groups is the use of aprotic solvents such as CFCs or perfluorocarbons, perfluoroalkyl sulfide and perfluorinated cyclic amines which however are too expensive and/or not environmentally friendly.

In this context a sustainable alternative to such VOCs is constituted by dense CO_2 as demonstrated by the great investments done in the last few years both in the USA (i.e. the cooperation of DuPont with University of North Carolina) and in the EU (European Research Grants Superpol and Ecopol) in the use of supercritical carbon dioxide for producing fluorinated polymers.

3.1 Homogenous synthesis of amorphous high molecular weight fluoropolymers

Due to the solubility of amorphous fluoropolymers in supercritical carbon dioxide, it has been possible to perform homogeneous polymerizations of fluorinated monomers with bulky substituent groups such as 1,1-dihydroperfluorooctyl acrylate (FOA) [11], p-perfluoroalkyl-ethyleneoxy-methylstyrene (STF) [12] and 1H,1H,5H-octafluoropentyl acrylate (OFP) [13] in carbon dioxide under relatively mild conditions (temperature about 60°C, initial pressure lower than 35 MPa) using AIBN as initiator. High molecular weight poly(FOA) and poly(STF) may be homogeneously synthesised with yields ranging from 65 to 85% and properties of the product similar to those of polymers prepared in CFCs [11,12]. Both FOA and STF monomer may be also homogeneously copolymerised with non-fluorinated vinyl monomers such as methyl methacrylate, butyl acrylate, styrene and ethylene: even if the hydrocarbon monomers lead to polymer unsoluble in supercritical CO_2 by this technique a random soluble copolymer is formed with a composition tuneable as a function of the comonomer feed ratio [11,12].

3.2 Heterogeneous polymerisations of fluorinated monomers in scCO_2.

High molecular weight PTFE may be synthesised using a heterogeneous hybrid carbon dioxide/aqueous medium [14]. By this route, with a proper selection of aqueous soluble perfluorinated initiator, a sort of compartmentalisation of monomer, polymer and initiator is achieved thus obtaining fast polymerisation kinetics and high values of yield and molecular weights (up to 90% and Mn = 900 Kg/mole) [14]. Moreover diluting the monomer with carbon dioxide moderates its reactivity, while the high heat capacity of water can be advantageous in controlling such exothermic polymerisations.

Processes carried out in carbon dioxide/water hybrid systems have been performed in batch reactors at 50-120 bar, using a water-soluble persulfate initiator, with and without a sodium or ammonium salt of perfluorooctanoic acid ($C_7F_{15}COO^-Na^+/NH_4^+$) as surfactant (to stabilize water-rich micelles). PTFE is collected as a "white spongy product", characterized by a morphology similar to conventional granular PTFE (particles of 0.5-3 μm) if no surfactant is added, while, with added surfactant, spherical particles of 100-200 nm diameters can be obtained [14].

It must be underlined that syntheses of polytetrafluoroethylene in carbon dioxide based systems are much safer as the highly dangerous handling monomer TFE [15] is always present as a mixture with carbon dioxide. In fact, TFE forms explosive mixtures with air and must be kept scrupulously free of oxygen even to prevent the formation of shock-sensitive polymeric peroxides; moreover, in the absence of air, it can disproportionate violently to form CF_4 and elemental carbon [15] ($CF_2{=}CF_2 \rightarrow C + CF_4$).

Poly(vinylidene fluoride) (PVDF) is the second most important thermoplastic within the fluoropolymer family after PTFE. Even if thermal and chemical stability of PVDF are somewhat lower with respect to PTFE the hydrogenated polymer can be easily processed on conventional equipment thus leading to advantageous compromise between quality and price. The synthesis of PVDF in a continuous stirred tank reactor has been recently investigated in supercritical CO_2 in order to model the rate of polymerisation in the case of a thermally initiated free radical chain process [16,17]. These studies demonstrate that vinylidene fluoride (VDF) can polymerise in dense carbon dioxide at 65-85°C, obtaining a polymer with properties similar to commercial samples prepared by conventional route. This polymerisation seems to follow a homogeneous chain growth kinetic law that, in particular, involves the absence of an acceleration effect and, consequently, an easier heat removal from the reactor [17].

The precipitation polymerisation of VDF in dense carbon dioxide initiated by γ-radiation has been object of investigation in the frame of the aforementioned Superpol EU research program [18]. The absence of chemical initiators, whose fragments would be finally incorporated in the polymer chains, together with the easy removal of volatile carbon dioxide and monomer, result in a "super clean" product obtained under very mild process conditions ($T \leq 40°C$, $p \leq 25MPa$).

Both TFE and VDF may be copolymerised with suitable comonomers in order to obtain polymers better matching their end-use application. Part of the investigation has been carried out with fluorinated comonomers such as perfluoropropyl-vinylether (PPVE) ($CF_3CF_2CF_2OCF{=}CF_2$): polymerisations performed under mild conditions (35°C and 90-110 bar) lead to the formation of polymer collected from the reactor under the form of "free flowing powder", with almost quantitative yields and copolymer composition tuneable by the molar ratio of comonomers in the feed. Quite interestingly chains have molecular weights higher than commercial product (ca. 10^3 Kg/mole, methanol can be used as a chain transfer to control its value) and are characterized by a concentration of carboxylic acid or acid fluoride end groups similar to those detected in polymers prepared in CFCs [9,19].

Similar results have been obtained in the copolymerisation of TFE and hexafluoropropene [9].

Of great relevance for the applications is the copolymerisation of TFE and functionalised perfluoro-vinyl-ether monomers, such as sulfonyl-vinyl fluoride ($CF_2{=}CFOCF_2CF(CF_3)O(CF_2)_2SO_2F$ (PSEVPE), for the production of perfluorinated ion exchange membranes. In view of its final utilisation the copolymer should have high mechanical strength and ionic conductivity. To satisfy this requirements chains must have high molecular weight coupled with a high concentration of ionic groups grafted onto the backbone. The main drawback for the synthesis of polymer with these joined properties is constituted by the β-scission of vinyl ether terminated macroradicals [19]. Experiments performed in supercritical CO_2 have shown that poly(TFE-*co*-PSEVPE) may be synthesized as a "tacky to gummy solid", with a concentration of ionic groups similar to those obtained in conventional routes but virtually in the absence of acid end groups as a result of a substantial suppression of β-scission reactions [19].

TFE can be copolymerised in dense carbon dioxide also with hydrocarbon vinyl monomers such as vinyl acetate (VAc); in this case it is possible to prepare an organic solvent-soluble fluoropolymers that might be functionalized in a further step, for example through hydrolysis (VAc groups will be converted into vinyl alcohol groups). P(TFE-*co*-VAc) can be synthesized in aqueous media (for example by emulsion) but through this route a branched structures is obtained due to hydrogen abstraction from VAc groups (as a consequence, a significant reduction in molecular weights is observed after hydrolysis) [20]. On the other hand copolymers produced in supercritical carbon dioxide (45 °C, 230-260 bar initial pressure) are predominantly linear: this suggests that hydrogen abstraction of radicals from VAc groups is negligible probably due to the swelling properties of carbon dioxide towards fluoropolymer matrices, that facilitates the diffusion of monomers into the polymer phase, thus increasing the rate of propagation (bimolecular process) relative to chain transfer (that can be considered an unimolecular process as the result of an intramolecular mechanism) [20]. Note that in the cases of both β-scission and hydrogen abstraction the plasticising effect of CO_2 promotes propagation reactions over side reactions, resulting in a product with high molecular weights.

4. FREE RADICAL POLYMERISATIONS OF HYDROCARBON MONOMERS IN $scCO_2$: A BRIEF OVERVIEW

Owing to the already mentioned limited solubility of most hydrocarbon polymers in supercritical CO_2, their polymerisations in such solvent are mainly run by heterogeneous techniques if high molecular weight must be achieved.

Precipitation processes (see Table 7.1 for selected examples) were the first to be investigated in the perspective of industrial utilisation of polymerisations in CO_2. This is the case of polymerisation of ethylene in CO_2 using both gamma radiation and AIBN as free radical initiators [21,22]. In the former case the high stability of carbon dioxide to ionising radiation was used as a further advantage of this unconventional solvent.

The results obtained in the precipitation polymerisation of acrylic acid (AA) [23] suggest that the performances of the process depend also on the kinetic competition between the propagation step and the nucleation step of the polymer particles. In the case of the synthesis of poly(acrylic acid) the extremely fast propagation rate allows high molecular weight polymer to form even though it demonstrates poor solubility in the continuous phase.

As, in principle, the relative rates of these stages can be changed by the tuneable physico-chemical properties of the supercritical fluid it could be possible to enlarge the field of utilisation of precipitation polymerisations.

Table 7.1 - Selected precipitation polymerisations in dense CO_2.

Monomer [ref.]	T (°C)	P (bar)	M_n (Kg/mol) (M_w/M_n)	Yield %
CH_2=CH(COOH) Acrylic acid (AA) [23]	62	125-345	2.9-149 (1.3-3.9)	>90
CH_2=CH_2 Ethylene [21,22]	20-45	50-400	14-258 (n.a.)	(n.a.)

M_w: weigh average molecular weight; M_n: number average molecular weight; n.a.: not available.

Other polymers that can be synthesized through free radical precipitation polymerization in dense CO_2 include PVC, polystyrene, polyacrylonitrile and polyvinylacetate. [24]

Owing to the good solubility of many monomers and initiators in CO_2, dispersion polymerisation is up to now the most investigated heterogeneous method for the syntheses of high molecular weight CO_2-insoluble polymers.

A central role for the performance of a successful dispersion or emulsion polymerisation is the surfactant. Since traditional surfactants, designed for use in aqueous or organic continuous phase, are completely insoluble in CO_2 [25], a relevant and challenging area of research is the design and synthesis of novel surfactants specific for carbon dioxide. In order to stabilise the polymer colloid in CO_2 special stabiliser compounds have been developed. All these molecules are characterized by a CO_2-philic portion showing high solubility in the continuous medium and a CO_2-phobic part which has high affinity to the polymer phase. The latter constitutes the anchoring portion of the surfactant which attaches to the surface of the polymer particle by either physical adsorption or chemical grafting depending on its nature. The CO_2-philic portion extends in the continuous phase and prevents the onset of flocculation of the particles by imparting long-range repulsion among them which must be strong enough to compensate the van der Waals attractions.

Surfactants used in CO_2 are either graft copolymers or homopolymers, such as poly(1,1-dihydroperfluorooctyl acrylate) (poly(FOA)) which has an acrylic like anchoring portion and fluorinated CO_2-philic side chains [26-32], or block copolymers constituted by a fluorinated or silicone portion which functions as the CO_2-philic part [31-41] and a suitable CO_2-phobic anchoring part. Another group of stabilisers successfully used are the polydimethylsiloxane macromonomers containing a reactive group such as methyl acrylate (PDMS-mMA) which grafts to the polymer thus forming "in situ" block copolymers, [31,42-44] and carboxylic acid-terminated perfluoropolyethers able to anchor via hydrogen bond formation between carbonyl and carboxylic acid groups. [45,46]

Many recent papers deal with the design and synthesis of surfactants for carbon dioxide [6,7,26-47] to define criteria for formulation and production of effective low cost compounds. This research field is of leading importance for the set-up of really exploitable polymerisation processes and has reached interesting results thanks to the contribution of Beckman *et al.* [6,7] who have shown how an intellectual pathway based on thermodynamic considerations can lead to the syntheses of new classes of CO_2-philic compounds potentially usable for the preparation of low cost non fluorinated amphipatic compounds which can exhibit self-assembly behaviour in CO_2 based polymerisation systems. The CO_2-philic parts of these new amphipatic surfactants may be either polyethers modified with acetate side groups or copolymers of epoxydes (such as propylene or cyclohexene oxide) and CO_2 (note that only carbon, hydrogen and oxygen atoms are present).

For what concern the polymerisation process in itself by far the most investigated has been the dispersion polymerisation of methyl methacrylate which leads to the formation of high molecular weight polymer with high reaction rate and almost quantitatively yields. The polymer product is recovered in the form of a dry free-flowing powder constituted of 1-4 μm spherical particles with a narrow particle size distribution. [26-28,38,42-46]

Similar results has been obtained with other vinyl monomers, provided that a good stabilising agent is used (Table 7.2).

Another example of dispersion free radical polymerization in supercritical CO_2 is the copolymerization of ethyl vinyl benzene (EVB) and divinyl benzene (DVB) to produce, in relatively mild conditions (65 °C, 310 bar) and with almost quantitative yields, a cross-linked polymer with a high surface area to be used for example as catalyst after further functionalisation [48].

Emulsion polymerisations have been scarcely investigated in $scCO_2$ up to now. Apart from the inverse emulsion polymerisation of TFE previously mentioned, in the literature it is reported the inverse emulsion polymerisation of acrylamide in CO_2/water systems at 345 bar and 60 °C both in the presence and in the absence of amide end-capped poly(hexafluoropropylene oxide) as stabilising agent [49]. In this case molecular weights are comparable to those obtained in a process using an alkane medium [49]. Interesting possibilities in this field have been opened by the discovery of surfactants which can form microemulsions of water and water soluble molecules in CO_2 [50,51,52] so an increased interest in this area could be expected.

Table 7.2 - Selected dispersion polymerisations in dense CO_2

Monomer [ref.]	T (°C)	P (bar)	M_n (Kg/mol) (M_w/M_n)	d_n (μm) (d_w/d_n)	Yield %
CH_2=C(CH_3)COOCH_3 MMA [26-28,38,42-46]	65	145-400	77-321 (2.2-5.3)	1.7-3.6 (1.02-1.17)	>90
CH_2=CH-Ph Styrene [30,35-37]	65	235-439	22.5-56.0 (3.1-4.3)	0.21-9.6 (1.03-1.30)	>90
CH_2=CH-N(CH_2)$_3$CO VP [29,34]	65,80	141-390	---	1.00-1.20 (1.10-1.16)	64-82
CH_2=CH-O-CO-CH_3 Vinyl acetate [31]	25,65	54-359	39-51 (2.3-4.3)	7.2-7.6 ---	>85
CH_2=CH-CN Acrylonitrile [39]	65	207-444	---	0.21-0.46 (1.02-1.05)	68-87
CH_2=C(CH_3)COO-E HEMA [41]	65	274-452	---	0.20-0.87 (1.01-1.04)	>85
CH_2=C(CH_3)COO-P GMA [40]	65	126-451	---	0.28-4.78 (1.01-1.08)	>90

M_w: weight average molecular weight; M_n: number average molecular weight; d_w: weight average particle diameter; d_n: number average particle diameter; MMA = Methyl methacrylate; VP =1-vynil-2-pyrrolidone; HEMA= 2-hydroxyethyl methacrylate; -E = -CH_2-CH_2-OH; GMA= Glycidyl methacrylate; -P = -CH_2-CH-CH_2O.

5. OTHER POLYMERISATIONS IN $scCO_2$

Dense CO_2 has been tested as a continuous phase also in cationic polymerisation reactions. In this class of chain reactions the chain carrier is a carbocation whose high reactivity results in high reaction rate but also leads to unwanted side reactions such as transfer to the monomer or termination. Indeed the propagating species exist under the form of ion couples whose intimacy is strongly dependent on the polarity of the solvent. The stabilisation of this ion pair is achieved through nucleophilic interaction between the carbocation and the counterion alone or associated to a suitable Lewis base to increase its nucleophilicity. In both cases the positive charge density of the carbocation is lowered and so is the acidity of the α-hydrogens thus leading to a more hindered chain transfer to the monomer. It has been found that solvent separated ion pairs are much less stable with respect to contact ion pairs and between them an equilibrium is established whose position is mainly dependent on the solvent. As the dielectric constant of a SCF can be changed with temperature and pressure it could be possible to control the "livingness" of the ion pair in order to form high molecular weight product.

It must be pointed out that, since cationic polymerisations are usually performed at very low temperature (-30, -70°C) to increase the selectivity, most of the researches have been carried out in liquid CO_2, even if recent investigations have shown that interesting results can be also obtained in supercritical conditions. Some example of cationic polymerisations in CO_2 are reported in Table 7.3, for more detailed information the reader is addressed to pertinent reviews [3,4].

Table 7.3 - Selected cationic polymerisations in dense CO_2.

Monomer [ref.]	Initiator	T (°C)	P (bar)	M_n (Kg/mol) (M_w/M_n)	Yield %
CH_2=CH-OiBu [53]	$CH_3CH(OiBu)COOCH_3/EtAlCl_2$	40	345	7-152 (1.2-4.0)	44-91
$CH_2=C(CH_3)_2$ [54]	$TiCl_4$, $AlCl_3$, BCl_3, $SnCl_4$, TMPCl/$TiCl_4$, TMPCl/$SnCl_2$	-50°C	---	30-500	---
$CH_2=C(CH_3)_2$ [55,56]	$TiCl_4$, $AlCl_3$, BCl_3, $SnCl_4$, TMPCl/$TiCl_4$, TMPCl/$SnCl_2$	32-42	< 150	0.7-2.5 (1.5-3.1)	4-40

iBu = $-CH_2CH(CH_3)_2$; TMPCl=2-chloro-2,4,4-trimethylpentane.

Also step growth polymerisations in the presence of supercritical CO_2 are becoming object of increasing interest even if until now they have been much less studied with respect to chain addition polymerisations. Many of these reactions are carried out in the melt phase in order to avoid the use of organic solvents. During the polymerisation low molecular weight molecules are co-produced, and they must be removed to shift the equilibrium towards the synthesis of high molecular weight polymer. The increasing melt viscosity lowers the diffusivity of such small molecules in the polymer phase and usually high vacuum is necessary to have acceptable removal rate.

On the other hand supercritical CO_2 can plasticize the polymer phase thus enhancing the kinetic of diffusion and the mobility of the polymer chain in the melt with a consequent improvement of reaction kinetic. This potentiality have been explored in the synthesis of polycarbonate by the melt phase transesterification of bisphenol A with diphenyl carbonate in the presence of supercritical CO_2. [57-60] Polycarbonate has been obtained with yields approaching 95% at about 270°C and 200-275 bar. The supercritical fluid offers a mean of efficient extraction of phenol co-produced during the polymerisation and lowers the viscosity of the molten polymer for easier handling. The swelling of the melt phase improves also the rate of the polymerisation by increasing the exchange surface available for condensation removal. Such synthetic approach has been tested also for the synthesis of poly(ethylene terephtalate) [59] and polyamides. [59,60]

6. CONCLUSIONS

The advantages that supercritical fluids offer as polymerisation media have been object of growing research interest in the last decade and supercritical CO_2 has emerged by far as the best choice as it joins favourable physical-chemical properties (tuneability of the solvent power, plasticizing effect of polymers, inertness to free chain reactions) with positive technical and environmental qualities (naturally occurring and abundant, inexpensive, non-toxic, non flammable).

In spite of all this bonanza, the real breakthrough for its utilisation as viable alternative to conventional polymerisation solvent has been the discovery of surfactants which are interfacially active in supercritical CO_2, thus solving the problems related to the low solubility of most hydrocarbon polymers in this compound.

This consideration clarify the decisive role of the design and synthesis of new classes of stabilising agents, less expensive than the fluorinated and polisiloxane based, for the effective scale-up of processes from laboratory scale to production plant.

This effort must be coupled with a more thorough understanding of the phase behaviour of the reaction system not only in its initial configuration but also as a function of the extent of the polymerisation. This aspect include the investigation of the distribution of the different species between the continuous phase and the dispersed one which could open interesting possibilities for the performance of new coating processes.

Moreover a more detailed investigation of the solvent effect on the kinetic of polymerisation reactions must be evaluated, to test the possibility of getting a better control on the polymer architecture; just as an example, it could be possible to tune the reactivity ratio of two comonomer having different activation volume by changing the density of the polymerisation medium.

From all these considerations it seems reasonable to foresee that the sector of polymerisation in supercritical CO_2 could be inserted among those of strategic interest for the western industrial chemistry which is living the gradual tendency to shift the productive attention from high volume low added value commodity chemicals to high speciality high performance products prepared under sustainable conditions.

Indeed as potential prerequisite for applicative advantages are clearly present the effective exploitation of such technology in industrial processes will be markedly dependent on effective co-operation among polymer scientists, physical-chemists, chemist and chemical engineers in order to manage simultaneously and synergistically all the interacting aspects of this very promising research field.

REFERENCES

1. Shaffer, K.A., DeSimone, J.M. "Chain Polymerizations in Inert Near- and Supercritical Fluids" Trends Polymer Science, 3, 146-153, 1995.
2. Cooper, A.I., DeSimone, J.M. "Polymer synthesis and characterization in liquid/supercritical carbon dioxide" Materials Science (Current Opinion in Solid State), 1, 761-768, 1996.
3. Canelas, D.A., DeSimone, J.M. "Polymerizations in Liquid and Supercritical Carbon Dioxide" Advances in Polymer Science, 133,103-140, 1997.
4. Kendall, J.L., Canelas, D.A., Young, J.L., DeSimone, J.M. "Polymerizations in Supercritical Carbon Dioxide" Chem. Rev., 99, 543-563, 1999.
5. Kirby, C.F., McHugh, M.A. "Phase Behaviour of Polymers in Supercritical Fluid Solvents" Chem. Rev., 99, 565-602, 1999.
6. Sarbu, T., Styranec, T., Beckman, E.J. "Non-fluorous polymers with very high solubility in supercritical CO_2 down to low pressures" Nature, 405, 165-168, 2000.
7. Sarbu, T., Styranec, T., Beckman, E.J. "Design and Synthesis of Low Cost, Sustainable CO_2-philes" Ind. Eng. Chem. Res., 39, 4678-4683, 2000.

8. DeSimone, J.M., Guan, Z., Elsbernd, C.S. "Synthesis of Fluoropolymers in Supercritical Carbon Dioxide" Science, 257, 945-947, 1992.
9. Romack, T.J., DeSimone, J.M., Treat, T.A. "Synthesis of Tetrafluoroethylene-Based Nonaqueous Fluoropolymers in Supercritical Carbon Dioxide" Macromol., 28, 8429-8431, 1995.
10. Dolbier, W.R.Jr. "Structure, Reactivity, and Chemistry of Fluoroalkyl Radicals" Chem. Rev., 96, 1557-1584, 1996.
11. Guan, Z., Elsbernd, C.S., DeSimone, J.M. "Homogeneous Free Radical Polymerizations of Fluorinated Monomers in Supercritical Carbon Dioxide" Polym. Prepr. (Americam Chemical Society), 33, 329-330, 1992.
12. Guan, Z., Combes, J.R., Elsbernd, C.S., DeSimone, J.M. "Synthesis of Fluoropolymers in Supercritical Carbon Dioxide" Polym. Prepr. (Americam Chemical Society), 34, 447-448, 1993.
13. Ryan, J., Erkey, C., Shaw, M. "Kinetics of Polymerization of 1H,1H,5H Octafluoropentyl Acrylate in Supercritical Carbon Dioxide. Polym. Prepr. (Americam Chemical Society), 38, 428-429, 1997.
14. Romack, T.J., Kipp, B.E., DeSimone, J.M. "Polymerization of Tetrafluoroethylene in a Hybrid Carbon Dioxide/Aqueous Medium" Macromol., 28, 8432-8434, 1995.
15. Romack, T.J., Combes, J.R., DeSimone, J.M. "Free-Radical Telomerization of Tetrafluoroethylene in Supercritical Carbon Dioxide" Macromol., 28, 1724-1726, 1995.
16. Charpentier, P.A., Kennedy, K.A., DeSimone, J.M., Roberts, G.W. "Continuous Polymerizations in Supercritical Carbon Dioxide: Chain-Growth Precipitation Polymerizations" Macromol., 32, 18, 5973-5975, 1999.
17. Charpentier, P.A., DeSimone, J.M., Roberts, G.W. "Continuous Precipitation Polymerization of Vinylidene Fluoride in Supercritical Carbon Dioxide: Modeling the Rate of Polymerization" Ind. Eng. Chem. Res., 39, 4588-4596, 2000.
18. Galia, A., Caputo, G., Spadaro, G., Filardo, G. "Gamma-radiation initiated polymerization of vinylidene fluoride in dense carbon dioxide" Ind. Eng. Chem. Res. 41(24), 5934-5940, 2002.
19. De Young, J.P., Romack, T.J., DeSimone, J.M. "Fluoroolefin Polymerization in Carbon Dioxide: Synthesis and Characterization of TFE Based Fluoropolymers" Polym. Prepr. (Americam Chemical Society), 38, 424-425, 1997.
20. Lousenberg, R.D., Shoichet, M.S. "Synthesis of Linear Poly(tetrafluoroethylene-co-vinyl acetate) in Carbon Dioxide" Macromol., 33, 1682-1685, 2000.
21. Hagiwara, M., Mitsui, H., Machi, S., Kagiya, T. "Liquid Carbon Dioxide as a Solvent for the Radiation Polymerization of Ethylene" J. Pol.ymer Science: Part A-1, 6, 603-608, 1968.
22. Hagiwara, M., Mitsui, H., Machi, S., Kagiya, T. "Two-stage Irradiation Study on Propagation and Termination in the γ-radiation-induced Polymerization of Ethylene in Liquid Carbon Dioxide" J. Polymer Science: Part A-1, 6, 721-728, 1968.
23. Romack, T.J., Maury, E.E., DeSimone, J.M. "Precipitation Polymerization of Acrylic Acid in Supercritical Carbon Dioxide" Macromol., 28, 912-915, 1995.
24. Fukui, K., Kagiya, T., Yokota, H., Toriuchi, Y., Kuniyoshi, F. US Patent 3,522,228, 1970.
25. Consani, K.A., Smith, R.D. "Observations on the Solubility of Surfactants and Related Molecules in Carbon Dioxide at 50°C" J. Sup. Fluids, 3, 51-65, 1990.
26. DeSimone, J.M., Maury, E.E., Mengeloglu, Y.Z., Mc Clain, J.B., Romack, T.J., Combes, J.R. "Dispersion Polymerizations in Supercritical Carbon Dioxide" Science, 265, 356-359, 1994.

27. Yu-Ling, H., Maury, E.E., DeSimone, J.M. "Dispersion Polymerization of Methyl Methacrylate Stabilized with Poly(1,1-dihydroperfluorooctyl acrylate) in Supercritical Carbon Dioxide" Macromol., 28, 8159-8166, 1995.
28. Lepilleur, C., Beckman, E.J. "Dispersion Polymerization of Methyl Methacrylate in Supercritical CO_2" Macromol., 30, 745-756, 1997.
29. Carson, T., Lizotte, J., DeSimone, J.M. "Dispersion Polymerization of 1-Vinyl-2-pyrrolidone n Supercritical Carbon Dioxide" Macromol., 33, 1917-1920, 2000.
30. Shiho, H., DeSimone, J.M. "Preparation of Micron-Size Particles in Supercritical Carbon Dioxide" J. Polymer Science: Part A: Polymer Chemistry, 27, 2429-2437, 1999.
31. Canelas, D.A., Betts, D.E., DeSimone, J.M., Yates, M.Z., Johnston, K.P. "Poly(vinyl acetate) and Poly(vinyl acetate-co-ethylene) Latexes via Dispersion Polymerizations in Carbon Dioxide" Macromol., 31, 6794-6805, 1998.
32. O'Neill, M.L., Yates, M.Z., Harrison, K.L., Johnston, K.P. "Emulsion Stabilization and Flocculation in CO_2. 1. Turbidimetry and Tensiometry" Macromol., 30, 5050-5059, 1997.
33. Mc Clain, J.B., Betts, D.E., Canelas, D.A., Samulski, E.T., DeSimone, J.M., Londono, J.D., Cochran, H.D., Wignall, G.D., Chillura-Martino, D., Triolo, R. "Design of Nonionic Surfactants for Supercritical Carbon Dioxide" Science, 274, 2049-2052, 1996.
34. Berger, T., Mc Ghee, B., Scherf, U., Steffen, W. "Polymerization of Vinylpyrrolidone in Supercritical Carbon Dioxide with a Diblock Copolymer Stabilizer" Macromol., 33, 3505-3507, 2000.
35. Canelas, D.A., Betts, D.E., DeSimone, J.M. "Dispersion Polymerization of Styrene in Supercritical Carbon Dioxide: Importance of Effective Surfactants" Macromol., 29, 2818-2821, 1996.
36. Canelas, D.A., DeSimone, J.M. "Dispersion Polymerization of Styrene in Carbon Dioxide Stabilized with Poly(styrene-b-dimethysiloxane)" Macromol., 30, 5673-5682, 1997.
37. Shiho, H., DeSimone, J.M. "Dispersion Polymerization of Styrene in Supercritical Carbon Dioxide Utilizing Random Copolymers Containing a Fluorinated Acrylate for Preparing Micron-Size Polystyrene Particles" J. Polymer Science: Part A: Polymer Chem., 38, 1146-1153, 2000.
38. Giles, M.R., O'Connor, S.J., Hay, N.J., Winder, R.J., Howdle, C.M. "Novel Graft Stabilizers for the Free Radical Polymerization of Methyl Methacrylate in Supercritical Carbon Dioxide" Macromol., 33, 1996-1999, 2000.
39. Shiho, H., DeSimone, J.M. "Dispersion Polymerization of Acrylonitrile in Supercritical Carbon Dioxide" Macromol., 33, 1565-1569, 2000.
40. Shiho, H., DeSimone, J.M. "Dispersion Polymerization of Glycidyl Methacrylate in Supercritical Carbon Dioxide" Macromol., 34, 1198-1203, 2001.
41. Shiho, H., DeSimone, J.M. "Dispersion Polymerization of 2-Hydroxyethyl Methacrylate in Supercritical Carbon Dioxide" J. Polymer Science: Part A: Polymer Chemistry, 38, 3783-3790, 2000.
42. Shaffer, K.A., Jones, T.A., Canelas, D.A., DeSimone, J.M. "Dispersion Polymerizations in Carbon Dioxide Using Siloxane-Based Stabilizers" Macromol., 29, 2704-2706, 1996.
43. O'Neill, M.L., Yates, M.Z., Johnston, K.P., Smith, C.D., Wilkinson, S.P. "Dispersion Polymerization in Supercritical CO_2 with a Siloxane-Based Macromonomer: 1. The Particle Growth Regime" Macromol., 31, 2838-2847, 1998.
44. O'Neill, M.L., Yates, M.Z., Johnston, K.P., Smith, C.D., Wilkinson, S.P. "Dispersion Polymerization in Supercritical CO_2 with a Siloxane-Based Macromonomer: 2. The Particle Formation Regime" Macromol., 31, 2848-2856, 1998.

45. Christian, P., Howdle, S.M., Irvine, D.J. "Dispersion Polymerization of Methyl Methacrylate in Supercritical Carbon Dioxide with a Monofunctional Pseudo-Graft Stabilizer" Macromol., 33, 237-239, 2000.
46. Christian, P., Giles, M.R., Griffiths, M.T., Irvine, D.J., Major, R.C., Howdle, S.M. "Free Radical Polymerization of Methyl Methacrylate in Supercritical Carbon Dioxide Using a Pseudo-Graft Stabilizer : Effect of Monomer, Initiator, and Stabilizer Concentrations" Macromol., 33, 9222-9227, 2000.
47. Meredith, J.C., Johnston, K.P. "Theory of Polymer Adsorption and Colloid Stabilization in Supercritical Fluids. 2. Copolymer and End-Grafted Stabilizers" Macromol., 31, 5518-5528, 1998.
48. Cooper, A.I., Hems, W.P., Holmes, A.B. "Synthesis of Highly Cross-Linked Polymers in Supercritical Carbon Dioxide by Heterogeneous Polymerization" Macromol., 32, 2156-2166, 1999.
49. Adamsky, F.A., Beckman, E.J. "Inverse Emulsion Polymerization of Acrylamide in Supercritical Carbon Dioxide" Macromol., 27, 312-314, 1994.
50. Lee, C.T., Psathas, P.A., Johnston, K.P., deGrazia, J., Randolph, T.W. "Water-in-Carbon Dioxide Emulsions : Formation and Stability" Langmuir, 15, 6781-6791, 1999.
51. Psathas, P.A., daRocha, S.R.P., Lee, C.T., Johnston, K.P., Lim, K.T., Webber, S. "Water-in-Carbon Dioxide Emulsions with Poly(dimethylsiloxane)-based Bolck Copolymer Ionomers" Ind. Eng. Chem. Res., 39, 2655-2664, 2000.
52. Psathas, P.A., Janowiak, M.L., Garcia-Rubio, L.H., Johnston, K.P. "Formation of Carbon Dioxide in Water Miniemulsions Using the Phase Inversion Temperature Method" Langmuir, 18, 3039-3046, 2002.
53. Clark, M.R., DeSimone, J.M. "Cationic Polymerization of Vinyl and Cyclic Ethers in Supercritical Carbon Dioxide" Macromol., 28, 3002, 1995.
54. Biddulph, R.H., Plesch, P.H. "The Two-temperature Polymerization of Isobutene. IV. Exploratory Experiments" J. Chem. Soc., 3913-3920, 1960.
55. Pernecker, T., Kennedy, J.P. "Carbocationic Polymerisation in Supercritical Carbon-Dioxide .1. Exploratory Experiments with Isobutylene" Polym. Bull. (Berlin), 32, 537-543, 1994.
56. Deäk, G., Pernecker, T., Kennedy, J.P. "Carbocationic Polymerization in Supercritical CO_2 .3. The Ceiling Temperature of and the Effect of Temperature on the Polymerization of Isobutylene" Polym. Bull. (Berlin), 33, 259-265, 1994.
57. Odell, P.G., Hamer, G.K. "Polycarbonates via Melt Transesterification in Supercritical Carbon Dioxide" Polym. Prepr. (Americam Chemical Society), 38, 470-471, 1997.
58. Odell, P.G., United States Patent 5,698,665. 1997.
59. Burke, A.L.C., Givens, R.D., Jikei, M., DeSimone, J.M. "Use of CO_2 in Step-Growth Polymerizations: From Plasticized Polymer Melts to Solid State Polymerizations" Polym. Prepr. (Americam Chemical Society), 38, 387-388, 1997.
60. Givens, R.D., Jikei, M., DeSimone, J.M. "Synthesis of Polyamides and Polycarbonates using Supercritical Carbon Dioxide" Polym. Prepr. (Americam Chemical Society), 38, 468-469, 1997.

Chapter 8

MODIFICATION OF POLYMERS IN SUPERCRITICAL CARBON DIOXIDE

Giuseppe Filardo, Alessandro Galia, Alberto Giaconia
Dipartimento di Ingegneria Chimica dei Processi e dei Materiali, Università di Palermo, Italy filardo@dicpm.unipa.it

1. INTRODUCTION

The interaction of scFluids and polymers are governed by the intermolecular forces between solvent-solvent, solvent-polymer segment, and polymer segment-segment pairs. Because of its symmetry, within reasonable pressure values, CO_2 does not have a dipole moment, but it does have a quadrupole moment significant over a much shorter distance than dipolar interactions. The quadrupole moment and the Lewis acidity of CO_2 imparts to the carbon dioxide the peculiarity to be a solvent for selected classes of polymers like perfluorinated polyacrylates, polysiloxanes and polyether-polycarbonate diblock copolymers [1-3].

Apart from the specific interactions due to the nature of its atoms towards the aforementioned classes of polymer, dependently on the low value of its polarizability, $scCO_2$ is not expected to be a good solvent for many macromolecular materials. Nevertheless supercritical CO_2 interacts deeply with polymers.

2. GENERAL CONSIDERATIONS

The action of $scCO_2$ on polymers is, as a rule, confined within the amorphous regions of the macromolecular structure and results in the modification of the internal forces among the chains with consequent

M. Aresta (ed.), Carbon Dioxide Recovery and Utilization, 197-207.

swelling and changes of the mechanical and physical properties of the polymers.

The whole effect, often called plasticization, is characterised by increased segmental and chain mobility and by an increase in the interchain distance.

The plasticization has a strong impact on many processing operations.

In so far as the viscosity reduces, this influences polymer extrusion and blending; additionally the enhancement of free volume in the polymer matrix allows a faster diffusion of additives inside the polymer for impregnation and extraction, or of the monomer in mass transfer controlled polymerisation processes (gel effect).

Plasticization also influences the foaming of polymers, and can lead to changes in polymer morphology due to induced crystallization as a consequence of a reduction in the kinetic barrier to the molecular rearrangement.

Depending on the nature of the macromolecule, this plasticizing effect can be the result of the ability of CO_2 molecules to interact with the basic sites in polymer molecules. To this regard the interaction between CO_2 and the carbonyl group in polymethylmethacrylate (PMMA) has been experimentally proved and related to the Lewis acid-base bond [4]. It has also been shown experimentally that such interactions between CO_2 and polymer functional groups result in a decreasing of the chain-chain interactions and the increasing of the mobility of polymer segments in strict analogy to the effect of the temperature on the polymers. [5]

The effect of the presence of CO_2 into the polymer matrices has been studied by a variety of methods as comphensively summarized on recent reviews [6, 7]: gas sorption and permeability, sorptive dilation, creep compliance, high-pressure differential scanning calorimetry (DSC); infrared spectroscopy; dielectric measurements, fluorescence; NMR spectroscopy; X-ray diffraction; dynamic mechanical response; chromatography.

3. EXTRACTION

Because of their peculiar properties supercritical fluids have been chosen as fluid process in a number of applications for their tuneable solvency, through the proper choice of pressure and/or temperature, coupled with their favourable transport properties. Based on these preconditions supercritical fluids have been widely investigated as solvent media for extractions (SFE).

The most common application of SFE is the use of $scCO_2$ for decaffeination of coffee, and in extraction processes of hops and spices. [8]

As a natural extension of the extraction processes from solid matrices, the extraction of low molecular compounds (residual solvent, unreacted monomer, oligomer, initiator decomposition products, additives) from polymers represent an interesting applicative field which can exploit the peculiar properties of supercritical fluids. Indeed the combination of favourable mass transport properties and perfect wettability with the plasticization and swelling of the polymer matrix allows one to enhance significantly the kinetics of extraction of aforementioned compounds from polymeric samples. As recent examples of application of SFE from polymeric materials it is worth to mention the extraction of copper ions using sc-CO_2 soluble chelating agents [9] and an industrial leather degreasing process [10].

4. IMPREGNATION

The action of scCO_2 on the macromolecular matrices allows not only the extraction process but also makes possible to infuse into them different species.

Once more the appeal of this technology lays on the wider spectrum of polymeric materials which can be prepared using the CO_2 induced acceleration of mass transfer kinetics inside amorphous polymers. A key factor for the success of this modification method is constituted by the intrinsic high diffusion coefficient of small carbon dioxide molecules inside the matrix so that this plasticizing agent can diffuse out promptly upon depressurisation thus leaving no solvent residue in the material and making possible the practical immobilisation of the infused compound as a consequence of rapid free volume decrease.

A relevant property to be considered in the design of impregnation experiments is the partitioning coefficient of the solute between the fluid and the swollen polymer phase. For additives with relevant solubility in the supercritical fluid high level of impregnation can be achieved even in the case of partitioning favourable to the gaseous phase, in the case of low solubility compounds high concentration of additives inside the matrix can be reached only if partitioning is strongly shifted towards the polymer phase owing to specific interactions between polymer chain and the infusing molecule.

Depending on the situation the contact time can be relevantly different.

Examples of CO_2-assisted impregnation are the production of polymer films with highly reflective metallised surfaces for various applications, obtained through infusion of organometallics or metal complexes into a polymer matrix [11-13]. Coating and impregnation of a fibrous polyethylene material with a non-ionic surfactant using sc-CO_2 is also reported [14]. Other interesting applications of SCF impregnation are polymer dyeing, grafting and reactive blending that will be described in the following sections.

Carbon dioxide induced plasticization of amorphous polymers offers new opportunities in processing polymeric based materials for optical applications. Thank to the expanded structure of the matrix it is possible, for example, to deliver azobenzene derivatives, that represent a class of molecules frequently used as dopants in non linear optical (NLO) devices for second harmonic generation, into PMMA and polystyrene (PS) [15,16]. In order to prepare doped polymeric materials having an NLO effect, the impregnated azobenzene chromophores must be oriented usually via poling, at temperatures near the glass transition temperature of the polymer matrix. The advantages of such mild approach is evident since long-lived NLO systems have to be based on highly rigid polymeric matrices with high Tg and, unfortunately, the high temperature needed for the alignment can cause degradation of the materials. Depending on the density of the $scCO_2$, by this approach, it is possible to reduce the Tg of PMMA and PS at values very close to the room temperature.

5. DYEING

Due to the environmental concerns linked to the dyeing technologies carried out with the conventional methodologies, dyeing of polymeric materials by means of supercritical fluids, and namely of $scCO_2$, represents a particular but very important case of environmental friendly technological advancement. The ease recovering of SCF after the dyeing of the fibres, achieved by simply depressurisation, and the possibility of recycling the SCF repeatedly, without needing to clean the solvent (since the dye will precipitate from SCF solvent once its density is significantly reduced) has been a winner argument for commercialising this process.

Applicative examples are the dyeing of PET [17] and of fibre textile [18] in $sc\text{-}CO_2$.

6. GRAFTING

A different way to modify a macromolecular material is based on the grafting of a specific group onto polymer backbones. Examples of this approach are constituted by the thermally initiated free-radical grafting of maleic anhydride onto poly(4-methyl-l-pentene) [19], and the gamma rays induced grafting of the same compatibilizer onto isotactic polypropylene [20], both carried out in an $scCO_2$ environment. The grafting of isocyanato-isopropyl groups onto semicrystalline poly(ethylene-co-vinylalcohol) using SCCO2 has also been achieved [21].

7. REACTIVE BLENDING

The possibility of swelling a polymer thus preparing the matrix to the infusion of a monomer different by the repeat union precursor of the preformed matrix is at the base of a quite relevant technique able to bypass limitations like the incompatibility of polymers. Interpenetrating networking can be obtained through the infusion of styrene and/or simple acrylates in the presence of suitable initiator into CO_2-swollen polymer matrices, (PCTFE, PMP, Polyethylene, Nylon 66, polyoxymethylene, polycarbonates, PVC, PTFE, poly(TFE-co-hexafluoroprolylene). The polymerisation of the infused monomer within the polymer matrix, [22-27], have been achieved through free-radical or anionic polymerizations both "in situ" during diffusion of the components or "ex-situ" after depressurisation of the polymer phase. By a proper selection of operative conditions it is possible to vary the penetration depth, thus modifying surface layers leaving unaltered the bulk of the polymer matrix. [22]

It should be noted that the properties of the polymer blends obtained by this route can be different from the properties of the same blends produced through conventional technologies. [7]

8. COATING

The appeal about the utilisation of $scCO_2$ as a fluid in coating applications is due to the intrinsic advantage of reducing or eliminating the emission of volatile organic during spray painting.

Thus the coating materials are dissolved in $scCO_2$ and subsequently released through a specially designed nozzle from the spraying head as an atomized spray. The dispersed entrained phase is constituted by droplets having narrow size distribution so that a well defined coated surface can be prepared. Since the solubility of many compounds used in coating systems is quite low in $scCO_2$ an organic cosolvent is usually added. As an alternative to the addition of cosolvents the use of stabilizers, similar to those employed in dispersion polymerisation reactions, has been proposed (section 3c).

Successful spray coatings of poly(2-ethyl hexylacrylates) dispersed in sc-CO_2 have been recently reported [28].

9. EXTRUSION AND BLENDING

The plasticizing action of supercritical CO_2 in polymer melts may be applied in extrusion and mould injection processes where, due to the reduced shear stresses, lower temperatures with respect to conventional processes may be used, thus preventing degradation of thermal sensitive polymers. Simultaneously it would be possible to take advantage of the polymer swelling to conduct extraction of contaminants from polymer matrices (useful, for example, in recycling), impregnation of additives or reactants into polymer matrices (useful, for example, in dyeing and in reactive extrusion processes, respectively), and production of microcellular polymer structures by foaming.

In all these processes rapid formation of a single-phase polymer-CO_2 mixture must be assured, while the pressure along the extrusion barrel should be high enough to prevent anticipated gaseous phase separation. The possibility of foaming of the molten polymer once it comes off the die must be considered as well; as discussed later foaming can lead to a material with a macrocellular or a microcellular structure.

Specific interaction between carbon dioxide and polymer functional groups, have broad implications for the blending of non-miscible polymers. In fact, if CO_2 interacts differently with components of the polymer blend which have different molecular structures, it should have different plasticising action on them. Consequently blending may be improved as the viscosities of these polymers approach each other. Moreover, an effect of CO_2 on the interfacial surface tension between polymeric phases may facilitate blending of non-miscible polymers.

Examples of CO_2-assisted polymer blending (a part those already discussed as examples of reactive blending) have been reported by Lee M. *et al.* [29-32] as they performed PE/PS blending using single and twin-screw extruders with different arrangements, claiming a decrease of size and a more even distribution of the dispersed PS domains in the blend by increasing the CO_2 dose rate [29,30]. Similar results have been attained for the blending PMMA/PS [31,32].

10. INDUCED CRYSTALLISATION

The increase of the chain mobility induced by the $scCO_2$ that plasticize glassy polymers has important implications especially for semicrystalline polymers. This occurs in some polymers when CO_2 induced plasticization allows the chains to rearrange into more energetically favoured ordered configurations, thus forming crystallites. This effect may be used to induce crystallisation at temperatures low enough to prevent the activation of unwanted side reactions. This effect has been particularly investigated in the case of poly(ethylene terephthalate) (PET) [33-35], whose cristallinity is relevant in determining the gas permeability in the soft-drink bottles and the mechanical properties in the case of fibers. Note that the morphology of PET processed with sc-CO_2 could be different from what obtained by heat or drawing [35,36]. Moreover sc-CO_2 can change the degree of crystallinity in various other polymers such as poly(phenylene sulfide) [37,38], polycarbonate [39], poly(aryl ether ketone) [40], syndiotactic PS [41,42]. No change in the degree of crystallinity upon treatment with dense CO_2 were observed in some other polymers like PVDF [43].

11. FOAMING

Supercritical carbon dioxide can be used as a foaming agent in polymer melt processing since, as pressure of a polymer plasticized with CO_2 is released, a plastic foam may be produced.

This route may enhance the production of polymers characterised by microcellular structures, i.e. materials with cell diameters smaller than 10 μm and cell densities greater than 10^8 cells/cm^3, with uniform size and distribution. These microcellular materials exhibit interesting thermal, electrical and mechanical properties (with respect to the same unfoamed bulk polymer) thus enhancing their field of utilisation for specific applications (e.g. the production of microfiltration membranes).

Microcellular processing is characterised by three basic steps: polymer-CO_2 solution formation, microcell nucleation and cell growth [44]. Several recent studies deal with the possibility of controlling nucleation and growth of cells during depressurisation of CO_2-saturated polymers using both batch and continuous processes [44-52]. It has been observed that the CO_2 content, the temperature, the pressure, and the rate of depressurisation have a basic role in determining the structure of the formed material. The rate of depressurisation is a crucial parameter because the achievement of microcellular structure requires a fast homogeneous nucleation, which, in turn, is determined by high supersaturation induced by a quick lowering of the solubility of CO_2 in the melt. Cooling due to Joule-Thomson effect and possibility of crystallisation during pressure release, as well as non-porous skin formation [48, 49] should be considered in these processes.

As already mentioned, polymer foams may result from CO_2-assisted polymer extrusion processes [29,30]. In this case the rapid formation of a single phase polymer-CO_2 inside the extruder must be achieved [52,53], while the rate of pressure drop will be determined by different parameters, including nozzle geometry and polymer melt flow rate. Other examples of microcellular polymers produced by this route are polycarbonates [46,47], polystyrene [51-53] and polypropylene [53].

Due to its relatively high solubility in polymers, carbon dioxide has proven a better foaming agent (in terms of higher cell densities) then other non-expensive atmospheric gases like nitrogen [51,53].

An example of commercially available process to produce foams of thermoplastics (including polystyrene, polyolefines and polyesters) is the MuCell® process developed by Trexel Inc.. In this process supercritical carbon dioxide is injected into the polymer melt in an extruder; the production of foams with superior properties than hydrocarbon-blown foams of similar densities is claimed.

12 CONCLUDING REMARKS

The use of $scCO_2$, which is an inexpensive environmentally friendly solvent, could open a whole new dimension for processing of polymeric materials when high diffusion rate inside the polymer matrix, increased chain mobility or reduced polymer melt viscosity are required preconditions for a successful modification.

A number of relevant papers as a source of basic information about CO_2-assisted polymer modification are present in the literature. Particularly, for more detailed information the authors suggest to refer to the useful reviews published in the recent open literature [1].

REFERENCES

1. DeSimone, J. M., Guan, Z., Elsbernd, C.S. "Synthesis of Fluoropolymers in Supercritical Carbon-Dioxide" Science, 257 (5072), 945-947, 1992.
2. Shaffer, K. A., Jones, T. A., Canelas, D. A., DeSimone, J. M. "Dispersion Polymerization in Carbon Dioxide Using Siloxane-Based Stabilizers" Macromol., 29, 2704, 1996.
3. Sarbu, T., Styranec, T., Beckman, E.J. "Design and Synthesis of Low Cost, Sustainable CO_2-philes" Ind. Eng. Chem. Res., 39, 4678-4683, 2000.
4. Kazarian, S.G., Vincent, M.F., Bright, F.V., Liotta, C.L., Eckert, C.A. "Specific intermolecular interaction of carbon dioxide with polymers" J. Am. Chem. Soc., 118 1729-1736, 1996.
5. Kazarian, S.G., Brantley, N.H., West, B.L., Vincent, M. F., Eckert, C.A. "In situ spectroscopy of polymers subjected to supercritical CO_2: Plasticization and dye impregnation" Appl. Spectrocopy, 51, 491-494, 1997.
6. Kendall, J.L., Canelas, D.A., Young, J.L., De Simone, J.M. "Polymerizations in Supercritical Carbon Dioxide" Chem. Rev., 99, 543-563, 1999.
7. Kazarian, S.G. "Polymer Processing with Supercritical Fluids" Polymer Science, 42, 78-101, 2000.

8. Hubert, P., Vitzthum, O.G. "Fluid Extraction of Hops, Spices, and Tobacco with Supercritical Gases" Angew. Chem. Int. Ed. Eng., 17, 710-715, 1978.
9. Zacharia, R.E., Simon, S.L., Beckman, E.J., Enick, R.M. "Improving the thermal stability of a polymer through liquid carbon dioxide extraction of a metal compound" Polymer Degradation and Stability., 63, 85-88, 1999.
10. Marsal, A., Celma, P.J., Cot, J., Cequier, M. "Supercritical CO_2 extraction as a clean degreasing process in the leather industry" J. Sup. Fluids., 16, 217-223, 2000.
11. Rosolovsky, J., Boggess, R.K., Rubira, A.F., Taylor, L.T., Stoakley, D.M., StClair, A.K. "Supercritical fluid infusion of silver into polyimide films of varying chemical composition" J. Mat. Res., 12, 3127-3133, 1997.
12. Boggess, R.K., Taylor, L.T., Stoakley, D.M., StClair, A.K. "Highly reflective polyimide films created by supercritical fluid infusion of a silver additive" J. Appl. Polymer Sci., 64, 1309-1317, 1997.
13. Said-Galiyev, E., Nikitin, L., Vinokur, R., Gallyamov, M., Kurykin, M., Petrova, O., Lokshin, B., Volkov, I., Khokhlov, A., Schaumburg, K. "New chelate complexes of copper and iron: Synthesis and impregnation into a polymer matrix from solution in supercritical carbon dioxide" Ind. Eng. Chem. Res., 39, 4891-4896, 2000.
14. Ma, X., Tomasko, D.L. "Coating and impregnation of a nonwoven fibrous polyethylene material with a nonionic surfactant using supercritical carbon dioxide" Ind. Eng. Chem. Res., 36, 1586-1597, 1997.
15. Barry, S.E., Soane, D.S. "Second Harmonic Generation in Carbon Dioxide Processed Thin Polymer Films" Macromol., 29, 3565-3573, 1996.
16. Chapman, B.R., Paulaitis, M.E., Gochanour, C.R. "Effect of CO_2 plasticization on azobenzene diffusion in glassy polystyrene near the glass transition" Macromol., 34, 340-342, 2001.
17. Bach, E., Cleve, E., Schollmeyer, E. "Dyeing of poly(ethylene terephthalate) fibers in supercritical carbon doxide". Rudolph von Rohr, Ph. And Trepp, C., Eds., Zurich: Elsevier Science B.V., 1996.
18. Saus, W., Knittel, D., Schollmeyer, E. "Dyeing of Textiles in Supercritical Carbon-Dioxide" Textile Research Journal., 63, 35, 1993.
19. Hayes, H.J., McCarthy, T.J. "Maleation of poly(4-methyl-1-pentene) using supercritical carbon dioxide" Macromol., 31, 4813-4819.1998.
20. Spadaro, G., De Gregorio, R., Galia, A., Valenza, A., Filardo, G. "Gamma radiation induced maleation of polypropylene using supercritical CO_2: preliminary results" Polymer., 41, 3491-3494, 2000.
21. Friedmann, G., Guilbert, Y., Catala, J.M. "Modification chimique de polymères en milieu CO_2 supercritique: Greffage de groupes isocyanato-isopropyle sur une chaîne de poly(éthylène-co-alcool vinylique)". Eur. Polymer J., 36, 13-20, 2000.
22. Muth, O., Hirth, T., Vogel, H. "Polymer modification by supercritical impregnation" J. Super. Fluids., 17, 65-72, 2000.
23. Watkins, J.J., McCarthy, T.J. "Polymerization in supercritical fluid-swollen polymers: a new route to polymer blends" Macromol., 27, 4845-4847, 1994.
24. Kung, E., Lesser, A.J., McCarthy, T.J. "Morphology and mechanical performance of polystyrene/polyethylene composites prepared in supercritical carbon dioxide" Macromol., 31, 4160-4169, 1998.
25. Arora, K.A., Lesser, A.J., McCarthy, T.J. "Synthesis, characterization, and expansion of poly(tetrafluoroethylene-co-hexafluoropropylene)/polystyrene blends processed in supercritical carbon dioxide" Macromol., 32, 2562-2568, 1999.

26. Kung, E., Lesser, A.J., McCarthy, T.J. "Composites prepared by the anionic polymerization of ethyl 2-cyanoacrylate within supercritical carbon dioxide-swollen poly(tetrafluoroethylene-co-hexafluoropropylene)" Macromol., 33, 8192-8199, 2000.
27. Li, D., Han, B. "Impregnation of polyethylene (PE) with styrene using supercritical CO_2 as the swelling agent and preparation of PE/polystyrene composites". Ind. Eng. Chem. Res., 39, 4506-4509, 2000.
28. Shim, J.J., Yates, M.Z., Johnston, K.P. "Polymer coatings by rapid expansion of suspensions in supercritical carbon dioxide" Ind. Eng. Chem. Res., 38, 3655-3662, 1999.
29. Lee, M., Tzoganakis, C., Park, C.B. "Effects of supercritical CO_2 on the viscosity and morphology of polymer blends" Adv. Polymer Tech., 19, 300-311, 2000.
30. Lee, M., Tzoganakis, C., Park, C.B. "Extrusion of PE/PS blends with supercritical carbon dioxide" Polymer Eng. Sci., 38, 1112-1120, 1998.
31. Elkovitch, M.D., Lee, L.J., Tomasko, D.L. ANTEC'98, p.2538-2541.
32. Elkovitch, M.D., Tomasko, D.L., Lee, L.J. "Supercritical carbon dioxide assisted blending of polystyrene and poly(methyl methyacrylate)" Polymer Eng. Sci., 39, 2075-2084, 1999.
33. Zhong, Z., Zheng, S., Mi,Y. "High-pressure DSC study of thermal transitions of a poly(ethylene terephthalate)/carbon dioxide system" Polymer., 40, 3829-3834, 1999.
34. Mensitieri, G., DelNobile, N.A., Guerra, G., Apicella, A., AlGhatta, H. "Low-Temperature Melting Behavior of CO_2 Crystallized Modified PETS" Polymer Eng. Sci., 35, 506, 1995.
35. Kazarian, S.G., Brantley, N.H., Eckert, C.A. "Applications of vibrational spectroscopy to characterize poly(ethylene terephthalate) processed with supercritical CO_2" Vibrat. Spectrosc., 19, 277-283, 1999.
36. Kazarian, S.G., Brantley, N.H., Eckert, C.A. "Dyeing to be clean: Use supercritical carbon dioxide" CHEMTECH, 29, 36-41, 1999.
37. Schultze, J.D., Bohning, M. "Sorption and Permeation Properties of Poly(P-Phenylene Sulfide) Crystallized in the Presence of Sorbed Gas Molecules" Springer J. Macromol. Chem., 194, 431-444, 1993.
38. Schultze, J.D., Bohning, M., "Enhanced Crystallization by Annealing of Poly(Phenylene Sulfide) in Presence of CO_2 and N_2O" Springer J. Macromol. Chem., 194, 339, 1993.
39. Gross, S.M., Roberts, G.W., Kiserov, D.J., DeSimone, J.M. "Crystallization and solid-state polymerization of poly(bisphenol A carbonate) facilitated by supercritical CO_2" Macromol., 33, 40-45, 2000.
40. Handa, Y.P., Capowski, S., O'Neill, M. "Compressed-Gas-Induced Plasticization of Polymers" Thermochim. Acta., 226, 177-185, 1993.
41. Kazarian, S.G., Briscoe, B.J., Lawrence, C.J. "Supercritical Enhanced Processing" Polymer Process Engineering'99, 28-36, 1999.
42. Handa, Y.P., Zhang, Z.Y., Wong, B. "Effect of compressed CO_2 on phase transitions and polymorphism in syndiotactic polystyrene" Macromol., 30, 8499-8504, 1997.
43. Briscoe, B.J., Lorge, O., Wajs, A., Dang, P. "Carbon dioxide poly(vinylidene fluoride) interactions at high pressure" J. Polymer Sci., Part B: Polymer Physics., 36, 2435-2447, 1998.
44. Park, C.B., Baldwin, D.F., Suh, N.P. "Effect of the pressure drop rate on cell nucleation in continuous processing of microcellular polymers" Polymer Eng. Sci., 35, 432-440, 1995.
45. Liang, M.T., Wang, C.M. Proceeding of the 6th Meeting on Supercritical Fluids: Chemistry and Materials, Nottingham (UK), p.151-156, 10-13 April 1999.
46. Huang, Q., Seibig, B., Paul, D. "Polycarbonate hollow fiber membranes by melt extrusion" J. Membr. Sci., 161, 287-291, 1999.
47. Seibig, B., Huang, Q., Paul, D. "Design of a novel extrusion system for manufacturing microcellular polymer" Cellular Polymers, 19, 93-102, 2000.

48. Goel, S.K., Beckman, E.J. "Generation of microcellular polymeric foams using supercritical carbon dioxide. I: Effect of pressure and temperature on nucleation" Polymer Eng. Sci., 34, 1137-1147, 1994.
49. Goel, S.K., Beckman, E.J. "Generation of microcellular polymeric foams using supercritical carbon dioxide. II: Cell growth and skin formation" Polymer Eng. Sci., 34, 14, 1148-1156, 1994.
50. Goel, S.K., Beckman, E.J. "Nucleation and growth of microcellular materials: supercritical CO_2 as foaming agent" AIChE Journal, 41, 357-366, 1995.
51. Park, C.B., Suh, N.P. "Filamentary extrusion of microcellular polymers using a rapid decompressive element" Polymer Eng. Sci., 36, 34-48, 1996.
52. Park, C.B., Behravesh, A.H., Venter, R.D. "Low density microcellular foam processing in extrusion using CO_2" Polymer Eng. Sci., 38, 1812-1823, 1998.
53. Park, C.B., Suh, N.P. "Rapid polymer/gas solution formation for continuous production of microcellular plastics" J. Manuf. Sci. Eng., 118, 639-645, 1996.

SECTION IV

CHEMICAL CONVERSION OF CARBON DIOXIDE

Chapters 9-13 in Section IV summarize the activity performed within WP5 – Part A of the BRRT-CT98-5089 RUCADI Project, "Recovery and Utilization of Carbon Dioxide"-Chemical Utilization of Carbon Dioxide by the following Partners:

Michele Aresta, *METEA Research Center*, University of Bari, Italy
Marco Ricci, *PolimeriEuropa*, Italy
Giuseppe Filardo, *University of Palermo*, Italy
Alessandro Galia, *University of Palermo*, Italy
Danielle Ballivet, *CNRS*, France
Claude Mirodatos, *CNRS*, France
Eckard Dinjus, *FZK*, Germany
Ragnar Larsson, *University of Lund*, Sweden
Marek Borowiak, ICRI, Poland
R.Kieffer, *ULPS Strasbourg,* France
A.Kiennemann, *ULPS Strasbourg,* France
Jan Augustynski, *University of Genève*, Switzerland
Julian Ross, *University of Limerick*, Ireland
D. Treacy, *University of Limerick*, Ireland
Mike Hayes, *University of Limerick*, Ireland
J.J Lehay *University of Limerick*, Ireland
Paul Sermon, *University of Surrey*, UK
Kulathu Seshan, *University of Twente*, The Netherlands
Giuliano Dolcetti, *University of Udine*, Italy
Séverine Camy, *ENSIACET*, France
Angela Dibenedetto, *METEA Research Center*, University of Bari, Italy
Julia Valla, *C.P.E.R.I. / C.E.R.T.H.*, Thessaloniki, Greece
Angeliki Lemonidou, *CPERI/CERTH*, Thessaloniki, Greece
Immacolata Tommasi, *NOWATECH*, Italy
Eugenio Quaranta, *NOWATECH*, Italy

Chapter 9

CARBON DIOXIDE FIXATION INTO ORGANIC COMPOUNDS

Michele Aresta and Angela Dibenedetto
Department of Chemistry, University of Bari, Campus Universitario, 70126, Bari – Italy
aresta@metea.uniba.it

1. INTRODUCTION

The utilisation of carbon dioxide in the synthesis of organic chemicals is an outstanding example of "carbon recycling" [1], that can effectively contribute to both develop clean synthetic methodologies and avoid CO_2 emission. A quite large number of ways of using CO_2 (Fig.9.1) are known today. Nevertheless, only few processes are presently on stream. (Table 9.1)

Table 9.1- Use of carbon dioxide in the Chemical Industry for the synthesis of organic compounds

Industrial processes that utilise CO_2 as raw material	World capacity per year	Amount of fixed carbon dioxide
Urea	95 Mt	54 Mt
Salicylic acid	70 kt	25 kt
Methanol	20 Mt	2 Mt
Cyclic carbonates	80 kt	*ca.* 40 kt
Poly (propylene carbonate)	70 kt	*ca.* 30 kt

In addition to the uses described in Table 9.1, that are discussed in detail in Chapters 10-11, ca. 30 Mt/y of carbon dioxide are used in the synthesis of inorganic carbonates for the glass industry or for the synthesis of inorganic pigments. Some more 15 Mt per year are used in technological applications, without conversion of the cumulene. This makes the total amount of carbon dioxide used in the industry equal to *ca.* 100 Mt per year. However, the

M. Aresta (ed.), Carbon Dioxide Recovery and Utilization, 211-260.

amount of carbon dioxide will be expanded to unforeseeable levels if the CO_2-application in the enhanced oil recovery (EOR) technology will be implemented on a large scale. Also, if abundant good-quality low-price carbon dioxide will be available and new catalysts are developed, the chemical use will considerably increase.

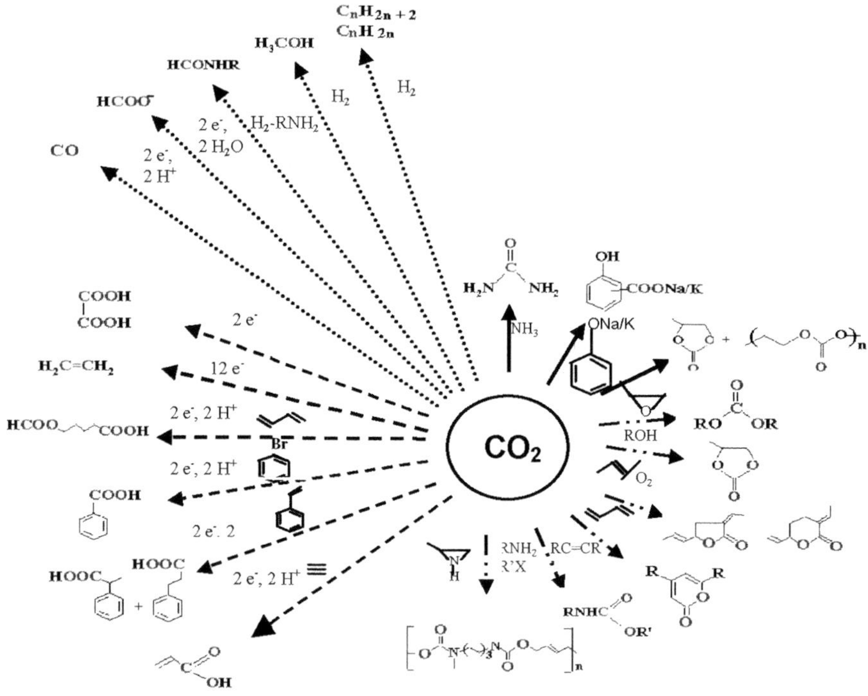

Figure 9.1 - Synthesis of chemicals from carbon dioxide

Of the processes mentioned in Table 9.1, two are exploited at the industrial level since more that one century, namely the synthesis of urea (that dates back to 1870), and the Kolbe-Schmitt reaction, (1890) used for the production of salicylic acid. Interestingly, the two processes mentioned above do not require any specially designed catalyst. The urea synthesis is a thermal process, and the salicylic acid synthesis is driven by Group 1 metal cations (essentially Na^+ and K^+). Despite these processes are on stream since long, several issues need to be addressed for a more efficient use of energy and materials. This is mostly true for the Kolbe-Schmitt reaction that uses one mol of Na^+ or K^+ per mol of acid produced with a selectivity that also

requires amelioration. These applications have been the only industrial processes based on carbon dioxide for decades. Other possible uses of CO_2 have been deemed so far by the prevalence of the economic- over the environmental-factor in decision making. In fact, either the use of carbon monoxide, or its derivatives like phosgene, as source of carbon, or the adoption of cheap processes, also if environmentally-harsh (e.g. non selective oxidation), has been a convenient solution for the synthesis of several chemicals, so far. The synthesis of carboxylic moieties by non-selective oxidation of hydrocarbon fragments or moieties is a clear example of low efficiency use of energy and atoms.
The new attitude towards environmental protection is now making economic- and environmental-issues equally important in decision making at the industrial level. Therefore, carbon dioxide utilisation is getting much attention as it responds to the principles of:

- recycling carbon while developing syntheses with atom economy,
- waste reduction at source,
- raw material diversification,
- dematerialization,
- and energy saving.

In fact, the use of carbon dioxide may allow to implement more direct and innovative synthetic pathways which respond to the concepts listed above. Noteworthy, the chemical utilization of carbon dioxide responds to the need of "avoiding carbon dioxide" not only through the amount of recycled and converted carbon dioxide (the amount directly fixed into the chemicals), but also through the amount of carbon dioxide not produced by implementing the atom efficiency and waste minimization principles [2].

It is worth to emphasise at this point, that all other reactions shown in Fig. 9.1, besides those discussed above, need the intervention of a very specific catalyst, that in general must be able to drive a nucleophile-electrophile, or acid-base, catalysis. The "transition-metal" catalyzed carbon dioxide conversion is very young and was mostly developed after the first example of carbon dioxide co-ordination to a metal centre appeared in the scientific literature in 1975 [3]. Since then, the knowledge on how to use carbon dioxide in synthetic processes has tremendously grown.

If one compares the CO_2-chemistry to the carbon monoxide chemistry, finds that the former is still in its infancy, with a gap of almost one century. The driving force in developing the CO-chemistry was the necessity to produce liquid fuels from coal in order to address specific needs in the 1913-1945 years. The driving force that is pushing the development of the CO_2 chemistry is the new attitude towards environmental problems and the need to sink CO_2 or to recycle it. It took roughly forty years for implementing at the industrial level the Syngas chemistry. The same time may be needed for bringing the new carbon dioxide chemistry from the "scientific curiosity" to the "industrial use" level. As a matter of fact, it is possible to foresee that in ten years from now the number of industrial processes based on carbon dioxide may grow significantly.

As it can be seen in Fig. 9.1, the reactions that use carbon dioxide can be divided into two classes:

- Reactions that use the entire molecule (fixation onto an organic substrate).
- Reactions that convert CO_2 into either another C_1 molecule or C_n molecules.

These two classes of reactions have different energy requirements and are relevant to products that have unlike market size, belonging to different industrial sectors, like the chemical industry (from fine chemicals to commodities) or the energy industry. Although some bulk chemicals can reach a market of tens Mt per year, this value is still enormously lower that the market of fuels. The market size of a product is that specific issue that is in general considered as a criterion for accepting or not a synthesis based on carbon dioxide, being it related to the amount of carbon dioxide that may be recycled or avoided. Nevertheless, it should not be considerd as the only driver, as technology innovation in the chemical industry may also be a key strategy for reducing the emission of carbon dioxide, contributing, thus, in a substantial mode to the control of the CO_2 emission.

2. ENERGETICS OF CARBON DIOXIDE UTILIZING REACTIONS

The utilization of carbon dioxide as a source of carbon in synthetic chemistry [4,5] appears very attractive in view of the fact that, if the separation and recovery of carbon dioxide will be adopted as a technology for controlling its immission into the atmosphere, large amounts will be available. The products that can be conveniently obtained from carbon dioxide can be categorised as:

i) *Fine- or commodity-chemicals*, i.e. molecules containing functionalities such as: -C(O)O- acids, esters, lactones; -O-C(O)O- organic carbonates; -N-C(O)O- carbamates; -N-C(O)- ureas, amides.

ii) *Products for the energy industry*, i.e. energy-rich C_1 molecules (mainly CO and CH_3OH) and C_n hydrocarbons or their derivatives.

Developing industrial processes in these areas would deeply innovate the chemical industry, with several advantages from the point of view of rational use of energy and materials, and waste minimization. The exploitation of such synthetic technology must overcome energetic and kinetic barriers. In fact carbon dioxide lays in a potential energy well (ΔG= -394.6 kJ mol^{-1}) that is not indeed unsurpassable: Nature uses carbon dioxide in a miryad of reactions. In fact, the carbon dioxide-utilizing reactions mentioned so far, can be divided into two main categories from the energetic point of view, as specified in Table 9.2.

Table 9.2 - Reactions involving CO_2 and their energetics

	Low energy processes Short term development	High energy processes Medium-term development
Source of energy	The substrate itself or heat (low-temperature processes)	Heat (high-temperature processes) - Dihydrogen - Electrons
Nature of the compounds prepared from carbon dioxide	Molecular compounds as: carboxylates, carbamates, carbonates, specialty chemicals. Polymers	Fuels Commodity chemicals
Industrial sector interested	Chemical industry	Energy and chemical industry

Those reactions in which the whole carbon dioxide molecule is fixed, may use the substrate (amines, unsaturated hydrocarbons) as the energy vector.[1] Therefore, the amount of extra energy, if required, is usually very low. These syntheses include the carboxylation reactions with the formation of a C-C or C-heteroatom bond (heteroatom = O, N, P, other element). Such low-energy processes occur at moderate temperature (from 240 to 400 K) and can be considered as short-term options. Conversely, the reduction reactions of CO_2 to C_1 or C_n species, use dihydrogen, electrons, or heat as energy source. Therefore, such processes are characterized by much energy input, occur at high temperature (600-1000 K) and can be considered as medium- to long-term options. Noteworthy, all the reactions mentioned above are common to, and very important in, biological systems (plants and bacteria). A considerable number of the enzymes involved in carbon dioxide utilisation are metal enzymes with the metal acting as the active site [1a-b,6].

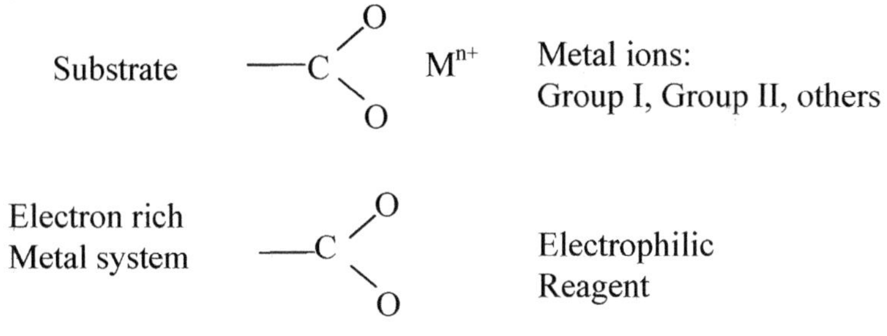

Scheme 9.1- Role of the metal in CO_2 fixation

As mentioned above, the carbon dioxide conversion requires in general an "acid-base", or in other terms, a "nucleophile-electrophile" catalysis. Metal centres may play a different role according to their oxidation state, which may produce an inversion of the role of the metal from electrophile to nucleophile. (Scheme 9.1) The latter type of interaction is typical of carbon dioxide complexes with transition metal systems. In general, the activation process of CO_2 involves a charge transfer from a nucleophile (the the transition metal, the catalyst, or the electron-rich substrate) into the lowest unoccupied $2\pi^*_u$ molecular orbital of CO_2, which causes the CO_2 molecule to bend in order to minimize its energy. The electronic structure of the bent molecule exhibits features which can be associated with a $CO_2^{\cdot -}$ species. In the bent radical-anion state, that precedes the dissociation $CO_{2,s} \rightarrow CO_s + O_s$, the O-C bond is longer (by 0.12 A) than in the ground state and weaker [3]. The additional electronic charge can be accommodated by the closed shell CO_2 molecule through the breaking of one (or two) C=O double bond(s) with subsequent formation of one (or two) additional oxygen lone-pair(s). This bond breaking prompts the carbon and the oxygen atoms to form new bonds. Such situation occurs, at a different extent, any time carbon dioxide interacts with an electrophile, should it be a metal complex, a metal surface, [7] or an electron rich organic substrate.

The data on the dissociative adsorption of CO_2 are rather controversial, especially those relevant to noble metals, for which factors causing or preventing the dissociation should be thoroughly analyzed prior to draw any conclusion. Useful information comes from the studies on the interaction of carbon dioxide with low temperature gas matrices [8].

A great emphasis has been put in the '70s and '80s on elucidating the possible role of CO_2-metal complexes [6,9], bearing the cumulene either η- or μ-bonded to one or more metal centres, in both the carbon dioxide reduction and the reaction with organic substrates like olefins. This issue is relevant to the more general question if the coordination of CO_2 to a metal is a necessary pre-requisite for the CO_2 fixation onto an organic substrate may occur. While the co-ordination to a metal system is essential for carbon dioxide reduction (C-O bond splitting [6,8] or protonation [10,11]), literature data do not show that such metal complexes have a role in the olefin-CO_2 coupling. In fact, either the olefin is activated first (a) or a three centers mechanism (b) may operate. (Scheme 9.2).

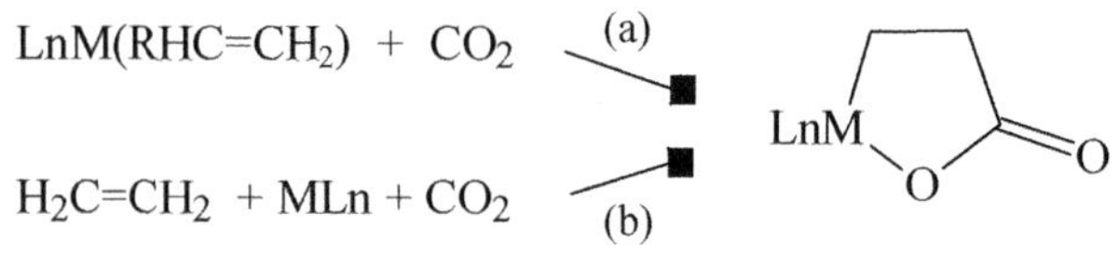

Scheme 9.2 – Carboxylation of olefins

The carboxylation reactions may have a key role in the industrial application of carbon dioxide. A number of CO_2-based synthetic approaches to carboxylates (acids, esters, lactones), carbamates, carbonates can be found in the literature and will be analyzed in the following paragraphs. They are characterised by a different level of development: from laboratory bench-scale, to various size (from a few kilos to hundred kilos) pilot plant.
The high energy processes for carbon dioxide reuse will be considered in a separate Chapter.

3. LOW-ENERGY ROUTES TO CHEMICALS FROM CARBON DIOXIDE: A COMPARISON OF EXISTING AND INNOVATIVE SYNTHETIC METHODOLOGIES

A selected number of chemicals that are or may be synthesised from carbon dioxide (*urea, formic acid, acetic acid, long chain carboxylic acids, oxalic acid, salicylic acid, 4-OH-benzoic acid, carbamates, lactones, linear esters, cyclic carbonates, linear carbonates)* will be considered in this Chapter, comparing the existing synthetic technologies with innovative ones based on carbon dioxide. The world capacity of these chemicals is presently around 120 Mt/y. A large increase is foreseeable if carbonates will be used as gasoline additives. This means that if the synthetic methodologies based on carbon dioxide were to be introduced, the amount of CO_2 used in the chemical industry would be increased by several tens Mt/y. Such analysis confirms the evaluation made some years ago of the potential of the chemical utilization for carbon dioxide mitigation [13]. Recycling carbon was estimated to contribute to cut 7-10% of the excess amount of carbon dioxide.

3.1 Industrial processes that utilize CO_2 as raw material. Existing technologies and perspectives

3.1.1 Urea: economics and general aspects

Urea, $(H_2N)_2CO$, is by far the chemical for which synthesis the largest amount of carbon dioxide is used (see Table 9.1). Urea world capacity has been estimated, in 1998, to be ca. 95 Mt/y with an operative index of *ca.* 89%. Assuming an average price of 180 Euro/t, the value of the urea production can be estimated to be *ca.* 15 E9 Euro. Urea plants are typically parts of integrated facilities, also comprehensive of ammonia plants. With 330 plants around the world, urea production is very spread. More than 54% of global capacity is located in Southeast Asia.

Innovative technologies and perspectives

No alternative technologies are available: probably, they are simply not needed: urea producers use, all over the world, the synthesis from carbon dioxide and ammonia as described above.

A net increase of the market would be possible if urea will be used as a "reactive carbon dioxide". In fact, urea could be the raw material for the synthesis of carbonates according to Eq. 4.

$$2\,ROH + H_2NCONH_2 \;=\; (RO)_2CO + 2\,NH_3 \qquad (1)$$

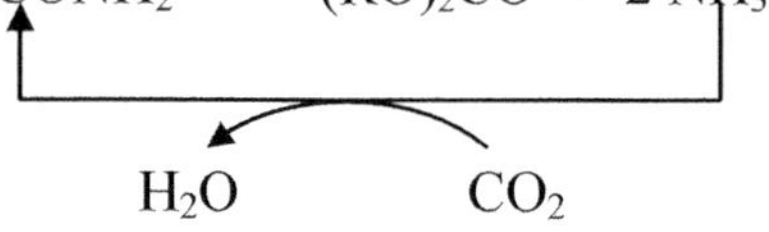

Ammonia formed in the reaction could be easily recycled for the synthesis of urea (Eq. 4), affording a "closed cycle" route to organic carbonates [14].

3.1.2 Salicylic acid: economics and general aspects

As mentioned above, the synthesis of salicylic acid is one of the few industrial processes based on carbon dioxide. Despite it is operated since more than one century, there is still a large interest in fully understanding the reaction mechanism in order to master the yield and selectivity issues as specified above. Technical–grade and USP-grade salicylic acid are produced in Europe by Rhone Poulenc, that has also plants in Brazil, and Lonza. In the United States, Dow Chemical Company is the main producer. Smaller plants around the world primarily supply regional demands. The world capacity of salicylic acid is approximately 70 kt per year. Salicylic acid is prepared on a industrial scale by the Kolbe-Schmitt reaction. (See Chapter 9). Also, air oxidation of o-cresolate at 230 °C in presence of a copper catalyst or copper benzoate (175-215 °C) can be used. Benzoyl-salicylic acid is formed in the synthesis of phenol from toluene by the Dow Chemicals process. Salicylic acid can also be obtained by fermentation of polycyclic aromatic compounds.

All these processes are not implemented on a large scale, the Kolbe-Schmitt reaction being the most largely used technology.

Innovative technologies and perspectives

There are no major changes foreseen from the technological point of view. The main expectations are on developing a catalytic cycle that may prevent the loss of one mol of sodium hydroxide per mol of product and may help to implement less hard operative conditions. The amelioration of the selectivity is also a key point, for avoiding the production of 4-OH-benzoic acid.

3.1.3 4-Hydroxybenzoic acid: economics and general aspects

4-Hydroxybenzoic acid is an intermediate in the manufacture of pharmaceuticals, plastics, pesticides and dyes. Important uses of 4-OH-benzoic acid are as an emulsifier and corrosion-protection agent. A more recent use of 4-hydroxybenzoic acid is in the preparation of polymers, either as a component in the manufacture of polyesters, or as a constituent of liquid crystals. The total market is over 250 kt per year, most of which is for captive use. "PHB-Ester", a mixture of the ethyl and *n*-propyl esters (60% ethyl and 40% propyl ester) of 4-hydroxybenzoic acid is used as food preservative: the sodium salts are antimicrobial agents with a high no-effect dose 1.2 g/kg body weight. Higher esters may be used for termite control.

Innovation in the synthetic strategy. The biotechnological way

Quite recently, the enzymatic synthesis of 4-hydroxybenzoic acid has been reported by Aresta et al. (Scheme 9.3). [15]

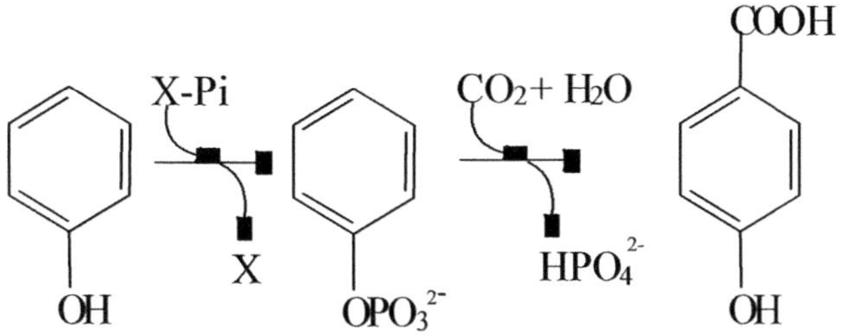

Scheme 9.3- Enzymatic conversion of phenol into 4-OH-benzoic acid

A phenylphosphate carboxylation enzyme extracted from *Thauera aromatica* and supported on low-melting agar, has been shown to be able to afford 4-hydroxybenzoic acid from phenol with 100% selectivity. This approach may be of interest as it cuts off the problem of separation and that of harsh reaction conditions. In fact, the enzymatic synthesis is performed at room temperature (mesophilic conditions) and sub atmospheric pressure of carbon dioxide.

3.1.4 *Organic carbonates: economics and general aspects*

Organic carbonates contain the OC(O)O functionality and can be categorized as follows. (Scheme 9.4)

i) Linear carbonates, either aliphatic or aromatic. The most used are dimethyl carbonate (DMC), diallyl carbonate (DAC), diethyl carbonate (DEC), diphenyl carbonate (DPC).

ii) Cyclic carbonates, such as ethylene carbonate (EC), propylene carbonate (PC), cyclohexene carbonate (CC) and styrene carbonate (SC).

iii) Polycarbonates, such as poly-(propylene carbonate) and bis-phenol-A-polycarbonate.

i) $(CH_3O)_2CO$ DMC; $(CH_2CH{=}CH_2O)CO$ DAC; $(EtO)_2CO$ DEC; $(PhO)_2CO$ DPC

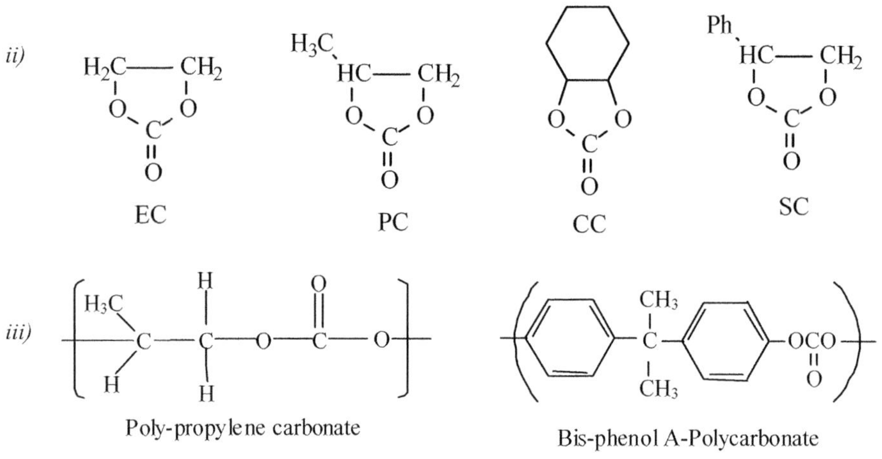

Scheme 9.4 - Molecular and polymeric carbonates

Molecular carbonates find a quite diversified use in the chemical industry as solvents, reagents (for alkylation or acylation reactions), monomers for polymers, component of special materials. The total market ranges around 1.8 Mt/y, each class having a different share, according to the use. BPA-PC is by far the carbonate that has the largest market (over 1.5 Mt/y), and its potential is far from a full exploitation [16]. Actually, the largest producers of BPA-PC are General Electric Plastics 0.6 Mt/y, Bayer 0.5 Mt/y, and Dow Chemicals 0.3 Mt/y.

Also for molecular carbonates there exists the potential of a great expansion of the market. This appears particularly true for DMC that could be used as gasoline additive, demanding, thus, amounts that would be impossible to produce with the existing synthetic technologies [17]. The current DMC production approaches 100 kt/y. Most of it is used in the synthesis of polycarbonates, while ca. 10-15% is on the market. Increasing the use of DMC as non toxic solvent and "green" reagent [18] and using DMC as gasoline additive, easily its market would reach the Mt/y size. The existing limitation is on the synthetic technologies, as the use of phosgene cannot be further expanded, as discussed above. A specific interest also exists for diallylcarbonate, used for the preparation of optical materials.

The major producers of cyclic carbonates are in the USA (Arco, Dow Chemicals, Huntsman, and Texaco) and EU (BASF). Huntsman Corp. (USA) is the largest world producer of cyclic carbonates with facilities in the USA, France and Germany. The total production is in the range of 100 kt/y, with Huntsman (ca. 40 kt/y) and BASF as leaders.

Almost the total amount of carbonates are produced from phosgene: this is today a major drawback. Only minor amounts (see Chapter 10) are produced by oxidative carbonylation of methanol or via the epoxide carboxylation (see below). There is a need for new technologies that may produce large amounts of carbonates using non toxic and cheap reagents. Aliphatic carbonates are used as solvents, green reagents for the synthesis of pharmaceuticals and agrochemicals, and monomers for polymers. DMC is an excellent substitute for dimethylsulphoxide in methylation reactions, and is also active as methoxycarbonylating agent [19]. Additionally, diphenylcarbonate [20] and methylphenylcarbonate [21] are good agents for the conversion of amines into carbamates (see below). Allylcarbonate [22] is used for the preparation of optical materials. Other carbonates have special uses. As reported above, the alkylcarbonate that may have an explosion of the market is DMC, if it will be used as combustion improver agent of gasoline. The great advantage of DMC, with respect to other additives used today, is that it has a high efficiency because of the high oxygen content and releases the lowest amount of carbon dioxid. What reported above explains the great interest in finding clean synthetic methodologies to DMC, and other cabonates, possibly based on carbon dioxide. Additionally, for its high versatility, DMC plays a key role in the carbonates panorama.

Cyclic carbonates (ethylene, propylene, and cyclohexene, mainly) are used as solvents for macromolecules and in the production of polymers. They are used in transesterification reactions, for example they afford DMC upon reaction with methanol. The reaction is catalysed by several agents [23]. They are also used in lithium batteries, as extractants and reagents.

BPA-PC, a versatile plastic, is second (27%) only to polyamides (35%) on the plastic market [24]. It is used in several industrial sectors the most relevant being the electrical and electronic (38%), followed by the construction, automotive, optical and information storage systems, medical, packaging [25]. The majority of carbonates are actually synthetised from phosgene, $COCl_2$. This technology is used since long for the synthesis of DMC [26], or other linear carbonates, cyclic carbonates [27] and BPA/PC. Chapter 11 is dedicated in this book to DMC and other aliphatic carbonates. Here, cyclic carbonates and BPA-PC will be treated.

Cyclic carbonates

Two synthetic routes to cyclic carbonates are commercially available:

* phosgenation
* carboxylation of epoxides using CO_2

The phosgenation of glycols has been for long the prominent technology [28]. It was exploited as early as 1970 by SNPE. This technology has several drawbacks, like the use of phosgene and the production of halogenated waste (HCl and organic solvents) that have a negative environmental impact. Currently, the carboxylation of epoxides, is a fast developing important industrial process.

It is worth to note that shifting from phosgene to carbon dioxide means to use from one hand a much safer species, from the other hand a much less reactive species. This requires the development of ad hoc catalysts for circumventing the kinetics of the reaction.

Carboxylation of epoxides

Cyclic carbonate formation from epoxides represents one of the few examples of recent successful syntheses based on carbon dioxide. IG Farben's discovery of the route based on ethylene oxide and CO_2 has led to the increased interest in the production of the alkylene carbonates [29]. A remarkable variety of active catalysts have been described [30].

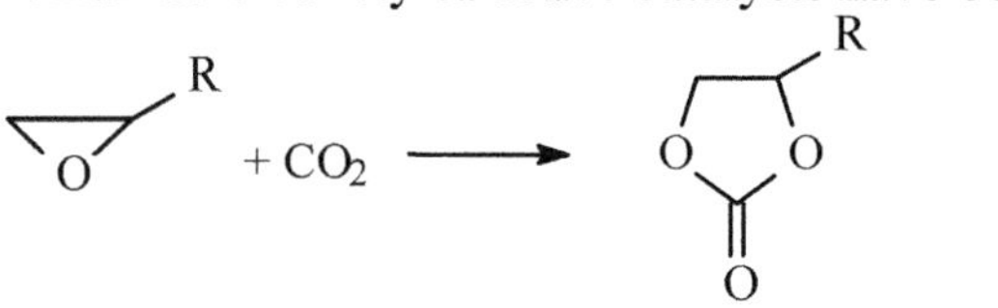

(2)

Among them, alkyl ammonium-, phosphonium- and alkali metal halides are extremely effective in providing carbonates with 90-99% yields. Such catalysts homo-polymerise epoxides, but do not produce copolymers with CO_2. A mechanism involving nucleophilic attack of the anion at the least-hindered carbon atom of the epoxide, CO_2 addition, intramolecular cyclization with release of the anion has been proposed.

Main group metal halide salts [31] (M_nX_m: M-metal of III-V groups; X-Cl, Br, I) have also been found to catalyse reaction (6). However, the systems require high concentration of catalyst. In general, these processes are run under high pressures of CO_2, 5 MPa, and temperatures ranging from 370 to 400 K. Several reports claim that main Group metal catalysts [32] afford high yield of cyclic carbonate at atmospheric CO_2 pressure. Organometallic halides R_nMX_m or R_nMX_m/Base [33], where R-Me, Et, Bu, Ph; M-Sn, Te, Sb, Bi, Ge, Si; X-Cl, Br, I, are good catalysts. Classical Lewis acids, organometallic complexes, e.g., $(Ph_3P)_2Ni$ [34], heteropoly-acids and metal phtalocyanines [35] (M = Co, Cr, Fe, Mn) have also been used as catalysts for reaction (6). Most recently, metal oxides (MgO [36], MgO/Al_2O_3 [37], Nb_2O_5 [38], others) have been used as heterogeneous catalysts. The latter have shown a considerable life-time and interesting turnover numbers (TON).

Recently, optically active carbonates have been synthesized using pure enantiomers of epoxides. Group 2 [36], 5 [39], and 13 [37] oxides have been successfully used. Also, an enantiomeric excess has been observed in the reaction of racemic mixtures of epoxides with carbon dioxide in presence of metal complexes with optically active ligands [39]. The *ee* depends on the stability of the metal complexes under the reaction conditions.

Interestingly, the same reagents epoxide and carbon dioxide may produce copolymers that find interesting applications [40]. Ethylene oxide, propylene oxide, cyclohexene oxide can be co-polymerized with carbon dioxide to afford high molecular mass polymers. The research in this area is in the direction of developing selective catalysts that may either co-polymerize or produce monomer carbonates, avoiding the formation of mixtures. Also, engineering catalysts that drive an alternate insertion of epoxide and carbon monoxide with respect to a random insertion of carbon dioxide into an epoxide polymer is important. This drives the load of carbon dioxide into the polymer and its quality.

As recalled above, cyclic carbonates can be used in transesterification reactions for the synthesis of linear carbonates (Eqs. 7, 8) in a two-step, equilibrium driven process [41]. Also in this area there is a need for more efficient catalysts.

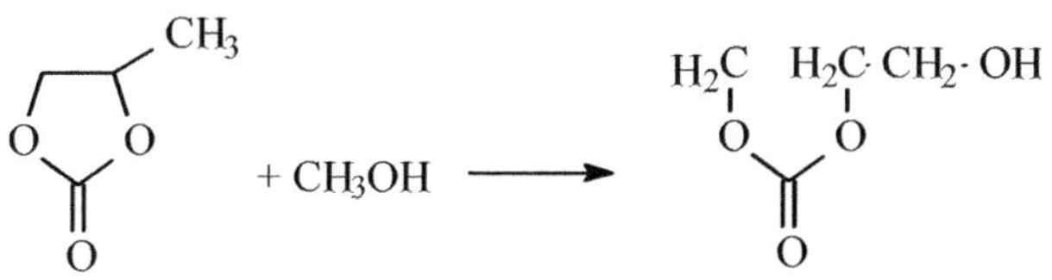

(3)

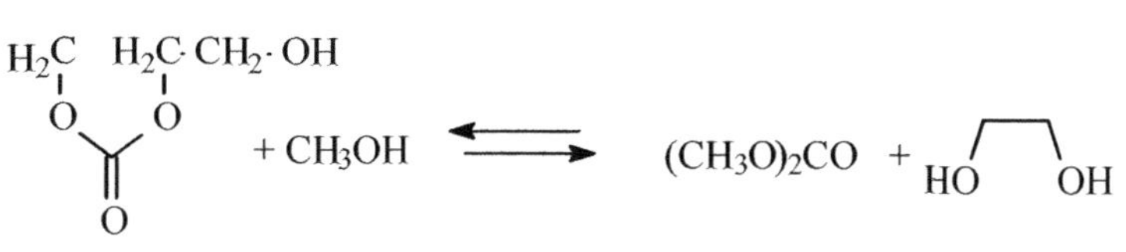

(4)

Oxidative carboxylation of olefins

The oxidative carboxylation of olefins, i.e. the reaction of olefins with carbon dioxide and dioxygen (Scheme 9.5), is a straight way to cyclic carbonates. All these reagents are cheap and easily available. The reaction is catalysed by homogeneous and heterogeneous catalysts [42]. This approach has a great potential. Finding catalysts that may assure high conversion and selectivity towards the carbonate is of great interest.

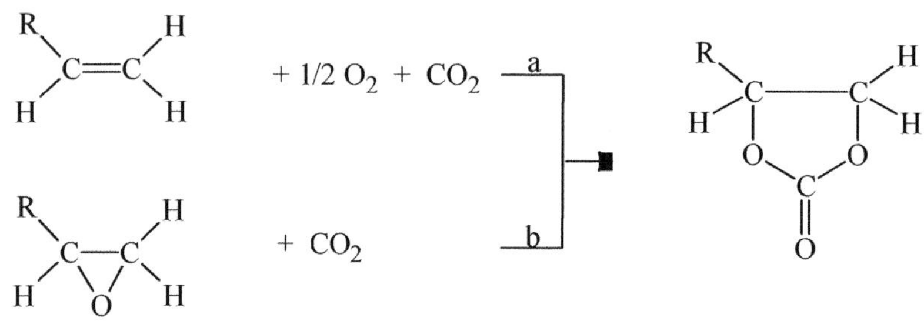

Scheme 9.5 - Synthetic ways to organic carbonates: oxidative carboxylation of olefins and carboxylation of epoxides

Reactions (a, b in Scheme 9.5) are someway correlated. As a matter of fact, reaction (a) avoids the synthesis of the epoxide that is, however, formed during the reaction [42]. The great advantage of reaction (a) is that it uses dioxygen as oxidant and affords directly the carbonate. Still research is needed for mastering the reaction and producing better yields and higher selectivity.

Bisphenol-A-polycarbonates

BPA-PC is commercially produced by the following technologies:

* reaction of BPA with phosgene using the interfacial polymerization
* melt transesterification.

The interfacial polymerization was almost simultaneously developed in the early 50s at Bayer and General Electric Plastics. The latter improved the technology and obtained high molecular mass polymers, with the synthesis of cyclic oligomers of the carbonate which were subsequently converted into high molecular mass polymers. The interfacial polymerization consists of three steps:

* Phosgenation of BPA to afford a bis-chloroformate.
* Cyclization of bischloroformate to carbonate oligomers.
* Ring condensation to produce BPA-PC.

In order to respond to the market request and to the continuously increase of the commercial interest in BPA-PC, a new synthetic route was developed, namely the melt transesterification of BPA. [43] Either DPC or DMC were used. The molecular carbonates react in presence of basic catalysts (oxides, hydroxides or phenolates of Group 1 and 2 elements), Lewis acids; organotin (dibutyltin oxide) and organotitanium compounds, phosphonium phenolate or borohydride, tetramethylammonium hydroxide. The whole process consists of several reaction steps that produce first a prepolymer formed at 450-530 K and 0.03-0.13 Mpa, then oligomers that are finally condensed at 550-570 K and 1.3 kPa to produce high molecular mass BPA-PC. In general the non phosgene routes afford a better quality polymers, with higher molecular mass, chlorine-free, more stable to heat. Also, such polymers are easily workable. A solvent free solid state polymerization has been developed by Bayer [44] that produces BPA-PC with molecular mass in the range 15-200 kDa.

As reported above, molecular carbonates may find use in the synthesis of polymers. Therefore, the development of carbon dioxide-based technologies for the production of molecular organic carbonates (either linear or cyclic) has a great importance because CO_2 could in this way eventually be fixed into long-lasting compounds that would represent a chemical sequestration of carbon dioxide [45].

Aliphatic polycarbonates

The formation of aliphatic-polycarbonates (al-PC) from oxiranes and CO_2 catalyzed by $ZnEt_2/H_2O$ was first discovered in 1969 [46] and currently PAC Polymers Inc. produces CO_2-epoxide copolymers by this method. The synthesis takes place at relatively mild temperatures (below 373 K) and pressures (2-8 MPa). Polymers with high molecular mass are obtained: 10^4 g/mol, from propylene oxide; 3.8 10^5 g/mol, from cyclohexene oxide. Yields are generally around 10-20%. In the reaction of co-polymerisation of CO_2 and propylene oxide, the monomer propylene carbonate is formed as a by-product and its production is increased with increasing the temperature. Together with the regular alternate insertion, avoiding formation of the monomer are key issues that deserve consideration and require research. Copolymer formation is suggested to proceed *via* a transition state.

During the recent years, a variety of catalysts, many of which based on zinc, have been shown to promote co-polymerisation. The most active catalysts are prepared by reaction of zinc hydroxide or zinc oxide with di-carboxylic acid [47]. A commercial process is based on these insoluble zinc catalysts. However, this method is plagued by low catalytic activity and the need to remove the metal from the polymer by an acid wash. Recently, soluble zinc complexes have been discovered which possess a significant catalytic activity that allows the production of high-molecular mass al-PCs [33]. The new catalysts are based on zinc complexes with phenoxide ligands bearing bulky substituents in the 2 and 6 positions: (2,6-diphenyl phenoxide)$_2$-Zn(THF)$_2$ is a typical representative of this group of catalysts.

3.1.5 Synthesis in-with supercritical-CO_2: $scCO_2$ as reactive-solvent

The use of supercritical carbon dioxide as solvent and reagent has been explored by several research groups. Such approach presents a large interest for its environmental implications as it may avoid the use of large amounts of organic solvents, allows a quite easy recovery of products at the end of the reaction, and combines two uses of CO_2 in one: solvent and reagent. A complete report on the use of $scCO_2$ is given in Section 3. Here we discuss the use of $scCO_2$ as reactive solvent.

To date, there is a limited number of well documented cases of this peculiar approach to the utilisation of CO_2. Namely: the synthesis of formic acid (Eq. 9) or its derivatives [48], the synthesis of lactones (Eq. 10) from dienes [49], the synthesis of linear carbonates (Eq. 11) from ortoformates [50] or aldols (Eq. 12) [51] and the conversion of cyclic ketals (Eq. 13) into cyclic carbonates [52].

$$\text{SC-CO}_2 + H_2 \rightarrow \text{HCOOH} \qquad (5)$$

(6)

$$CH_3C(OR)_3 + CO_2 \xrightarrow[\text{up to 9000 atm}]{\text{338-348 K}} CH_3C(=O)OR + (RO)_2CO \qquad (7)$$

$$(H_3C)_2C(OR)_2 + CO_2 \xrightarrow[\text{2000 atm}]{\text{453 K}} (H_3C)_2C{=}O + (RO)_2CO \qquad (8)$$

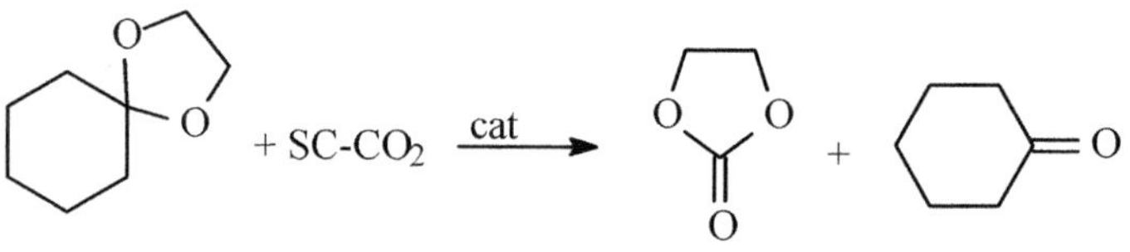

(9)

Reactions 11, 12, 13 are a particular way to the synthesis of linear and cyclic carbonates in which, starting formally from an alcohol, water is formed in a step in which carbonate itself is not generated. This is a quite important issue as water may limit the conversion yield in the synthesis of carbonates from alcohols and carbon dioxide. Water may also destroy the catalyst, rducing the efficiency of the catalytic system. Therefore, synthetic pathways in which water is not formed or is formed in a separate step may represent a better alternative. Reactions (11), and (12) require drastic conditions with pressures ranging from 200 to 900 Mpa. Reaction (13) has been carried out under much less drastic conditions.
The use of $scCO_2$ as reactive solvent results to be quite interesting for the low environmental impact and easy separation conditions. It requires the development of new catalysts, soluble under the reaction conditions, that may address the synthesis with high yield and selectivity.

3.1.6 Methanol

Carbon dioxide is used since long as additive to CO in the synthesis of methanol. Large amounts of CO_2 are used (2 Mt/y) to this end. A detailed analysis of the reaction conditions and mechanisms can be found in Chapters 11 and 13.

3.2 Perspective use of carbon dioxide in the synthesis of carboxylates and carbamates

In the following paragraphs the actual synthetic methodologies for a selected number of organic acids carboxylic and carbamic acids derivatives that eventually synthesised from CO_2 are reported. The actual market is indicated, that gives the perspective of the expansion of carbon dioxide utilisation if CO_2-based technologies will be eventually implemented.

3.2.1 Formic acid: economics and market

Worldwide capacity of formic acid, HCOOH, production is located for more than two thirds in Europe, with BASF as leader industry. Formic acid finds a large use in the leather and textile industries: skins and hides are acidified with a mixture of formic acid, sulphuric acid, and sodium chloride before chrome tanning. It is also used for adjusting the pH in dyeing of natural and synthetic fibres, in the coagulation of rubber latex, in the manufacture of pharmaceuticals, or as crop-protection agent. In Europe, most of the formic acid is used as a silage agent or in the fermentation of lactic acid as suppresser of the formation of butyric acid.

Recently, formic acid has been used as an additive for cleaning agents (replacement of mineral acid for environmental reasons), in the synthesis of aspartame and as pH stabiliser in the desulphurisation of flue gases in the Saarberg-Holter process. A large potential market for formic acid is in the Leonard Process in steel pickling and in the paper industry.

Formic acid is also considered as an hydrogen carrier in a hydrogen-based economy, as it can be easiily decomposed to CO_2 and H_2. The worldwide installed capacity for producing formic acid is over 400 kt/y.

Existing synthetic technologies

The installed formic acid processes can be classified in four groups

i) Oxidation of hydrocarbons has been used for long as the main technology. Today it covers less than 20% of the market.

$$C_4H_{10} \text{ (or naphta)} \rightarrow CH_3COOH + HCOOH + \text{other chemicals and tars} \quad (10)$$

ii) Hydrolysis of formamide had until the '70s a considerable share of the market (up to 33%). However, the large production of ammonium sulphate has made this process economically and environmentally not desirable and today it covers less than 6% of the market.

$$HCONH_2 + H_2SO_4 + H_2O \rightarrow HCOOH + NH_4HSO_4 \quad (11)$$

iii) Methylformate hydrolysis is the preferred technology today: it covers ca. 50% of the production. Methylformate is obtained by carbonylation of methanol, that is made from Syngas. Methanol formed as byproduct is easily recycled. Drawback of this technology is the Syngas production that is energy intensive.

$$CH_3OH + CO \rightarrow HCOOMe \quad (12)$$

$$HCOOMe + H_2O \rightarrow HCOOH + MeOH \quad (13)$$

iv) Formic acid can be also prepared from formates through an acid hydrolyis. This technology covers roughly 24% of the market.

$$NaOH + CO \rightarrow HCOONa \quad (14)$$

$$HCOONa + HX \rightarrow HCOOH + NaX \quad (15)$$

In principle, formic acid could be prepared from carbon monoxide and water:

$H_2O + CO \rightarrow HCOOH$ (16)

The reaction is exothermic, but below 420 K the reaction rate is very low. The equilibrium is reached rapidly at higher temperature, but it is shifted to the left. The pressure must be increased to obtain the acceptable formic acid yield. This process had no significant practical application so far.

Direct synthesis from carbon dioxide

Recent developments in carbon dioxide chemistry have made available new formic acid synthetic methodologies that implement the principle of "atom economy". In fact, the direct combination of carbon dioxide with dihydrogen produces formic acid with 100% atom utilization.

$CO_2 + H_2 \rightarrow HCOOH$ (17)

Zinc selenide and zinc telluride have been used as catalyst for the hydrogenation of carbon dioxide [53]. Homogenous transition-metal catalysts can also be employed. Using alcohol as solvent, the corresponding ester is produced. Ruthenium and palladium complexes have proved particularly active; organic bases (e.g., aliphatic tertiary amines) can be used as the alkaline material.

BP Chemicals [54] has developed a process in which formic acid is produced from carbon dioxide and hydrogen in a multistage process. First, a nitrogen base (triethylamine) reacts with carbon dioxide and hydrogen in presence of a ruthenium complex to yield the ammonium formate. The formate is then separated from the catalyst and the low-boiling constituent. The ammonium formate reacts with a high boiling base, for example 1-(n-butyl)imidazole, to yield a formate that can be thermally decomposed. The low-boiling base is liberated and distilled. Finally, the formate is thermally decomposed, formic acid distilled, and the high-boiling base recovered.

A recent development uses supercritical carbon dioxide as solvent and reagent [48]. The hydrogenation of $scCO_2$ in the presence of amines allows the conversion of the cumulene into formates under controlled conditions. In presence of Ru catalysts, $Ru(PMe_3)_4H_2$ or $Ru(PMe_3)_4Cl_2$, CO_2 itself can be hydrogenated to afford formic acid or formic acid derivatives [55]. Because of the unfavourable thermodynamics of the CO_2 hydrogenation, triethylamine was added as base to produce the stable formic acid-NEt_3 azeothrope. Very interesting results were obtained using co-catalysts or co-solvents such as water, MeOH or DMSO [56]. The observed catalytic efficiency of 7200 TON and initial TOF of 1400 per hour were greater than any previously reported in conventional solvents.

The formic acid synthesis can be coupled with subsequent reactions to obtain formic acid derivatives. In presence of alcohols or secondary amines the above systems provide an interesting "one pot" route to formate esters or formamides [56-58]. In particular, using dimethylamine as base, DMF was produced at 373 K in 99% yield and 99% selectivity (with TON as high as 420000) [59].

The formate moiety has also been formed in other reactions based on carbon dioxide. The efficient formation of DMF has been reported by Baiker using $Ru(dppe)_2Cl_2$ or an heterogenized silica-grafted Ru complex [60,61].

The catalyst $Ru(PMe_3)_4H_2$ promotes the hydrosilylation of sc-CO_2 with $Me_2(Et)SiH$ to give the corresponding silylformate, although in very low yield (2% vs silane, after 63 h) [62].

3.2.2 Acetic acid: economics and market

Acetic acid, CH_3COOH, has been for long produced by fermentation. This route has only a marginal interest today. The production of acetic acid is presently close to 6 Mt/y. It almost doubled in the last fifteen years. Production leader are USA (more than 1 Mt/y). Germany (close to ca. 0.5 Mt/y) is the larger producer in Europe.

In the USA about one-half of the acetic acid is consumed in the manufacture of vinyl acetate, whereas 20% goes into the manufacture of cellulose acetate. The remaining 30% is used in the manufacture of terephthalic acid and dimethyl terephthalate, esters of acetic acid, acetic anhydride (other than that used for cellulose acetate manufacture), and other minor uses.

Existing synthetic methodologies

Synthesis of acetic acid by liquid phase oxidation (LPO) of linear hydrocarbons (butane or naphta are most used) represents ca. 10% of the total production. The reaction temperature is in the range 420-470 K (the critical temperature of *n*-butane is 325 K). The reactor pressure is around 5.6 MPa and Mn, Co, Ni, and Cr are used as catalysts. Non-catalytic operations are also used. This technology has some drawbacks due to the environmental impact and separation costs. The oxidation of acetaldehyde to acetic acid proceeds through a free-radical chain, which produces per-acetic acid as an intermediate. Per-acetic acid can then react with acetaldehyde to generate acetaldehyde monoperacetate (AMP) that decomposes efficiently to acetic acid by a hydride shift in a Baeyer-Villiger rearrangement.

$$CH_3CHO + O_2 \rightarrow CH_3CO^{\cdot}OOH \rightarrow CH_3C(O)OOH \quad (18)$$

$$CH_3C(O)OOH + CH_3CHO \rightarrow CH_3CH(OH)OOC(O)CH_3 \rightarrow \rightarrow 2\,CH_3COOH \text{ or } 2\,HCOOCH_3 \quad (19)$$

Methyl formate can also be produced via a methyl migration.

There is little doubt that the process that will drive the market in the near future is the homologation of methanol, introduced by Monsanto [63] in 1968.

$$CH_3OH + CO \rightarrow CH_3COOH \quad (20)$$

The Monsanto technology has been implemented by other Companies in several Countries and represents today the major source of acetic acid.

Innovative technologies: process based on carbon dioxide

In Nature acetogenic bacteria are able to synthesise acetic acid from carbon dioxide (Scheme 9.6).

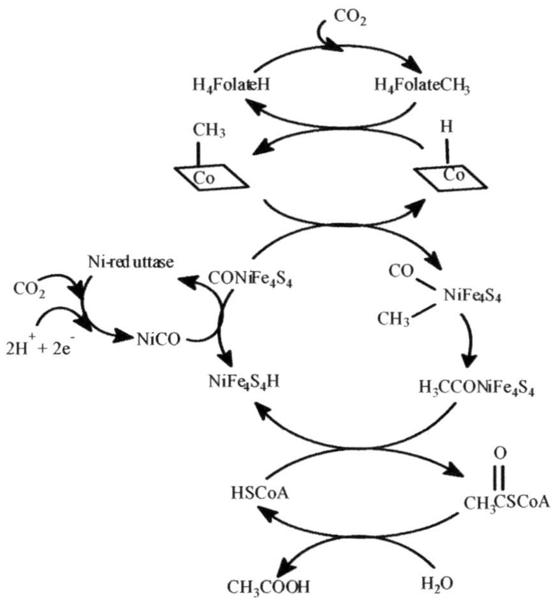

Scheme 9.6 - Formation of acetic acid

One mol of carbon dioxide is reduced to CO by Carbon monoxide dehydrogenase (CODH) enzyme, while one mol is first reduced to formate by Formate dehydrogenase (FDH) enzyme, then the formate is converted into methyl-group by the Tetrahydrofolate enzyme (H_4F) or a Tetrahydromethanepterin (H_4MPT). The methyl group and CO so generated are then coupled by a Fe_4S_4-Ni protein (Acetate synthase, ACS) with the co-operation of a Co-based enzyme (B12) to afford the acetyl group that is converted into a thioester (Acetyl coenzime A) AcS-$COCH_3$ and finally hydrolysed to acetic acid. This natural process is sometimes called "the biological Monsanto" for the similarity to the above mentioned Monsanto synthesis of acetic acid.

This reaction has been mimicked quite recently using Ni-systems [10, 11]. An olefin and carbon dioxide have been first converted into a thioester in the presence of thiophenol [64]. The thioester has been then hydrolysed to a carboxylic acid.

A reaction that is quite interesting is the *direct combination of methane and carbon dioxide* to afford acetic acid:

$$CH_4 + CO_2 \rightarrow CH_3COOH. \qquad (21)$$

Such reaction has been performed using V-based catalysts under moderate reaction conditions [65]. The key step is the activation of the C-H bond. Although the yield is still very low and far from practical application, this reaction remains of great interest as a process based on it would allow the conversion of natural gas, that usually contains 65-70 % of CH_4 and 35-30 % of CO_2, into acetic acid directly at the LNG extraction site. Research is still at the laboratory scale in this specific application. The same reaction can be carried out in a cold plasma [66]. The selectivity is not very high, in fact the yield of acetic acid is around 5.2 %. This process is very interesting from the atom economy point of view.

3.2.3 Oxalic acid: economic aspects and market

The capacity of the oxalic acid, $(COOH)_2$ world market was in 1992 around 190 kt per year, with the European Industry covering 10% of the total production and consumption. Five processes are in use world-wide: the oxidation of carbohydrates (Eq. 26), the oxidation of ethylene glycol (Eq. 27), the oxidation of propene (Eq. 28), the dialkyl oxalate process (Eq. 29), and the sodium formate process (Eqs. 30-31).

$$\text{Carbohydrate} \xrightarrow{NO_2} (COOH)_2 + \text{other products} \qquad (22)$$

$$(CH_2OH)_2 \xrightarrow{NO_2} (COOH)_2 \qquad (23)$$

$$CH_3CH{=}CH_2 \xrightarrow{NO_2} (COOH)_2 \qquad (24)$$

$$2\,ROH + 2\,CO \xrightarrow{H^+,\,H_2O} (COOR)_2 \rightarrow (COOH)_2 + 2\,ROH \qquad (25)$$

$$CO + NaOH \rightarrow HCOONa \qquad (26)$$

$$2\,HCOONa \xrightarrow{H_2SO_4,\,Ca^{2+}} (COOH)_2 \qquad (27)$$

Each of these processes has some drawbacks. The sodium formate process (Eq. 30) is too energy intensive, uses sulphuric acid and produces calcium sulphate as byproduct: it is no longer economic and has been substituted by the dialkyl oxalate process.

Oxalic acid is used in the recovery/separation of heavy or rare-earth elements, in textile treatment, in metal treatment, as bleaching agent, in leather tanning, in agrochemical/pharmaceutical production, in marble polishing. It also finds use for anodic aluminium coating, for millet jelly production, and for cobalt production. A larger market is foreseeable, supposed that low price oxalic acid will be made available.

Innovative technologies: carbon dioxide based synthesis

The electrochemical synthesis of oxalic acid from CO_2 is possible [67,68]. The synthesis is carried out in non aqueous media that are good solvents for carbon dioxide, often better than water. A drawback is the low conductivity of organic media that only permits a low current density.

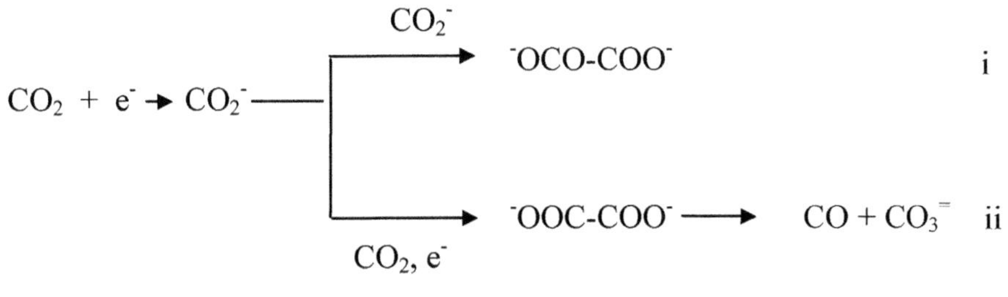

Scheme 9.7 - Modes of homocoupling of CO_2^-

As shown in Scheme 9.7, the first step is the transfer of one electron to carbon dioxide with formation of the radical anion $CO_2^{\cdot-}$. Once this species is

formed, two reactions compete for its use: the coupling of two radical anions, that originates the oxalate dianion upon further transfer of two electrons, and the head-to-tail coupling of the radical anion with free carbon dioxide. This coupling generates carbonate and carbon monoxide.
The yield and selectivity of the process depend on the current density, solvent, carbon dioxide pressure. A sacrificial anode is used, so that the oxalate is recovered as zinc oxalate. The current yield of oxalate is higher than 90%, that has justified the development of a small scale pilot plant [69].

3.2.4 Long chain aliphatic carboxylic acids, and aromatic mono- and di-carboxylic acids: economics and market

Carboxylic acids find a diversified use according to their molecular structure. Their market is in a continuous expansion.

Long chain aliphatic carboxylic acids $CH_3(CH_2)COOH$ find several uses in the chemical industry. Among others, their use as surfactants is an expanding market because of environmental issues that make carboxylic acids preferable to sulphonic acids. The world market of long chain aliphatic acids is around 10 Mt/y, with a continuous expansion for the reason given above.

Aromatic mono- (C_6H_5COOH) and di-carboxylic acids ($HOOCC_6H_4COOH$) find a large use in the polymer industry in the formation of esters and /or other derivatives.

Existing synthetic methodologies

The carboxylic acids are presently obtained by either oxidation of the corresponding alcohols (Eq. 32) or by hydrolysis of cyano-derivatives (Eq. 33).

$$RCH_2OH \rightarrow RCOOH \quad (28)$$

$$RCN \rightarrow RCONH_2 \rightarrow RCOOH + NH_3 \quad (29)$$

Aromatic carboxylic acids are produced by oxidation of either parts of a cyclic system such as cyclic olefins (Eq. 34), or aromatic rings (Eq. 35), or aliphatic substituents (Eq. 36) in an aromatic ring [70].

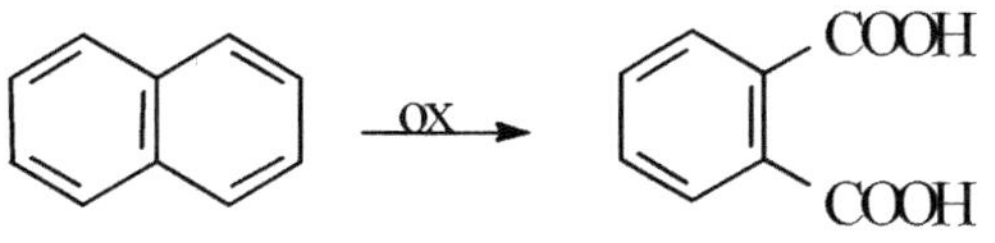

(30)

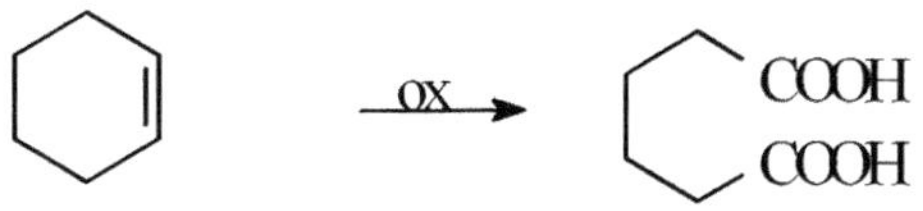

(31)

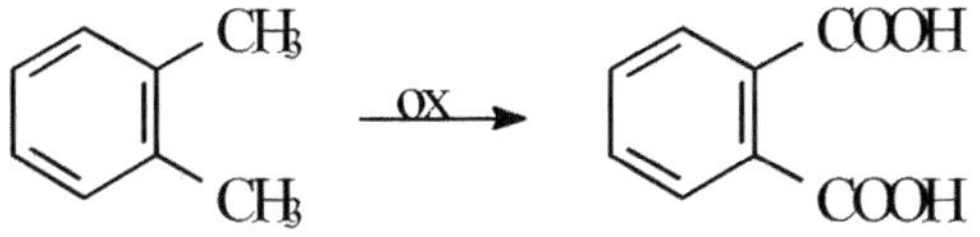

(32)

The above mentioned processes have a high environmental impact and a low atom efficiency. Developing alternative syntheses based on the direct carboxylation of organic substrates is of great interest from both the environmental and economic point of view.

Innovative technologies based on carbon dioxide

The transition metal mediated coupling of di-olefins, such as butadiene, and carbon dioxide leads, depending on the reaction conditions, to the formation of unsaturated carboxylic acids (Scheme 9.8).

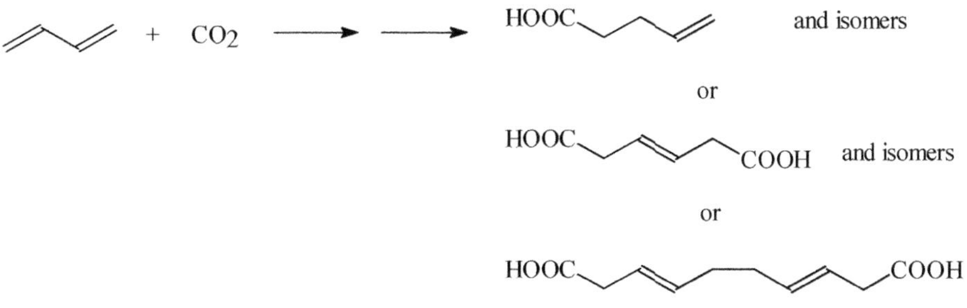

Scheme 9.8 – Products observed in the butadiene-carbon dioxide coupling reaction.

The carboxylation of unsaturated hydrocarbons (alkenes, alkynes, butadienes, allenes) [71] has been recently developed with interesting yield and selectivity (>90%) using transition metal complexes as catalysts. In all cases the homo coupling of substrates is a concurrent process with the hetero-coupling substrate-CO_2. As homo-coupling reaction the coupling of two identical moieties either two "olefin" or two "carbon dioxide" molecules, is meant. As a consequence, the formation of carboxylated products from olefins and carbon dioxide (hetero-coupling) is very often accompanied by homo-coupling products which, according to the experimental conditions, can also represent the most abundant fraction. The preliminary oligomerization of dienes is not a negative fact, as it can bear to the synthesis of long-chain or ring skeletons (Scheme 9.9) starting from C_3-C_4 monomers.

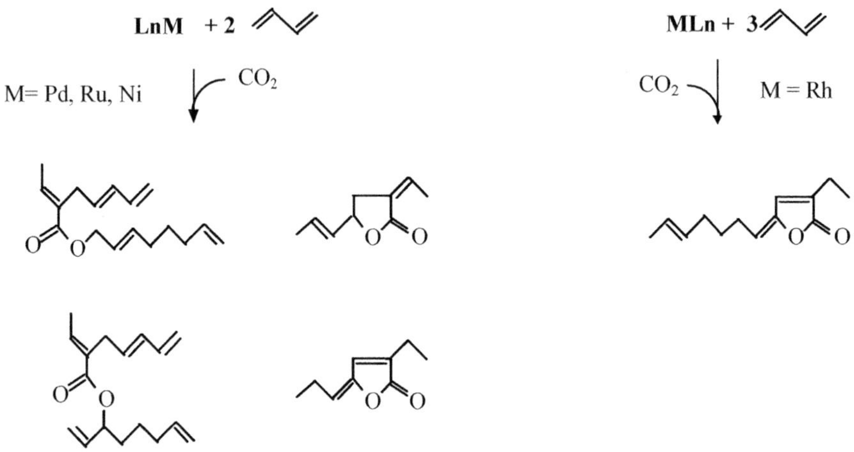

Scheme 9.9 - Metal catalysis in butadiene-CO_2 chemistry

To master the coupling issue is of fundamental importance in order to develop processes which may have an industrial application.
The analysis of the literature data indicates that the formation of either linear carboxylates or cyclic-lactones with high TON through the carbon dioxide-unsaturated hydrocarbon coupling is driven by one of the following conditions: i) low energy of the M-O bond that is formed in the process; ii) β-hydrogen shift from the hydrocarbon moiety to the metal, that can eventually produce an O-H bond in place of the O-M bond; iii) existence or generation of an allyl group that may assist the elimination through a σ-π interconversion. The simple reductive elimination from a metalla carboxylate has been observed in rare cases [6] and is not operative in processes with high TON.
The direct carboxylation of aromatic compounds is also possible. Naphtalene can be converted into compounds A or B at room temperature via one-electron

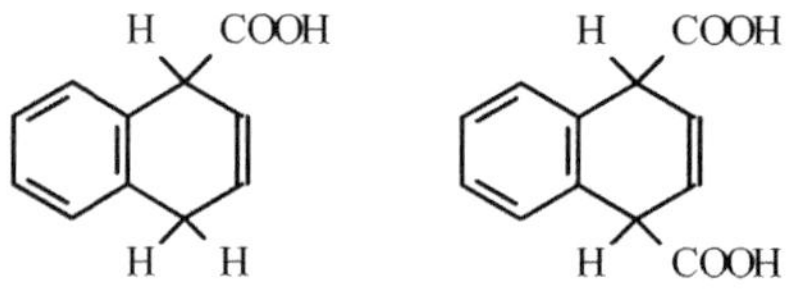

A B

transfer using Group 1 metals. The formation of either compounds, that can be synthesized also *via* the electrochemical carboxylation of naphtalene, depends on the metal/naphthalene ratio [1a].

In addition to the cases discussed above, it would be of great industrial interest to develop catalysts which produce acrylic acid from CO_2 and ethylene. The ethylene (or other olefins)-carbon dioxide coupling is mediated by Ni(0)-complexes. Stable intermediate metallorganic species are formed [71i]. The acidolysis of the metallo-carboxylates produces the carboxylic acids. The metal is oxidized and cannot be reused as catalyst. This is a case in which the electro-catalyzed synthesis works quite well. In fact, if the coupling reaction is carried out electrochemically in presence of sacrificial anodes, the metal salts of the carboxylic acids are formed with an interesting yield and selectivity [72]. The general application of this reaction towards different diolefines opens a possible pathway for producing carboxylic acids in a large variety [73].

Also the synthesis of four membered lactones from olefins and carbon dioxide would represent an interesting application of the new synthetic technology. To date only one example of such syntheses can be found in the literature [74] in which allene and CO_2 are coupled through a formal "2+2" addition to afford α-methylene-β-oxiethanone.

An interesting reaction would be the insertion of carbon dioxide into a C-H bond, that would convert an hydrocarbon into an acid.

$$RCH_2COCH_3 + CO_2 \rightarrow RCH_2COCH_2COOH \qquad (33)$$

The reaction occurs with activated C-H bonds and is promoted by several agents, like metal systems, [75] and Group 1-phenoxides [76]. Recently, it has been shown that the zwitterions 2-methyl-5-butyl-imidazolium-2-carboxylate is able to carboxylate active C-H and E-H bonds (E=O, N) at room temperature [77]. Such carboxylation reactions have a great potential for the synthesis of carboxylates.

3.2.5 Carbamates and isocyanates: economics and market

Carbamates have a quite diversified utilization, in pharmacology [78] as inhibitors and sedative, in agriculture [79] as herbicides insecticides and fungicides, and in the chemical industry [80] as precursors of carbonates or isocyanates. Methyl- or phenyl-carbamates of primary aliphatic- and aromatic-amines can be advantageously used as precursors of isocyanates. The latter find an extended industrial use for the preparation of speciality chemicals and polymers.

Carbamates and carbonates can be easily interconverted as shown in Scheme 9.10. Carbamates, carbonates and isocyanates are three classes of compounds of industrial relevance that are strictly interconnected and share some synthetic aspects.

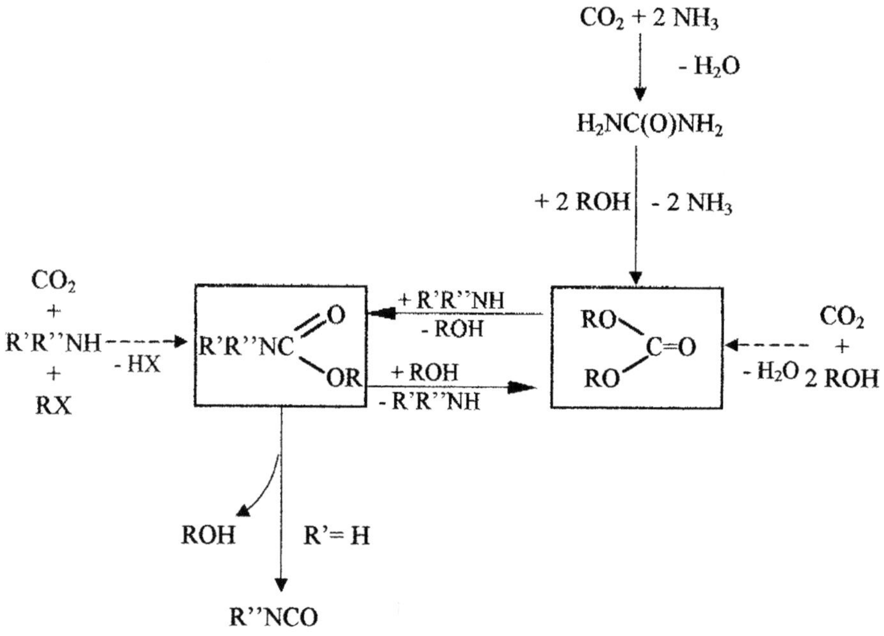

Scheme 9.10 - Relationship between carbamates, carbonates and isocyanates

It is quite difficult to give a global figure for the world capacity and production of carbamates and isocyanates as the total production is segmented in many different sectorial uses. Moreover, several compounds are synthesized for captive use. In general, as specialty chemical carbamates may have a market of few kt per year. Conversely, if they are used as precursors of isocyanates and polymers, the market may reach the Mt/y size. As an example, several carbamates are used in agriculture as herbicides, insecticides, and fungicides. About 10 carbamates are used as insecticides. In 1994 carbamate insecticides had an estimate end-user sale of about \$ 1.5 billion. Herbicide properties are presented by both N-aryl- (isopropyl N-phenylcarbamate) and N-alkyl or N,N-dialkyl-carbamates. Benzimydazol carbamates (carbendazim, benomyl) have found wide application as fungicides [81].

Carbamate esters also show interesting pharmacological properties. Ethyl carbamate inhibits mitosis, thus slowing cellular growth, especially of bone narrow cells, and inhibits enzymes such as cholinesterase. Its inhibition of cellular growth has prompted the use of ethyl carbamate as an antileukemic agent. It has also found use as an antidote for poisoning by central nervous system stimulants such as strychnine, picrotoxin. Other major therapeutic applications of carbamates are as:

- tranquilizers (meprobamate)
- muscle relaxants (carisoprodol, metocarbamol)
- sedative (mepentamate).

However, although so far carbamates have found application in fine chemicals production, they also show significant potential as intermediates in the synthesis of commodities. In particular, aromatic carbamates could become extremely important as intermediates in phosgene-free routes to TDI (toluene diisocyanate) and MDI (methyl diphenyl diisocyanate), provided that both convenient syntheses and suitable conversion processes are developed.

Existing methodologies for the synthesis of carbamates

Despite the diversified uses of carbamates, only very few synthetic methodologies have reached commercial relevance.

A) Phosgene-based route.

The phosgene-based synthetic routes still represent the most general way to carbamate esters. They include:

i) the reaction of phosgene with an alcohol followed by the aminolysis of the intermediate chloroformate;

$$R''OH + COCl_2 + Base \longrightarrow R''OC(O)Cl + (BaseH)Cl \quad (34)$$

$$R''OC(O)Cl + 2\ RR'NH \longrightarrow R''OC(O)NRR' + RR'NH_2Cl \quad (35)$$

ii) the reaction of isocyanate with alcohol.

$$RNH_2 + COCl_2 \longrightarrow RNCO + 2HCl \quad (36)$$

$$RNCO + R'OH \longrightarrow RNHC(O)OR' \quad (37)$$

Method (i) is very versatile and can be applied to several alcohols.
The reaction of alcohols with isocyanate, method (ii), is rapid and quantitative. Phenols react more slowly, but, in this case, the reaction can be catalysed by tertiary amines.

Isocyanates are, on turn, obtained by phosgenation of primary amines, a route that is not without significant drawbacks due to the forcing conditions usually required for the generation of the isocyanate (typically > 450 K). As a result, the amine can bear only a limited number of functionalities prior to generation of isocyanate with a serious limitation to the nature of carbamates generated. A further limitation comes from the presence of chloride containing impurities, such as unreacted carbamoyl chloride.

Worker safety, corrosion, and waste disposal suggest the development of alternative routes.

B) Synthesis of carbamates by reacting alcohols with urea.

This method is the preferred commercial route to methyl or ethyl urethane:

$$H_2NC(O)NH_2 + ROH \longrightarrow H_2NC(O)OR + NH_3 \quad (38)$$

In fact, it does not use phosgene and for methyl- and ethyl-urethane, $H_2NC(O)OR$ (R = Me, Et), it is economically more favourable than ammonolysis of methyl- or ethyl-chloroformate, as isolation of the product and recycling of ammonia is less costly than separating the carbamate from ammonium chloride and recovering ammonia. The reaction is carried out at temperatures around 430-490 K for reaching the optimum reaction rate, and ammonia has to be eliminated from the reaction medium, for better yield.

Butyl carbamate has been produced on a commercial scale by using a small excess of butanol and catalytic quantities of cupric acetate, at high temperatures and atmospheric pressures. Aliphatic carbamates have been prepared by reaction of urea or substituted ureas with aliphatic alcohols in the presence of an alkali metal (or an alkali metal compound) as catalyst [82].

The reaction of urea with alcohols to form carbamates fails when tertiary alcohols, phenols, or urea-reactive groups are present. Since urea is industrially prepared from carbon dioxide, this synthetic methodology is better considered as an indirect CO_2-based synthetic route to carbamate esters.

C) Transesterification with high boiling alcohols.

The transesterification reaction is promoted, at suitable temperatures, by Al or Sn catalysts [83].

$$RNHC(O)OR' + R''OH \longrightarrow RNHC(O)OR'' + R'OH \quad (39)$$

N-alkyl carbamates, as well as unsubstituted carbamates, can be used as interchange components. Strong bases, such as Group 1 alkoxides are ineffective as catalysts, as they react with the starting carbamate ester to form alkali cyanates. The reaction is driven to completion by removing the lower boiling alcohol (usually MeOH or EtOH) by distillation. Tertiary alcohols or phenols do not react under these conditions.

Recently, the transesterification reaction has been achieved using Ti(IV) alkoxides as promoters. The potentially low cost of manufacture of methyl- or ethyl-carbamate and their ability to undergo transesterification have made economically feasible the preparation of mono- and di-carbamates.

Innovative synthetic routes to carbamates based on carbon dioxide

Recently, a number of syntheses of carbamates based on the direct reaction of amines with carbon dioxide and an alkylating agent have been developed.

Reaction of amines with carbon dioxide and alkylating agents.

The reaction of CO_2 with amines is a direct route for building-up the carbamic moiety in the form of ionic alkylammonium carbamates:

$$RR'NH + CO_2 \rightarrow RR'NH{\cdot}CO_2 \qquad (40)$$

$$RR'NH + RR'NH{\cdot}CO_2 \rightarrow RR'NH_2^{+-}O_2CNRR' \qquad (41)$$

R' = H, Alkyl

This reaction does not proceed with aromatic amines, probably because of their poor basicity. Carbamic salts may be sparingly soluble. Alkylammonium carbamates undergo complex equilibria in solution (Eq. 46-48).

$$[RNH_3][O_2CNHR] \rightarrow RNHC(O)OH + RNH_2 \qquad (42)$$

$$RNHC(O)OH \rightarrow CO_2 + RNH_2 \qquad (43)$$

$$2\ RNHC(O)OH \rightarrow [RNHC(O)OH]_2 \qquad (44)$$

Depending on the experimental conditions, carbamic acid [RR'NC(O)OH] can undergo decarboxylation (47) or self- association through hydrogen bonding (48) leading to the formation of a dimer. Strong organic bases such as amidine or guanidine affect the equilibria of these reactions, and the formation of asymmetric salts is likely to occur.

$$RR'NH + CO_2 + Base \rightarrow (BaseH)(O_2CNRR') \qquad (45)$$

Carbamic acid is stable only in the dimer form [84] shown in Eq. 48. The dimmer is stavle in the solid state, while in solution easily looses CO_2 and affords the ionic carbamate.

Metal carbamates.

Metal carbamates have been obtained by interacting CO_2 with amines in the presence of metal systems or by formally inserting CO_2 into metal amides, while alkali carbamates $[M(O_2CNH_2)]$ can be synthesised from CO_2 [85a] and an alkali metal salt (nitrate or perchlorate) soluble in liquid ammonia. Under harsh conditions, the reaction of primary and secondary amines with CO_2 in presence of the metal can produce alkali carbamates. N-aryl-carbamates are made by reacting the corresponding aromatic isocyanate with sodium hydroxide in organic solvents [85b]. Metal carbamates can be used for the synthesis of metal-oxides and other materials[86].

Recently, a versatile route [87] to aliphatic and aromatic amine (primary or secondary) alkali-metal carbamates, has been reported, based on the reaction of CO_2 with amines in presence of commercially available alkali tetraphenylborate salts:

$$RR'NH + Base + MBPh_4 + CO_2 \rightarrow MO_2CNRR' + (BaseH)BPh_4 \qquad (46)$$

M = Na, Li, (K); R' = H.
If R = Alkyl, then Base = RR'NH.
If R = Aryl, then Base = NR''_3, where R" = Alkyl

Alkylammonium tetraphenylborate salts are co-products, that can directly be used for industrial applications or recycled to afford the metal-BPh_4.

Insertion of carbon dioxide into the E-N bond of p-block amides.

p-Block amides of the type $L_mE(NR_2)_n$ (E = B, Si, Ge, Sn, As, Sb, or P) also react with CO_2. The formal insertion of heterocumulene in the E-N bond [88] generates p-block carbamates $[L_mE(NR_2)_{n-x}(O_2CNR_2)_x]$. After the carbamic moiety is generated, it can be reacted with an electrophile to give the carbamic ester. Several reports of the carbamation of epoxides, alkynes, activated olefins, orthoesters, alkylating and acylating agents (e.g. alkyl halides and organic carbonates) are available.

Alkylating metal carbamates to produce carbamic esters is not a straightforward process. Carbamates present two nucleophilic sites: the nitrogen and oxygen atoms. (Scheme 9.11)

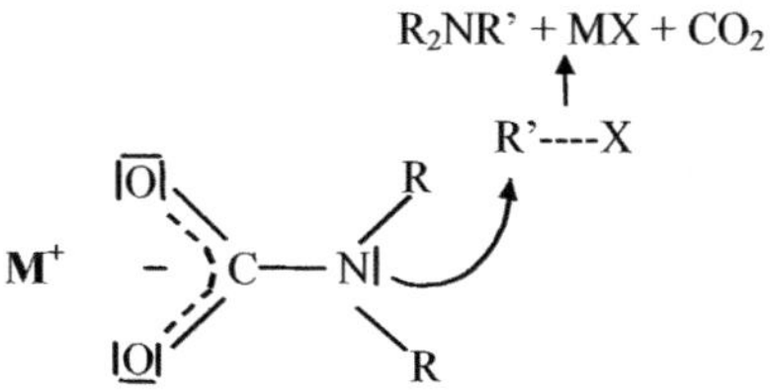

Scheme 9.11- Oxygen and nitrogen atom as nucleophilic sites

The interaction of an electrophile with nitrogen, which is favoured any time the oxygen atoms are engaged in a strong interaction with an electrophile such as a metal ion, causes N-alkylation with the loss of CO_2 and the generation of by-products. Such side reactions can be controlled using species that increase the electron density at the oxygen atom or that may stabilise the free carbamic moiety. Strong organic bases [89] and crown ethers [90] which capture the primary electrophiles (Scheme 9.12) have been used for this purpose.

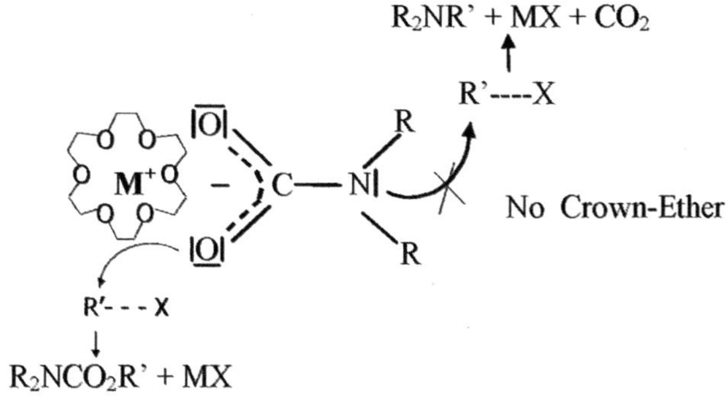

Scheme 9.12- Influence of the crown ether

Several element carbamate have been used for synthetic purposes. Phosphocarbamates [88a] easily prepared from readily available phosphorus precursors, are good sources of the carbamic moiety. The synthesis of organic carbamates can achieve yield close to 100% when phosphocarbamates are treated with alkyl halides in the presence of an alkali metal halide (MY) and a suitable crown ether (L). (Eq. 51).

$$P(NR_2)_{3-x}(O_2CNR_2)_x + xR'X + xMY + xL \longrightarrow xR_2N\text{-}C(O)O\text{-}R' + P(NR_2)_{3-x}Y_x + x(M{\cdot}L)X \quad (47)$$

Ammonium carbamates ($RR'NH_2^{+-}O_2CNRR'$; R = alkyl, R' = alkyl or aryl) have been alkylated using a variety of agents, like orthoesters and orthocarbonates, tosylate, dialkylsulfates, and alkyl halides. Recent examples of this alkylation include the formation of oxazolidinones and oxazinanones [91] by reaction of allylamines and homoallylamines with CO_2 and I_2 and the synthesis of 2-oxazolidones from ethanolamine derivatives and CO_2 in the presence of tertiary phosphines and diethylazodicarboxylate.

The synthesis of carbamic esters (52a) by direct reaction of amines, alkyl halides, and CO_2 can easily afford the alkylated amine as major products (52 b):

$$R_2NH_2^{+-}O_2CNR_2 + R'X \longrightarrow R_2NC(O)OR' + R_2NH_2^+X^- \quad (48a)$$

$$R_2MR' + R_2NH_2X + CO_2 \quad (48b)$$

In order to synthesize carbamic esters from primary amines, CO_2, and R'X, the $[RNH_3][O_2CNHR]$ reactivity must be modified by using either very strong organic bases [89] or polyethers [90] that can complex the ammonium cation (53) and increase the O-nucleophilicity:

$$2\ RNH_2 + CO_2 + R'X + L \longrightarrow RNHC(O)OR' + [RNH_3{\cdot}L]X \quad (49)$$

A number of alkali-carbamates have been used in the same conditions to afford organic carbamates in high yield [88].

Synthesis of carbamates by aminolysis of organic carbonates.

Synthesis of carbamate esters by aminolysis of organic carbonate has received poor attention as carbonic acid diesters have been traditionally synthesized from phosgene.

$$RR'NH + (R''O)_2C{=}O \longrightarrow RR'NC(O)OR'' + R''OH \quad (50)$$

To date, the interest for this synthetic route to carbamate esters is very high because new, non phosgene-based, synthetic methodologies are available on industrial scale for the synthesis of organic carbonates (see the carbonate section).

Among them, dimethylcarbonate (DMC) and diphenylcarbonate (DPC) are receiving growing attention as "carbonyl active species" in place of phosgene. In fact, their use [92] allows to eliminate handling and safety problems because of their low toxicity, while the absence of polluting wastes in DMC- or DPC-based processes, permits to eliminate problems connected with waste treatment and disposal.

Synthesis of N-alkyl carbamates.

Basic amines smoothly react with DPC or other organic carbonates having fairly good leaving groups (4-nitro-, 2,4-nitro-, 2,3,5-thrichloro-, 2,4,6-tricloro-, pentachloro-, 4-acetylphenoxy; $H_2C{=}C(Me)O$-; $H_2C{=}C(OMe)O$-). However, carboalkoxylation of aliphatic amines by unactivated carbonic acid diesters usually requires a suitable catalyst in order to observe good conversion rate and selectivity. Lewis acids, such as $AlCl_3$, $SnCl_2$, $ZnCl_2$, $Zn(O_2CCH_3)_2$, $FeCl_3$, or metal (Rh, Ru) complexes can catalytically promote the carboethoxylation of n-propylamine with diethylcarbonate a reaction that may occur also in absence of catalyst [93]. More recently, it has been shown that CO_2 itself is an effective catalyst for the synthesis of N-alkylmethylcarbamates from aliphatic primary amines and DMC [94]. This reaction involves the transfer of the carbamic group from alkylammonium carbamates, $[RNH_3][O_2CHNR]$, prepared in situ from aliphatic primary amines and CO_2, to the carbonyl carbon atom of DMC to give a carbonic-carbamic mixed anhydride, RNHC(O)OC(O)OMe, that, under the working conditions, decarboxylates to yield the N-alkylmethylcarbamate according to the following equations:

$$[RNH_3][O_2CHNR] + (CH_3O)_2CO \longrightarrow RNHC(O)OC(O)OCH_3 + RNH_2 + CH_3OH \qquad (51)$$

$$RNHC(O)OC(O)OCH_3 \longrightarrow RNHC(O)OCH_3 + CO_2 \qquad (52)$$

R = benzyl, cyclohexyl, allyl

This reaction is carried out in common organic solvents (aromatics, THF, and dichloromethane), but the organic carbonate itself may be used as the reaction solvent. Satisfactory reaction rates and good selectivities can be attained by heating the reaction mixture to at least 343 K under CO_2 atmosphere (P_{CO2} = 0.1-0.2 MPa). Ureas, N,N-substituted carbamates, and secondary and tertiary amines are formed in very low amount. The use of the heterocumulene as catalyst of the process is very appealing, as carbon dioxide presents no regeneration or recycling problems.

Synthesis of N-aryl carbamates.

Carbossialkoxylation or carbossaroylation of anilines and other aromatic amines can be achieved using catalysts such as zinc, cobalt, tin, aluminium, or titanium derivatives or by using strong bases under severe conditions [95]. These processes seem to involve either activation of the aromatic amine by a base or activation of the organic carbonate by a Lewis acid. One major problem related to these processes is selectivity, because side reactions generate N-alkylation products or substituted ureas (i.e., ArNHC(O)NHAr). However, the reaction of diaminotoluene with dimethylcarbonate to get the compounding dicarbonates, precursors of TDI, is being scaled up to the pilot plant level.

Organophosphoric acids ($Ph_2P(O)OH$, $(PhO)_2P(O)OH$, $(BuO)_2P(O)OH$, $(BuO)P(O)(OH)_2$) have been successfully used to promote the reaction of aromatic amines with dimethylcarbonate or diphenylcarbonate, or methylphenilcarbonate to generate N-arylcarbamates or N-alkylcarbonate [96]. At 393 K, in the presence of a slight excess of DPC, the conversion of $PhNH_2$ into the carbamate is almost quantitative in 15h, and selectivities are >99%. Diphenylphosphinic acid also catalyse the synthesis of N-1-naphtyl phenylcarbamate (used as an herbicide) from 1-aminonaphtalene and DPC [92a]. Because of the lower reactivity of 1-aminonaphtalene with respect to $PhNH_2$ using the former results in lower carbamate yields (30-40%). The catalyst can be recovered easily and quantitatively as arylammonium salt. The carbamate is not contaminated by phosphorus. This reaction is the first catalytic method for the synthesis of carbamates that does not use a metal-based catalyst. The acids mentioned above are also active in the reaction of arylamines with DMC that is less reactive. Although the conversion rate is lower, the high selectivity (almost 100%) makes this class of catalysts very attractive because the amine methylation is suppressed. The role of the organophosphorous acid has been investigated. Interestingly, these carbamation reactions take place through a mechanism that mimics the mechanism of formation of the carbamate anion in living systems, from ammonia and hydrogencarbonate, in the presence of carbamoylphosphate synthetase enzyme [97]. More recently [98], this process has been extended to the carbamation of industrially relevant aromatic diamines, such as 4,4'methylenediniline (MDA) and 2,4-diaminotoluene (TDA), with DPC or methylphenylcarbonate, in the presence of diphenylphosphinic acid as the catalyst of the carbamation process.

The carbamation of amines bearing silyl/groups in the aliphatic chain has been recently performed under mild conditions [99-100].

Synthesis of isocyanates

Isocyanates can be prepared by thermolysis of primary amines carbamates.

$$RHNCOOR' \rightarrow RNCO + R'OH \qquad (53)$$

$R' = CH_3$, Ph

This technology may be applied to MDA and TDA di-carbamates that are converted in MDI and TDI, precursors of polymers. Therefore, the development of a carbon dioxide based synthetic methodology of aromatic dicarbamates may bring to fix carbon dioxide into polyamides, that are the major plastics on the market.

Among isocyanates, MDI (methylenediphenyldiisocyanate) and TDI (toluenediisocyanate) occupy a top position in the market share. The global production of TDI has been, in 1997, 1.10 Mt. Western Europe plays a pivotal role in this business, accounting for 33% of the worldwide production. Global demand for MDI is 1.8 Mt/y and is expected to continue growing at around 8%/y, although, in Europe, the actual growth will be slightly below this figure. Again, Western Europe plays a crucial role in the business, with a capacity able to satisfy 50% of the overall MDI demand.

3.2.5 Esters and other derivatives (lactones, amides)

Coupling carbon dioxide with unsaturated hydrocarbons is a way to the synthesis of several compounds which contain the carboxylic function. Terminal and internal mono-enes, alkynes, dienes (both conjugated dienes and cumulenes) have been used to this end. Such reaction uses the internal energy of the unsaturated hydrocarbon to react carbon dioxide.

As already shown (Scheme 9.9) the reaction of butadiene with CO_2, catalysed by Pd- Rh- and Ru-complexes [101], is quite efficient for the synthesis of five-membered lactones bearing aliphatic chains as ring substituents (Scheme 9.8).

The metal can drive the length of the side-chain, as it determines the extension of the oligomerization of the diene prior to carbon dioxide coupling. This synthetic methodology results to be quite interesting as the ring can be opened to afford long-chain carboxylic acids and derivatives that find a large use in the chemical industry.

3.2.6 Insertion of CO_2 into C-C Bonds

The reaction of strained rings with CO_2 promoted by transition metal systems is of interest as it would be a way to synthesise lactones from cyclic hydrocarbons (or their derivatives) through a formal CO_2 insertion into a C-C bond. Rings of different size have been reacted with CO_2 in the presence of catalysts, both in thermal and light-driven reactions [101]. The size of the ring affects the reactivity. Cycloalkanes and cycloalkenes have different reactivity. As an example, the UV-visible irradiation of a solution of

$RhCl(PMe_2Ph)_3$ at low temperature in presence of cyclopropane and carbon dioxide under pressure, produces butyrrolactone (Scheme 9.13).

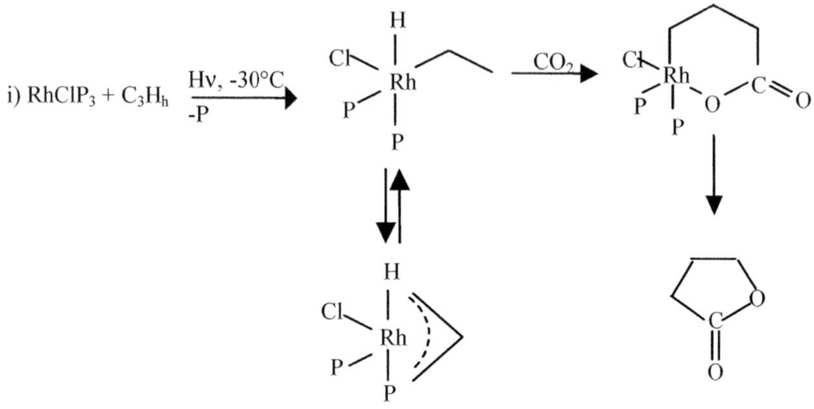

ii) "$RhClP_2$" + CO_2 + P=O $\xrightarrow[\text{room } \theta]{P}$ $RhCl(CO)P_2$

Scheme 9.13 - Reactivity of cycloalkanes towards carbon dioxide

It is interesting to note that, as evidentiated in Scheme 9.13, light can promote the reaction (*i*) in a different mode than heat (*ii*), which promotes phosphane oxidation with CO release, which changes the catalyst.

3.2.7 Electrochemical reactions that use CO_2

Tyssee et al. introduced [102] in 1972 the term electro-carboxylation to define numerous electrochemical processes by which one or more carboxylic groups are introduced into suitable organic starting materials. Electrocarboxylations, making use of carbon dioxide, offer a valuable alternative to the existing carboxylation procedures, many of which make use of dangerous reagents such as phosgene or cyanides, or involve systems very sensitive to contamination, such as those of the Grignard type.

These innovative synthetic routes are aimed at generating fine chemicals, for which energetic costs may not be so important as atom economy (yield and selectivity) and waste reduction. The versatility of this method is demonstrated by the wide choice of organic starting materials and by the corresponding variety of carboxylic acids obtained from them.

The electrocarboxylation processes can be grouped as:

a) Addition reactions, taking place on starting materials bearing different kinds of unsaturations, such as alkynes, alkenes, polyenes, ketones, aldehydes, aromatics, imines, unsaturated heterocycles and so forth.
b) Substitution reactions, taking place with starting materials bearing good leaving groups (X), such as halides, thioethers, thioacetals, acetates which are substituted by the carboxylate anion.

The reaction mechanism of electrocarboxylation in aprotic media is determined by the difference between the reduction potential of carbon dioxide and that of the organic compound to be carboxylated (S). Different pathways are possible that imply the generation of radical anions of the substrate or carbon dioxide.

Electro-catalysis by metal complexes

Uncatalysed electrocarboxylations in conventional diaphragm systems allow moderate product yields in several cases because of the occurrence of side or follow-up reactions. In order to improve yields and selectivities catalytic systems based on transition metal coordination compounds have been proposed. Among them nickel complexes appear particularly versatile and have been largely investigated, as it is possible to induce large modification in the reactivity of several systems, just modifying the type of ancillary ligands. Mechanistic aspects into the reaction of organic halides catalysed by Ni complexes have been investigated in detail by Amatore while synthetic and applicative aspects have been explored focusing the attention on the carboxylation of some benzyl halides which are precursors of alpha-arylpropionic acids used as nonsteroidal anti-inflammatory agents. Also complexes of Pd, Co, and Fe have shown interesting catalytic properties towards the electrocarboxylation of various organic compounds.

Considerable improvement in the performance of several electrocarboxylation processes has been obtained performing the syntheses in undivided cells equipped with suitable sacrificial anodes. The anodic dissolution of the metal electrode leads to the formation of solvated cations in stoichiometric balance with the carboxylated species $R\text{-}COO^-$ which can react with the metal ions to form complex salts that are very less reactive than the corresponding acids and are insoluble in the reaction medium, or can be made insoluble by addition of appropriate co-solvent, making easier the isolation procedures. This drawback can be overcome by the use of suitable homogeneous charge transfer catalysts which however during the electrolyses undergo decomposition, which extent can be limited by a proper selection of the operative conditions.

Electrochemical reduction of CO_2

Electrochemical reduction of carbon dioxide has widely been explored since long, stimulated by the similarity of CO_2 reduction to photosynthesis, the latter being one of the most important reactions in Nature as well as the basis of all living activity. Since carbon dioxide is easily available in Nature and huge amounts are generated by human activities, it constitutes an almost infinite source of carbon for the chemical industry. Its electrochemical or photochemical conversion to such compounds as formic acid, hydrocarbons, alcohols or aldehydes may be foreseen as a practical means of solar energy storage through the production of fuels.

A large number of papers have been published in recent years dealing with the reduction of carbon dioxide in aqueous and non-aqueous electrolytes. Several reviews provide detailed listing of the work performed so far [103].

The different aspects of the electroreduction of CO_2 investigated more recently can be grouped as follows :

(i) reduction in aqueous media
(ii) reduction in non-aqueous/organic media
(iii) surface studies on single crystals
(iv) use of gas-diffusion electrodes
(v) effects of pressure, temperature, ultrasonic treatment.

The central question arising from the analysis of a large amount of the research work devoted to the electrochemical carbon dioxide reduction is about the possible future role of such a technique as a means for converting CO_2 into value added chemicals. First, the choice of the electrochemical method can find a justification only when the electrical energy used in the process originates from other than fossil-fuels. However, even if a cheap and abundant electrical energy, provided by nuclear or solar power stations, were available, the direct electrochemical CO_2 reduction process would face the comparison with an indirect route involving hydrogen production by water electrolysis to feed a conventional heterogeneous catalytic process.

However, unexpectedly enough, the electrochemical CO_2 reduction may reach, till now on the laboratory scale, much higher specific rates (expressed with respect to the electrode/catalyst area) than an equivalent conventional catalytic process. This brings a confirmation of the fundamental superiority of the electric potential (over high temperatures/pressures) in enhancing the rates of reduction/oxidation reactions.

Taking into account all the above mentioned drawbacks of the direct electrochemical CO_2 reduction method, i.e., the fact of being yet an unproven and energy demanding technology, involving rather high investment cost, one has now try to identify prospects (if any) for the future practical applications of the method.

Storage of the electrical energy produced by solar or nuclear power stations appears as the most promising option, provided the elctroreduction of CO_2 results in a selective formation of a liquid C_1 (or C_2) product such as formic acid, methanol or ethanol. The latter compounds are, in fact, much easier to store than is hydrogen formed through the electrolysis of water. In this connection, the two-electron CO_2 reduction product formic acid appears as particularly attractive, due to its easy catalytic decomposition into hydrogen and CO_2. Formic acid may be formed with high faradaic efficiency, exceeding 90%, when using tin or indium as cathode materials (more problematic appears here the use of palladium, quite an efficient but prohibitively expensive electrocatalyst).

In contrast with formic acid, no efficient electrocatalyst leading to the CO_2 reduction to methanol has been found so far. On the other hand, ethanol is formed with faradaic efficiencies of ca. 20% and among other products in the course of the electroreduction at a copper cathode.Consequently, the production of formic acid remains, for the moment, the only viable option in view of the storage of electrical energy via CO_2 electroreduction.

A special Chapter is dedicated to the electrochemical fixation and reduction of carbon dioxide, where a more detailed information on this topic can be found.

3.2.8 Gamma-rays induced syntheses of intermediates and fine chemicals

Owing to the high stability of carbon dioxide its reduction is a high energy-demanding process. For this reason extensive investigations have been aimed at setting-up of reduction routes driven by one energy different from thermal. In this contest, it is well known that ionising radiation often stimulates chemical reactions and has been proved that CO_2 reduction can be initiated and maintained simply by the aid of radiation exposure.

So, the ionising radiation from ^{60}Co and ^{137}Cs sources and that produced by particle accelerator could be utilised for decreasing CO_2 produced by fossil fuels fired power plants. Moreover the gamma radiation from spent fossil fuels as well as radioactive waste could be preferably used, thus allowing a sort of recycle of the radioactive sources that must be isolated completely and safely from the biosphere [104]. Several attempts have been made to use γ-ray for CO_2 reduction or for its use in carboxylation reactions. Recently, polymerisation through ionising radiation in CO_2-based system was investigated as a new field of research.

REFERENCES

1. a) "Carbon Dioxide as a Source of Carbon", Aresta, M. and Forti, G. Eds., NATO-ASI Series C, 1987; b) "Enzymatic and Model Reaction for Carbon Dioxide Carboxylation and Reduction Reactions", Aresta, M. and Schloss, J.V. Eds., Elsevier Publ.,1990; c) "CO_2 conversion and Utilization", Song, C., Gaffney, A.M. and Fujimoto K. Eds, ACS Symp Series 809, 2002; d) "Utilization of Greenhouse Gases", Liu, C.J., Mallinson, R.G. and Aresta, M. Eds, ACS Symp Series, 2003; e) "Environmental Challenges and Greenhouse Gas Control for Fossil Fuel Utilization in the 21st Century" Maroto Valer et al. Eds, Kluwer Academic/Plenum Publisher, New York, 2002; f) Arakawa H., Aresta M., Armor JN, et al. "Catalysis research of relevance to carbon management: Progress, challenges, and opportunities" Chem. Rev. 101 (4): 953-996, 2001; g) Aresta M., Tommasi I. "Carbon dioxide utilisation in the chemical industry" Energy Convers. Mgmt. 38, S373-S378, 1997.
2. Aresta, M., Dibenedetto, A. "Key Issues in Carbon Dioxide Utilisation as a Building Block for Molecular Organic Compounds in the Chemical Industry." ACS Book on CO_2 Conversion and Utilisation, C. Song, A.M. Gaffney, and K. Fujimoto Eds, ACS Symp Series 809, 54-70, 2002.
3. Aresta, M., Nobile, C.F., Albano, V.G., Forni, E., Manassero, M. "New nickel-carbon dioxide complex. Synthesis, properties, and crystallographic characterization of (carbon dioxide)bis(tricyclohexylphosphine)nickel" J. Chem. Soc., Chem. Commun. (15), 636-7, 1975.
4. a) Aresta, M., Dibenedetto, A., Tommasi, I. "Developing Innovative Synthetic Technologies of Industrial Relevance based on Carbon Dioxide as Raw material" Energy&Fuels 15, 2, 269-273, 2001; b) Aresta, M. "The fixation of carbon dioxide in inorganic and organic chemicals" Energy Convers. Mgmt. 34 (9-11), 745-52, 1993.
5. Aresta, M. "Carbon Dioxide Utilization: Greening both the Energy and Chemical Industry. An Overview" Book on Utilization of Greenhouse Gases, Chang-jun Liu, R. G. Mallinson, M. Aresta Eds, ACS Symp Series, 2003.
6. Aresta, M., Quaranta, E., Tommasi, I., Giannoccaro, P., Ciccarese, A. "Enzymatic versus chemical carbon dioxide utilization. 1.The role of metal centres in carboxylation reactions" Gazz. Chim. Ital. 125 (11): 509-538, 1995.
7. Solymosi F. "The bonding, structure and reactions of CO_2 adsorbed on clean and promoted metal-surfaces" J. Mol. Catal. 65 (3): 337-358, 1991.
8. Galan, F., Fouassier, M., Mascetti, J., Papai, I. "CO_2 coordination and cooperative effects in ligand binding to transition metal atoms: matrix isolation and density functional studies" International Conference on Low Temperature Chemistry, 2nd, Kansas City, Mo., Aug. 4-9, 59-62, 1996.
9. Gibson, D.H. "The organometallic chemistry of carbon dioxide" Chem. Rev. 96 (6): 2063-2095, 1996.
10. Aresta, M., Quaranta, E., Tommasi, I. "The role of metal centers in reduction and carboxylation reactions utilizing carbon-dioxide" New J. of Chem. 18 (1): 133-142, 1994.
11. Tommasi, I., Aresta, M., Giannoccaro, P., Quaranta, E., Fragale, C. "Bioinorganic chemistry of nickel and carbon dioxide: a Ni complex behaving as a model system for carbon monoxide dehydrogenase enzyme" Inorg. Chim. Acta 272 (1-2): 38-42, 1998.
12. Aresta, M., Quaranta, E., Tommasi, I. "Prospects for the utilization of carbon-dioxide" Energ. Conv. Mgmt. 33 (5-8): 495-504, 1992.

13. a) Aresta, M., Quaranta, E., Tommasi, I. "Utilization of carbon dioxide: strategy for the control of its level I the atmoshere. The role of the photoconversion technology" in "Photochemical Conversion and Storage of Solar Energy", Pelizzetti, E., Schiavello, M. Eds. Kluwer Academic Publishers, 517-550, 1991; b) Solar photoconversion processes for recycling carbon dioxide from the atmosphere: an assessment of Energy, Research and industrial applications, Colorado Springs, March 13-16, 1990.
14. Aresta, M., Quaranta, E. "Carbon dioxide: a substitute for phosgene" ChemTech 27 (3), 32-40, 1997.
15. a) Aresta, M., Quaranta, E., Liberio, R., Dileo, C., Tommasi, I. "Enzymatic synthesis of 4-OH-benzoic acid from phenol and CO_2: The first example of a biotechnological application of a carboxylase enzyme" Tetrahedron 54 (30): 8841-8846, 1998; b) Aresta, M., Dibenedetto, A. "Development of environmentally friendly syntheses: use of enzymes and biomimetic systems for the direct carboxylation of organic substrates" Reviews in Molecular Biotechnology, 90(2), 113-128, 2002.
16. Chemical Economics Handbook (CEH) Report (Abs) Polycarbonates resins 1997, www-cmrc.sri.com
17. Pacheco, M.A., Marshall, C.L. "Review of Dimethyl Carbonate (DMC) Manufacture and Its Characteristics as a Fuel Additive" Energy & Fuels, 11(1), 2-29, 1997.
18. Delledonne, D., Rivetti, F., Romano, U. "Developments in the production and application of dimethyl carbonate" Appl. Catal., A: General, 221(1-2), 241-251, 2001.
19. Aresta, M., Dibenedetto, A. "Mixed anhydrides: key intermediates in carbamates forming processes of industrial interest" Chemistry-A Eur. J., 8, 685-690, 2002.Aresta, M., Dibenedetto, A., Quaranta, E. "Reaction of aromatic diamines with diphenyl carbonate catalyzed by phosphorus acids: a new clean synthetic route to mono- and dicarbamate", Tetrahedron 54(46), 14145-14156, 1998; b) Aresta, M., Bosetti, A., Quaranta, E. Ital. Pat. Appl. 002202, 1996.
21. Aresta, M., Dibenedetto A., Quaranta E. "Selective carbomethoxylation of aromatic diamines with mixed carbonic acid diesters in the presence of phosphorous acids" Green Chemistry, 237-242, 1999.
22. Fujio, Y., Matsukuma, K., Nishimoto, T. "Allyl benzoate-glycol bis(allyl carbonate)-diallyl benzenedicarboxylate copolymers for optical materials", Eur. Pat. Appl., 6 pp. EP 312091 A2 19890419, CAN 111:215663, 1989.
23. a) Tatsumi, T., Watanabe, Y., Koyano, K. A. "Synthesis of dimethyl carbonate from ethylene carbonate and methanol using TS-1 as solid base catalyst" Chem. Commun. 2281-2282, 1996; b) Watanabe, Y., Tatsumi, T. "Hydrotalcite-type materials as catalysts for the synthesis of dimethyl carbonate from ethylene carbonate and methanol" Microp. Mesop. Mat. 22: 399-407, 1998; c) Chang, C.D., Jiang, Z., LaPierre, R.B., Schramm, S.E. (Exxon Mobil) "Process for co-production of dialkyl carbonate and alkanediol" US 6365767, 2002
24. Serini, V. "Polycarbonates", Ullmann's Enciclopedia of Industrial Chemistry, 5th Ed. 1992, A21, 207.
25. Evans, T.L., Brunelle, D.J., Salem, A.J., Stewart, K.R. "Developments in the chemistry of oligocyclic carbonates for use in structural composites" Polymer Preprints (ACS, Division of Polymer Chemistry), 32(2), 176-7, 1991; Serini, V. "Polycarbonates" Industrial Polymers Handbook, 1, 291-304, 2001.
26. "Continuous manufacture of alkyl carbonates" Fr. Patent, 7 pp. FR 2163884 19730831, SNPE, 1973.
27. Damle, S. B. "Carbonic and carbonochloridic esters", Kirk-Othmer, Encyclopedia of Chemical Technology, 4th. Ed. 1993, 5, 77.

28. Buysch, H.-J. "Carbonic acid esters", Ullmann's Encyclopedia of Industrial Chemistry, 5th, Ed. 1986, A5, 197.
29. a) Limura, N., Takagi, M., Iwane, H., Ookago, J. "Preparation of propylene carbonate" Jpn. Kokai Tokkyo Koho, 4 pp. JP 07267944 A2 19951017 Heisei. CAN 124:146137, 1995; b) Inaba, M., Hasegawa, K., Nagaoka, H. "Process for preparation of alkylene carbonates by reacting alkylene oxides with carbon dioxide" Jpn. Kokai Tokkyo Koho , 5 pp. JP 09067365 A2 19970311 Heisei. CAN 126:277459 1997; c) Inoe, K., Ookubo, H. "Method for producing alkylene carbonate by cycloaddition of alkylene oxide with carbon dioxide" Jpn. Kokai Tokkyo Koho, 6 pp. JP 07206846 A2 19950808 Heisei. CAN 123:256691, 1995; d) Inoe, K., Ookubo, H., "Method for producing alkylene carbonate by addition of carbon dioxide to alkylene oxide" Jpn. Kokai Tokkyo Koho, 5 pp. JP 07206848 A2 19950808 Heisei. CAN 123:256692, 1995; e) Ichikawa, S., Iwane, H. "Preparation of alkylene carbonates" Jpn. Kokai Tokkyo Koho, 6 pp. JP 09235252 A2 19970909 Heisei. CAN 127:262668, 1997; f) Inoe, K., Ookubo, H. "Method for producing alkylene carbonate by addition of carbon dioxide to alkylene oxide" Jpn. Kokai Tokkyo Koho, 5 pp. JP 07206847 A2 19950808 Heisei. CAN 124:29749, 1995.
30. a) Marquis, E.T., Sanderson, J.R. "Process for preparing high-purity alkylene carbonates from epoxides and carbon dioxide using metal phthalocyanine catalysts" U.S. Patent, 5 pp. US 5283356 A, CAN 120:217649, 1994; b) Kuran, W., Listos, T. "Initiation and propagation reactions in the copolymerization of epoxide with carbon dioxide by catalysts based on diethylzinc and polyhydric phenol" Macromol. Chem. Phys. 195(3), 977-84, 1994; c) Tojo, M. Fukuoka, S. JP Patent 03120270, 1991 (Asahi Chem. Ind. Co. Ltd); d) Mais, F. J., Buysch, H.J., Mendoza-Frohn, C., Klausener, A. EP Patent 543 249, 1993, (Bayer); e) Bobyleva, L.I., Kryukov, S.I., Bobylev, B.N., Liakumovich, A.G., Surovtsev, A.A., Karpov, O.P., Akhmedyanova, R.A., Koneva, S.A. SU Patent 1 781 218, 1992 (Yaroslavskij Polit. Institut).
31. Sakai, T., Kihara, N., Endo, T. "Polymer Reaction of Epoxide and Carbon Dioxide. Incorporation of Carbon Dioxide into Epoxide Polymers" Macromol. 28(13), 4701-6, 1995.
32. Sone, M., Sako, T., Kamisawa, C. "Organic carbonates and catalysts and production methods therefore" Jpn. Kokai Tokkyo Koho JP 11335372, 8 pp., 1999.
33. Darensbourg, D.J., Holtcamp, M.W. "Catalysts for the reactions of epoxides and carbon dioxide" Coordination Chemistry Reviews, 153, 155-174, 1996.
34. De Pasquale, R.J "Unusual catalysis with Nickel(0) complexes" J. Chem. Soc., Chem. Commun., 157-158 , 1973.
35. Ji, D.F., Lu X.B., He R. "Syntheses of cyclic carbonates from carbon dioxide and epoxides with metal phthalocyanines as catalyst" Applied Catalysis A: General, 203(2), 329-333, 2000.
36. Yano, T., Matsui, H., Koike, T., Ishiguro, H., Fujihara, H., Yoshihara, M., Maeshima, T. "Magnesium oxide-catalysed reaction of carbon dioxide with an epoxide with retention of configuration" Chem. Commun. 1129-30, 1997.
37. Yamaguchi, K., Ebitani, K., Yoshida, T., Yoshida, H., Kaneda, K. "Mg-Al mixed oxides as highly active acid-base catalysts for cycloaddition of carbon dioxide to epoxides" J. Am. Chem. Soc. 121, 4526-27, 1999.
38. Aresta, M., Dibenedetto, A., Gianfrate, L., "Nb(V) Compounds as Epoxides Carboxylation Catalysts: The Role of the Solvent" J. Mol. Catal. 2003, in press
39. Aresta, M., Dibenedetto A., Gianfrate L. "Enantioselective synthesis of organic carbonates promoted by Nb(IV) and Nb(V) catalysts" Applied Catalysis A: General, 2003 in press.

40. Darensbourg, D.J., Zimmer, M.S. "Copolymerization and Terpolymerization of CO_2 and Epoxides Using a Soluble Zinc Crotonate Catalyst Precursor", Macromol., 32(7), 2137-2140, 1999.
41. Frevel, L. K.; Gilpin, J. A. "Carbonate synthesis from alkylene carbonates". US Patent 3 642 858, 3 pp., CAN 76:99140, 1972 (Dow); Urano, Y.; Kirishiki, M.; Onda, Y.; Tsuneki, H. "Preparing dialkyl carbonates from alkylene carbonate and alcohol using rare earth oxide" Eur. Pat. Appl. EP599287, 12 pp., CAN 121:258494, 1994; Kondo, T., Okada, Y., Tanaka, F., Asaoka, S., Yamamoto, S. "Preparation of dialkyl carbonates with ion-exchanged zeolite catalysts" Jpn. Kokai Tokkyo Koho JP 07048319, 8 pp. CAN 123:9072, 1995; Inoe, K.; Ookubo, H. "Simultaneous manufacture of dialkyl carbonates and glycols" Jpn. Kokai Tokkyo Koho JP Patent 06 343 870, 7 pp., 1994, (Mitsui Toatsu Chem.); Inoe, K.; Ookubo, H. "Manufacture of dimethyl carbonate and ethylene glycol using a heterogeneous catalyst" Jpn. Kokai Tokkyo Koho (JP 06343871, 6 pp., CAN 122:217155, 1994, (Mitsui Toatsu Chem.).
42. a) Aresta, M., Dibenedetto, A., Tommasi I. "Direct synthesis of organic carbonates by oxidative carboxylation of olefins catalysed by metal oxides: developing green chemistry based on carbon dioxide", Appl. Organometal. Chem., 14, 799-802, 2000; b) Aresta, M., Dibenedetto, A. "Carbon Dioxide as Building Block for the Synthesis of Organic Carbonates: Behavior of Homogeneous and Heterogeneous Catalysts in the Oxidative Carboxylation of Olefins"J. Mol. Catal. 182-183, 399-409, 2002.
43. Kim, W.B., Lee J.S. "Comparison of polycarbonate precursors synthesized from catalytic reactions of bisphenol-A with diphenyl carbonate, dimethyl carbonate, or carbon monoxide" J. Appl.Polym. Sci., 86, 937–947, 2002.
44. Kuehling, S., Pakull, R., Grigo, U., Tacke, P., Freitag, D., Alewelt, W. "Preparation of solvent-free polycarbonates by solid-state polycondensation of oligomers" Eur. Pat. Appl., 11 pp., EP 481296, 1992.
45. Wang, S.J., Du, L.C., Zhao, X.S., Meng, Y.Z., Tjong, S.C. "Synthesis and characterization of alternating copolymer from carbon dioxide and propylene oxide" J. Appl. Polym. Sci., 85(11), 2327-2334, 2002.
46. Inoue, S., Koinuma, H., Tsuruta, T. "Copolymerization of carbon dioxide and epoxide" J. Polym. Sci., Part B, 7 (4), 287-92, 1969.
47. Super, M., Berluche, E., Costello, C., Beckman, E. "Copolymerization of 1,2-epoxycyclohexane and carbon dioxide using carbon dioxide as both reactant and solvent" Macromolecules 30 (3), 368-372, 1997.
48. Jessop, P.G., Leitner, W. "Chemical Synthesis Using Supercritical Fluids", Wiley-VCH, 1999.
49. E. Dinjus "Organometallic reactions with CO_2-catalyst design and mechanism" ICCDU IV September 7-11, 1997, Kyoto, Japan.
50. Sakakura, T., Choi, J.-C., Saito, Y., Sako, T. "Synthesis of dimethyl carbonate from carbon dioxide: catalysis and mechanism", Polyhedron, 19, 573-76, 2000.
51. Isaacs, N.S., O'Sullivan, B. "High pressure routes to dimethyl carbonate from supercritical carbon dioxide", Tetrahedron, 55, 11949-56, 1999.
52. Aresta, M., Dibenedetto, A., Dileo, C., Tommasi I., Amodio E. "The First Synthesis of a Cyclic Carbonate from a Ketal in SC-CO_2" J. Sup. Fluid, 25/2, 177-182, 2003.
53. a) Kirovskaya, I.A., Pimenova, L.N., Votyakova, I.A., Sharangovich, N.N. "Reaction of hydrogen with carbon dioxide on the surface of diamondlike semiconductors" Zhurnal Fizicheskoi Khimii, 52(9), 2356-60, 1978; b) Kirovskaya, I.A., Zeleva, G.M., Votyakova, I.V., Chernega, T.S. "Catalytic hydrogenation of carbon dioxide on the surface of A3B5 and A2B6 compounds" Zhurnal Fizicheskoi Khimii, 57(4), 993-5, 1983.

54. a) Anderson, J.J., Drury, D.J., Hamlin, J.E., Kent, A.G. "Formic acid", BP Chemicals Ltd., UK, PCT Int. Appl., WO 8602066, 15 pp., CAN 105:210757, 1986; b) Anderson, J.J., Hamlin, J.E. "Formic acid", BP Chemicals Ltd., UK, Eur. Pat. Appl. 126524, 10 pp., CAN 102:95259, 1984.
55. Jessop, P.G., Ikariya, T., Noyori, R. "Homogeneous hydrogenation of carbon-dioxide" Chem. Rev. 95 (2), 259-272, 1995.
56. Jessop, P.G., Hsiao, Y., Ikariya, T., Noyori, R. "Homogeneous catalysis in supercritical fluids: Hydrogenation of supercritical carbon dioxide to formic acid, alkyl formates, and formamides" J. Am. Chem. Soc. 118 (2), 344-355, 1996.
57. Jessop, P.G., Hsiao, Y., Ikariya, T., Noyori, R. "Methyl formate synthesis by hydrogenation of supercritical carbon-dioxide in the presence of methanol" J. Chem. Soc., Chem. Commun. (6): 707-708, 1995.
58. Jessop, P.G., Hsiao, Y., Ikariya, T., Noyori, R. "Catalytic production of dimethylformamide from supercritical carbon-dioxide" J. Am. Chem. Soc. 116 (19): 8851-8852, 1994.
59. Jessop, P.G., Linehan, J.C., Main, A.D., Pitts, J.M. "Recent development in the hydrogenation of CO_2 catalyzed by ruthenium phosphine complexes" VI International Conference on Carbon Dioxide Uilization, Breckenridge – Colorado, p. 31, September 9-14, 2000.
60. Krocher, O., Koppel, R.A., Baiker, A. "Highly active ruthenium complexes with bidentate phosphine ligands for the solvent-free catalytic synthesis of N,N-dimethylformamide and methyl formate" Chem. Commun. (5), 453-454, 1997.
61. Kroecher, O., Koeppel, R.A., Baiker, A. "Highly active ruthenium complexes with bidentate phosphine ligands for the solvent-free catalytic synthesis of N,N-dimethylformamide and methyl formate" Chem. Commun. (5), 453-454, 1997.
62. Jessop, P.G. "Homogeneously-catalyzed syntheses in supercritical fluids" Topics in Catalysis 5 (1-4): 95-103, 1998.
63. Paulik, F.E., Roth, J.F. "Catalysts for the low-pressure carbonylation of methanol to acetic acid", Cent. Res. Dep., Monsanto Co., St. Louis, MO, USA, Chem. Commun., 24, 1578, 1968.
64. Tommasi, I., Aresta, M., Giannoccaro, P., Quaranta, E., Fragale, C. "Bioinorganic chemistry of nickel and carbon dioxide: an Ni complex behaving as a model system for carbon monoxide dehydrogenase enzyme" Inorg. Chim. Acta, 272(1,2), 38-42, 1998.
65. Taniguchi, Y., Kitamura, T., Fujiwara, Y. "Vanadium-catalyzed acetic acid synthesis from methane and carbon dioxide" ICCDU IV, Kyoto, Japan, September 5-11, 1997, P-30.
66. He, M.-X., Zou, J.-J., Li, Y., Eliasson, B., Liu, C.-J. "Oxygenates synthesis from methane and carbon dioxide using dielectric-barrier discharges" 223th ACS National Meeting, Orlando, FL, Fuel Chemistry Division Preprints 47(1), 328-29, 2002 He, M.-X., Zou, J.-J., Li, Y., Eliasson, B., Liu, C.-J. "Oxygenates synthesis from methane and carbon dioxide using dielectric-barrier discharges" 223th ACS National Meeting, Orlando, FL, Fuel Chemistry Division Preprints 47(1), 328-29, 2002.
67. Gressin, J.C., Michelet, D., Nadjo, L., Saveant, J.M. "Electrochemical reduction of carbon dioxide in weakly protic medium", Nouveau Journal de Chimie, 3(8-9), 545-54, 1979.
68. Vianello, E., Isse, A.A., Gennaro, A., Severin, M.G. "Carbon dioxide reduction by heterogeneous and homogeneous electrocatalysis" Proceedings International Conference on Carbon Dioxide Utilisation Bari, Italy, September 26-30, 1993, 287- 294.
69. Fisher, J., Lehmann T., Heitz, E. "The production of oxalic acid from carbon dioxide and water" J. Appl. Electrochem., 11(6), 743-50, 1981.

70. a) Abrams, K.J., Marsh, D.W., Doll, B.B. "Oxidation catalyst recovery during manufacture of aromatic carboxylic acids in a continuous liquid-phase process", US 4769489, 5 pp., CAN 110:10086, 1989.; b) Maruyama, K., Arakawa, M., Kubota, T. "Preparation of aromatic or heterocyclic carboxylic acids" Jpn. Kokai Tokkyo Koho, 3 pp., JP 2002363125, CAN 138:13960, 2002.
71. a) Walther, D., Dinjus, E. "Activation of carbon dioxide on transition metal centers: metal ring closure reaction between carbon dioxide and 1,3-dienes on the electron-rich nickel(0) complex moiety", Zeitschrift fuer Chemie, 22(6), 228-9, 1982; b) Dinjus, E., Walther, D., Schuetz, H., Schade, W. "Regioselective synthesis of 3-unsaturated carboxylic acids as a subsequent step of a nickel-induced coupling of conjugated dienes with carbon dioxide", Zeitschrift fuer Chemie, 23(8), 303-4, 1983; c) Hoberg, H., Schaefer, D., Oster, B.W. "Diene carboxylic acids from 1,3-dienes and carbon dioxide by carbon-carbon bonding to nickel(0)",J.Organomet. Chem., 266(3), 313-20, 1984; d) Walther, D., Dinjus, E. "Activation of carbon dioxide at transition metal centers: reactivity of the metallacycle formed between carbon dioxide and 2,3-dimethyl-1,3-butadiene on the complex center (TMED)Ni(0)" Zeitschrift fuer Chemie, 24(2), 63, 1984; e) Hoberg, H., Apotecher, B. "α,ω-Diacids from butadiene and carbon dioxide with nickel(0)". J. Organomet. Chem., 270(1), C15-C17, 1984; f) Walther, D., Dinjus, E., Goerls, H., Sieler, J., Lindqvist, O., Andersen, L. "Metallocyclics from carbon dioxide and dimethylbutadiene at the complex moiety L(Ni)°: structure and reactivity as a function of neutral ligands L", J. Organomet. Chem., 286(1), 103-14, 1985; g) Behr, A., Kanne, U. "Nickel-complex-induced C-C linkage of carbon dioxide with trienes", J. Organomet. Chem., 317(3), C41-C44, 1986; h) Fischer, R., Walther, D., Braunlich, G., Undeutsch, B., Ludwig, W., Bandmann, H. "Nickelalactones for synthesis - sonochemical and bimetallic activation of cross-coupling reactions with alkyl halides" J. Organomet. Chem., 427 (3), 395-407, 1992; i) Hoberg, H., Schäfer, D., Oster, B.W. "Diene carboxylic acids from 1,3-dienes and carbon dioxide by carbon-carbon bonding to nickel(0)", J. Organomet. Chem., 266(3), 313-20, 1994.
72. Derien, S., Dunach, E., Perichon, J. „From stoichiometry to catalysis: electroreductive coupling of alkynes and carbon dioxide with nickel-bipyridine complexes. Magnesium ions as the key for catalysis" J. Am. Chem. Soc., 113(22), 8447-54, 1991.
73. a) Derien, S., Clinet, J.C., Dunach, E., Perichon, J. „New carbon-carbon bond formation through the nickel-catalyzed electrochemical coupling of 1,3-enynes and carbon dioxide" J. Organomet. Chem., 424(2), 213-24, 1992; b) Derien, S., Clinet, J. C., Dunach, E., Perichon, J. "Electrochemical incorporation of carbon dioxide into alkenes by nickel complexes" Tetrahedron, 48(25), 5235-48, 1992.
74. Aresta, M., Dibenedetto, A., Papai, I., Gabor, S. "Unprecedented formal 2+2 addition of allene to CO_2 promoted by $[RhCl(C_2H_4)(PiPr_3)]_2$: direct synthesis of the four membered lactone α-methylene-β-oxiethanone. Theoretical aspects and experiments", Inorg. Chim. Acta, 334, 294-300, 2002.
75. Bringmann, J., Dinjus, E. "Electrochemical synthesis of carboxylic acids from alkenes using various nickel-organic mediators: CO_2 as C1-synthon" Applied Organometallic Chemistry 15 (2): 135-140, 2001.
76. a) Bottaccio, G., Marchi, M., Chiusoli, G.P. "Carboxylation of organic substrates with carbon dioxide in hydrocarbon media" Gazz. Chim. Ital., 107(9-10), 499-500, 1977; b) Bottaccio, G., Chiusoli, G.P., Alneri, E., Marchi, M., Lana, G. "Carboxylation of organic substrates with carbon dioxide" Ger. Offen., 16 pp., DE 2612577, CAN 86:5180, 1977.
77. a) Tommasi, I., Aresta, M., Tkatchenko I. "Unprecedented synthesis of 1,3-dialkylimidazolium-2-carboxylate: a carbon dioxide transfer agent to active C-H bonds" Division of Industrial and Engineering Chemistry - Ionic liquids as green Solvents:

Progress and Prospects", 224th ACS National Meeting, Boston, MA, August 18-22, 2002; b) Tommasi, I., Aresta, M., Tkatchenko I. "Unprecedented synthesis of 1,3-dialkylimidazolium-2-carboxylate: a carbon dioxide transfer agent to active C-H bonds" International Conference "Green Solvents for Catalysis – Environmentally Benign Reaction Media" October 13 – 16, 2002, Bruchsal/Germany.

78. Barthelemy, J. "Carbamates", Lyon Phar., 37(5), 249-63, 1986.
79. Wu, T.-T., Huang, J., Arrington, N.D., Dill, G.M. "Synthesis and herbicidal activity of □-heterocyclic carbinol carbamates", Journal of Agric. Food Chem., 35(5), 817-23, 1987; (c) Suzuki, K., Kato, T., Takahashi, J., Kamoshita, K. "Mode of action of methyl N-(3,5-dichlorophenyl)carbamate in the benzimidazole-resistant isolate of Botrytis cinerea" Nippon Noyaku Gakkaishi (1984), 9(3), 497-501, CAN 102:180714, 1985.
80. Romano, U., Rivetti, F., Sasselli, G. "Alkyl isocyanates" Eur. Pat. Appl. (1984), 15 pp., EP 125726, CAN 102:95246, 1985.
81. Piccardi, P. "The role of synthetic chemicals in insect pest control: present and future" La Chimica e l'Industria (Milan, Italy), 68(11), 108-17, 1986.
82. Ito, M. "Method for the preparation of aliphatic carbonic acid diesters and carbamates" Jpn. Kokai Tokkyo Koho, 10 pp. JP 09255630 A2 19970930 Heisei. CAN 127:307168 1997.
83. Volodarskaya, Yu.I., Sal'nikova, G.A., Shmidt, Ya.A. "Nucleophilic substitution of urethanes with hydroxyl-containing compounds in the presence of catalysts" Sintez i Fiziko-Khimiya Polimerov, 12, 12-17, 1973.
84. Aresta, M., Ballivet-Tkatchenko, D., Dell'Amico, D.B., Boschi, D., Calderazzo, F., Labella, L., Bonnet, M.C., Faure, R., Marchetti, F. "Isolation and structural determination of two derivatives of the elusive carbamic acid" Chem. Commun., (13), 1099-1100, 2000.
85. a) Bernard, M.A., Borel, M.M., Ledesert, M.A. "Metallic thioacetates. I. Preparation and thermal stability of trihydrated barium thioacetate and of two new compounds, hydrated calcium and strontium acetate thioacetates" Bull. Soc. Chim. Fran., (7-8)(Pt. 1), 2194-8, 1973; b) Johnson SL, Morrison D.L. "Kinetics and mechanism of decarboxylation of N-arylcarbamates. Evidence for kinetically important zwitterionic carbamic acid species of short lifetime" J. Am. Chem. Soc., 94(4), 1323-34, 1972.
86. Belli Dell' Amico, D., Calderazzo, F., Labella, L., Marchetti, F., Pampaloni, G., "N,N-dialkylcarbamato metal complexes, molecular inorganic precursors to functionalized inorganic matrices" Inorg. Chem. Commun., 5(9), 733-745, 2002.
87. a)Aresta, M., Dibenedetto, A., Quaranta, E. "Reaction of alkali-metal tetraphenylborates with amines in the presence of CO_2: a new easy way to aliphatic and aromatic alkali-metal carbamates" J. Chem. Soc. Dalton Trans., (20), 3359-63, 1995; b) Aresta, M., Quaranta, E. "Novel, CO_2-promoted synthesis of anhydrous alkylammonium tetraphenylborates: a study on their reactivity as intra- and inter-molecular proton-transfer agents" J. Organomet.Chem., , 488 (1-2), 211-22, 1995.
88. a) Aresta, M.; Quaranta, E. "Reactivity of phosphacarbamates: transfer of the carbamate group promoted by metal-assisted electrophilic attack at the carbon dioxide moiety" J. Org. Chem., 53(17), 4153-4, 1988; b) Aresta, M., Quaranta, E "Alkali-metal-assisted transfer of the carbamate group from phosphocarbamates to alkyl halides: a new easy way to alkali-metal carbamates and to carbamate esters", J. Chem. Soc., Dalton Trans. (12), 1893-9, 1992.
89. McGhee, W.D., Riley, D.P. "Palladium-mediated synthesis of urethanes from amines, carbon dioxide, and cyclic diolefins", Organometallics, 11(2), 900-7, 1992.

90. Aresta, M., Quaranta, E. "Role of the macrocyclic polyether in the synthesis of N-alkylcarbamate esters from primary amines, carbon dioxide and alkyl halides in the presence of crown ethers", Tetrahedron, 48(8), 1515-30, 1992.
91. Costa, M., Arisi, A. "1,3-Oxazolidin-2-one ring formation from carbon dioxide and N-alkyl-γ-aminocrotonate under homogeneous and supercritical conditions" ICCDU VI, Breckenridge, Colorato, USA, September 9-14, 2001.
92. a) Aresta, M., Berloco, C., Quaranta, E. "Biomimetic building-up of the carbamic moiety: the intermediacy of carboxyphosphate analogs in the synthesis of N-aryl carbamate esters from arylamines and organic carbonates promoted by phosphorus acids", Tetrahedron, 51(29), 8073-8, 1995; b) Aresta, M., Bosetti, A., Quaranta, E. "Procedimento per la produzione di dicarbammati aromatici" Ital. Pat. Appl. 002202, 1996.
93. Kanbe, H., Shimazaki, Y., Sotozono, M. "Preparation of carbamate esters" Jpn. Kokai Tokkyo Koho, 9 pp. JP 06298717, CAN 122:160120, 1994.
94. Aresta, M., Quaranta, E. "Mechanistic studies on the role of carbon dioxide in the synthesis of methyl carbamates from amines and dimethyl carbonate in the presence of carbon dioxide", Tetrahedron, 47(45), 9489-502, 1991.
95. Cesti, P., Bosetti, A., Mizia, F., Notari, M., Ricci, M., Rivetti, F., Romano, U. "Integrated process for the preparation of aromatic isocyanates by the reaction of aromatic amine and organic carbonate in presence of catalyst and procedures for effecting the relative intermediate phase" PCT Int. Appl., 57 pp., WO 0156977, CAN 135:154353, 2001.
96. Aresta, M., Dibenedetto A., Quaranta, E. "Reaction of aromatic diamines with diphenyl carbonate catalyzed by phosphorus acids: a new clean synthetic route to mono- and dicarbamates", Tetrahedron, 54(46), 14145-14156, 1998.
97. Aresta M., Dibenedetto A. "Mixed anhydrides: key intermediates in carbamates forming processes of industrial interest" Chemistry-A Eur. J., 8, 685-690, 2002.
98. a) Aresta, M., Dibenedetto, A., Quaranta, E. "Selective carbomethoxylation of aromatic diamines" Green Chemistry, 1(5), 237-242, 1999; b) Aresta, M., Quaranta . "Procedimento per la preparazione di carbammati e prodotti così ottenuti" Ital. Pat. 1198206, 1988; c) Aresta, M., Quaranta E. "Sintesi di carbammati da ammine e carbonati organici via attivazione del diossido di carbonio" Ital. Pat. 1237207, 1993; d) Aresta, M., Quaranta, E. "Sintesi di carbammati organici da ammine, diossido di carbonio e alogenuri alchilici in presenza di eteri corona" Ital. Pat. 1237208, 1993.
99. Aresta, M., Quaranta, E., Dibenedetto, A., Tommasi I., Marciniec B. "CO_2 catalysed carbamation of aminofunctional silanes" Appl. Organometal. Chem., 14, 871-873, 2000.
100. Dibenedetto, A., Narracci, M., Aresta M., Fragale C. "Reaction of silyl-mono and di-amines with carbon dioxide: evidence of formation of inter- and intra-molecular ammonium carbamates and their conversion into organic carbamates of industrial interest by trans-esterification of carbonates under carbon dioxide catalysis" Green Chemistry, 4 (05), 439-443, 2002.
101. a) Aresta, M., Quaranta, E., Ciccarese, A. "Head to head and head to tail coupling of allene and co-condensation with carbon dioxide promoted by 1,2-bis(diphenylphosphino)ethane (η6-tetraphenylborate)rhodium" C1 Mol. Chem., 1(4), 283-95, 1985; b) Behr, A.; Thelen, G., "Carbon-carbon coupling of carbon dioxide, cyclic hydrocarbons and nickel(0)." C1 Molecule Chemistry, 1(2), 137-53, 1984.
102. Tyssee, D.A., Wagenknecht, J.H., Baizer M.M.,. Chruma, J.L. "Cathodic organic syntheses involving carbon dioxide" Tetrahedron Lett., 47, 4809-12, 1972.
103. a) Frese, K.W., Jr. (1993). "Electrochemical and Electrocatalytic Reactions of CO_2 ".K.Christ, E. Guard and B.P. Sullivan, Elsevier:148. b) Sammells, A.F. and R.L.Cook

(1993). "Electrochemical and Electrocatalytic Reactions of CO_2 " K.Christ, E. Guard and B.P. Sullivan, Elsevier:245.

104. a) Fujita, N., Fukuda, Y., Matsuura, C., Saigo, K. "Changes in pH and redox potential during radiation-induced CO2 reduction in an aqueous solution containing iron powder" Radiat. Phys Chem., 47(4), 543-549, 1996; b) Fujita, N., Morita, H., Matsuura, C., Hiroishi, D., "Radiation-induced CO_2 reduction in an aqueous-medium suspended with iron-powder", Radiat. Phys Chem. 44 (4), 349-357, 1994.

Chapter 10

LINEAR ORGANIC CARBONATES

Danielle Ballivet-Tkatchenko and Svetlana Sorokina
Laboratoire Laboratoire de Synthèse et Electrosynthèse Organométalliques, Université de Bourgogne, UMR CNRS 5632, 9 av A. Savary, BP 47870, 21078-Dijon cedex, France. ballivet@u-bourgogne.fr
and Institut de Recherches sur la Catalyse, UPR CNRS 5401, 2 av. A. Einstein, 69626-Villeurbanne cedex, France

1. INTRODUCTION

The most common and successful method for producing organic carbonates, including polycarbonates, has been based for long time on the reaction between an alcohol or phenol and phosgene. The existing strict environmental regulations and the expanding market for organic carbonates do not assure that this technology can continue to have a leading role in this area, due to the inherent riskfulness of the processes. The market expansion of dimethyl carbonate (DMC), diphenyl carbonate (DPC), and bisphenol A polycarbonate (BPA-PC) is a reality, finding its driving force from the industrialization of alternative routes by EniChem (Italy) and UBE (Japan). This favorable situation can be foreseen for other carbonates.

The scope of this review is to present the industrially important processes for the most widely used open chain organic carbonates on the basis of data collected from the open literature and patents. Alternative technologies are described with special emphasis on those which involve carbon dioxide as a C1 synthon for introducing the carbonate function. The limited number of commercialization is certainly indicative of unfavorable economics. However, the need for environmental improvements and sustainable development in the chemical industry may reverse the situation by stimulating research programs from both the academic and industrial communities. As a matter of fact, during the last year of the RUCADI

M. Aresta (ed.), Carbon Dioxide Recovery and Utilization, 261-277.

Network, there was a significant increase in the number of publications and patents reporting new catalysts and process technologies to improve the performances of product oriented syntheses.

2. PROPERTIES AND USES

Open chain organic carbonates can be classified into two groups, namely, dialkyl (diaryl, alkyl-aryl) carbonates and polycarbonates. The molecular compounds find applications either as intermediates in the pharmaceutical, agrochemical, polymer industries or as solvents for batteries, extractive separation, additives in hydraulic fluids.[1] Among them, DMC and DPC are those which are expected to have an increasing demand connected with "green chemistry" and more environmentally acceptable processes. The areas of potential utilization are numerous. DMC is a methylating agent replacing dimethyl sulfate and methyl halides, with the advantages of generating less wastes (no salt formation) and avoiding carcinogen risks. [2] It is also a carbonylating agent, like DPC, finding new applications in place of toxic phosgene in polycarbonate and polyurethane productions.[3] Moreover, carbonates are among the combustion-improving agents those with the lowest emission of carbon dioxide. Should this use be implemented, the market will expand to several Mt/y. DMC as an oxygenate in reformulated gasoline is positively considered, but its current price should dramatically decrease for such application.[4] Polycarbonates have a broad spectrum of commercial applications. The aromatic polycarbonate, BPA-PC, is the most widely used as an engineering thermoplastic. It exhibits an exceptional combination of properties such as optical clarity, heat and impact resistance, and processing characteristics. Its ability to be either blended with other polymers, copolymerized, or functionalized enhances special performance requirements for a market in expansion. Applications concern a variety of economic sectors: electrical/electronic (36%), building/construction (24%), automobile (15%), and optical information storage (13%). Other important areas include medical, packaging, photographic, and optical applications.[5] The aliphatic polycarbonates, poly(ethylene or propylene carbonate), are primarily used as sacrificial binders for metals and ceramics in the electronics industry due to their low decomposition temperature (<300°C) and very low ash content.[6]

3. ECONOMIC ASPECTS

Large amounts of molecular carbonates do not reach the market because they are utilized as intermediates in different processes. For example, the current DMC production is close to 80 kt/y. On the free market, there are only 16-17 kt/y of DMC produced by two european companies EniChem (8 kt/y) and SNPE (ca. 4 kt/y), and by UBE in Japan (ca. 5 kt/y). Each company has its own technology. Assuming an average price of 1 Euro/kg, the value of DMC production for the free market is about 16-17 MEuros. However, the importance of DMC is mainly connected to its use as an oxygenated solvent with low toxicity, in polycarbonate production, and, even more important, as a fuel additive.[3] An increase of at least one order of magnitude in its market size should be expected. On the contrary, a large use of DMC as a phosgene substitute in the synthesis of aromatic di-isocyanates would only have consequences on its captive uses.

BPA-PC is the carbonate that has the largest market (1.94 Mt in 2000). It is one of the top three engineering polymers, behind polyamide and acrylonitrile-butadiene-styrene. The market potential is far from having been fully exploited. World demand is forecasted to grow at 9 % annually through 2003.[7] The dominant producers are General Electric Plastics, Bayer, and Dow Chemicals, accounting for 84% of global capacity with 30% arising from Western Europe. The fastest growth is expected in Asia due to investment of the major players coupled with the increase in the number of asian producers. In less than 15 years, optical information storage market experienced one of the largest growth. Market expansion is foreseen for aliphatic polycarbonates due to their potential applications in drug delivery systems and as bioresorbable materials.[8]

4. INDUSTRIALIZED REACTIONS

Phosgene-based technology has a leading role in the production of a variety of organics including carbonates and polycarbonate plastics. The expanding carbonate market do not assure that this technology can continue to dominate. Nevertheless, the stronger demand by the fine chemicals industry increases phosgene production with improved processes driven by environmental issues.[9] In the field of carbonates, alternative routes have been proposed during the last 30 years. However, in the absence of environmental constraints, only a limited number have been commercialized due to unfavorable economics arising from limiting factors such as thermodynamics and kinetics.

4.1 The phosgene route

The high reactivity and versatility of phosgene have made its fortune up to date, giving high yields under mild conditions.[10] Schematically, by reacting phosgene with mono- and polyhydric alcohols chloroformates are formed in a first step, leading to carbonates upon further addition of the alcohols (Eqs. 1,2).

$$ROH + COCl_2 \rightarrow ROC(O)Cl + HCl \quad (1)$$

$$ROC(O)Cl + ROH \rightarrow (RO)_2CO + HCl \quad (2)$$

The two-stage process requires special equipment to avoid environmental risk and corrosion. Dry alcohols and anhydrous $COCl_2$ are required to obtained the carbonates. Moreover, spent halogenated solvents need post-treatment and HCl, the co-product, is neutralized, then disposed as it contains impurities unsuitable for recycling in other applications. The overall process, including phosgene synthesis, is highly energy demanding.

The production of dialkyl carbonates is mainly based on phosgene alcoholysis.[11] The technology was developped in France by SNPE in the early 1970's.[12] Unlike the majority of alcohols, phenols are relatively inert to phosgene. For example, the direct conversion of phenol into diphenyl carbonate is best achieved at high temperature (160-180°C) with catalytic amounts of onium salts or tertiary amines.

Most of the phosgene technology is associated with the production of di-isocyanates as starting materials of polyurethane chemistry. The next largest consumption is for BPA-PC plastics either directly or indirectly.[13] The direct method is the interfacial polycondensation between phosgene, dissolved in methylene chloride, and bisphenol A, dissolved in aqueous alkali solution to enhance its reactiviy. Bayer was the first company to industrialize this route, in the late 1950s in Germany. Besides safety concerns, there are difficulties in separating and removing chloride-containing residues leading to a lowering in quality of the polymer fatal for optical applications. The melt transesterification between DPC and bisphenol A is the indirect method offering advantages of a solvent-free and phosgene-free route. Conceptually, the overall process scheme is simple since the polymer is obtained in undiluted form and is directly pelletized.

4.2 Non-phosgene routes

Synthetic routes based on phosgenation are being replaced, in some instances, by innovative reaction pathways, *all of them being catalytic*

(Scheme 10.1). The corresponding most significant processes involve the production of:

- DMC via the oxidative carbonylation of methanol,
- DPC by the transesterification between DMC and phenol,
- BPA-PC through the transesterification between BPA and DPC,
- poly(alkylene carbonate)s via the carbonation of ethylene or propylene oxide; cyclic alkylene carbonates can also be obtained under appropriate conditions.

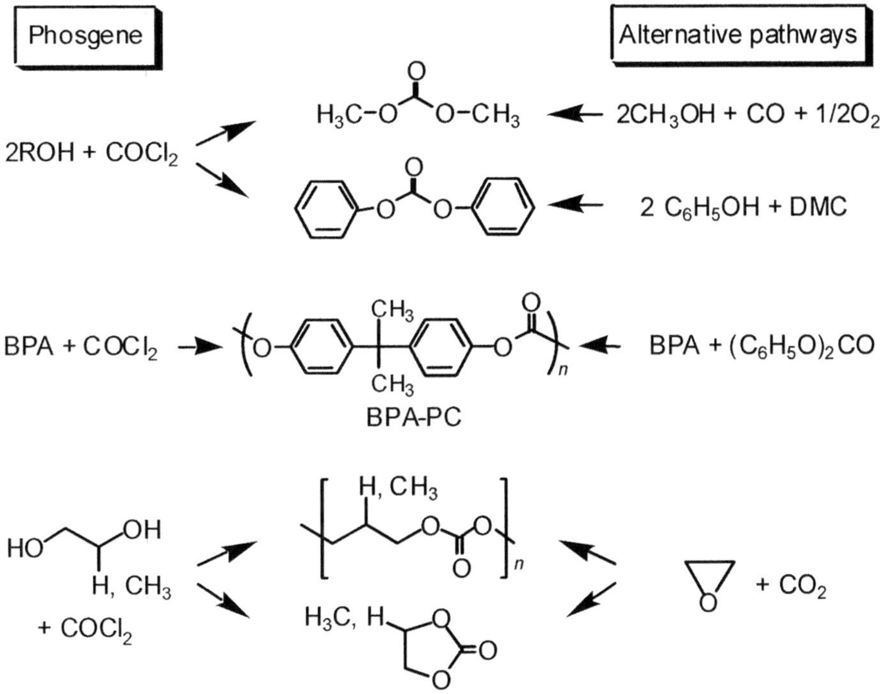

Scheme 10.1. Innovative reaction pathways

4.2.1 Dimethyl and diphenyl carbonates

The catalytic oxidative carbonylation of methanol is the innovative process for DMC production (Eq 3).

$$2CH_3OH + CO + 1/2O_2 \rightarrow (CH_3O)_2CO + H_2O \quad (3)$$

It was introduced on an industrial scale in the early 1980's by EniChem in Italy.[14] The liquid-phase reactor run under reasonable conditions (100-

130°C, 2-3 MPa) with copper chlorides as catalyst precursors. The selectivity towards DMC is higher than 95%, the main by-products being methyl chloride, dimethyl ether, and CO_2. The recovery of pure DMC involves azeotropic separation from methanol.[3] The productivity (135-250 $gL^{-1}h^{-1}$) is fairly constant, but conversion per pass is limited due, in part, to catalyst deactivation by water (Eq 3).[15] Additional constraints are (*i*) the presence of chloride, which requires corrosion-resistant reactor, and (*ii*) the nature of reactants, CO and O_2, which implies a good control of the kinetics. In principle, the use of solid catalysts should overcome the corrosion problems and improve product recovery. This approach was described by Dow Chemicals but no process has yet been established.[16]

A gas phase technology was successfully implemented by UBE in the early 1990's.[17] The process also runs under reasonable conditions (110-150°C, 0.1-2 MPa) with an heterogeneous $PdCl_2$-based catalyst supported on active carbon according to Eq 4.

$$2\ CH_3ONO + CO \rightarrow (CH_3O)_2CO + 2\ NO \qquad (4)$$

Methyl nitrite and carbon monoxide react to form DMC and nitric oxide. The selectivity for DMC lies in the range 90-95% based on CO and MeONO consumptions. Methyl nitrite is regenerated in a separate reactor by the uncatalyzed reaction between methanol and NO at ~50°C (Eq 5).

$$2\ CH_3OH + 2NO + 1/2O_2 \rightarrow 2\ CH_3ONO + H_2O \qquad (5)$$

The main by-products are dimethyl oxalate, methyl formate, and methylal. The gas phase process is based on two successive stoichiometric reactions; one is DMC synthesis and the other methyl nitrite synthesis. As the formation of DMC is not accompanied by water generation (Eq 4), catalyst stability and efficiency are higher than in the liquid-phase process; DMC recovery is also made easier. However, in both cases the catalysts contain chloride; leaching is observed for the heterogeneous system but corrosion seems less crucial than in the liquid-phase process. Nevertheless, there are some concerns about the toxicity and handling of methyl nitrite and NO.

DPC production according to EniChem technology involves to successive reactions.[3] At first, the transesterification reaction between DMC and phenol, catalyzed by titanium alkoxides, leads to methylphenyl carbonate (Eq 6). Then disproportionation of MPC forms DPC and DMC (Eq 7).

$$C_6H_5OH + (CH_3O)_2CO \rightarrow (CH_3O)(C_6H_5O)CO + 2\ CH_3OH \qquad (6)$$

$$2\,(CH_3O)(C_6H_5O)CO \rightarrow (C_6H_5O)_2CO + (CH_3O)_2CO \quad (7)$$

This option was chosen to overcome overcome unfavorable thermodynamics of the direct transesterification. Optimized reactions conditions and engineering design allows a productivity fitting with the economics.

4.2.2 Polycarbonates

For BPA-PC, the melt transesterification between bisphenol A and DPC is the method offering advantages of a solvent-free and phosgene-free route. In Europe, the technology was introduced by General Electric Plastics in the early 1990s in Spain; DPC is produced according to the EniChem technology, which relies upon the DMC/DPC sequence.[18] The reaction run in the melt state at temperatures ≤310°C, in the presence of basic catalysts such as lithium, sodium, potassium, or tetraalkylammonium hydroxides or carbonates. As transesterification is reversible, phenol should be distilled off continuously under vacuum to facilitate the forward chain growth reaction. Sophisticated equipment is needed to control temperature and reduced pressure sequences. The removal of volatiles and catalysts becomes difficult during the procedure due to melt polycarbonate viscosity.[19] Moreover, high temperature conditions induce undesirable chain branching that can be substantially reduced through catalyst optimization.[20] Molecular weights are lower than in the phosgene technology. Higher molecular weights have been obtained by a solid state polymerization below the melting point (210-220°C). The three-stage process includes prepolymerization of BPA with DPC, crystallization, and then solid state polymerization.[21]

The discovery of metal-catalyzed alternating copolymerization of ethylene oxide, or propylene oxide, and CO_2 that yields the corresponding poly(alkylene carbonate)s led to the first polymer production involving the direct use of carbon dioxide.[22] *It is one of a few successful examples of exploitation of carbon dioxide in the chemical industry.*[23] The polymer chains are composed predominantly of carbonate linkages with some polyether chains (Eq 8).

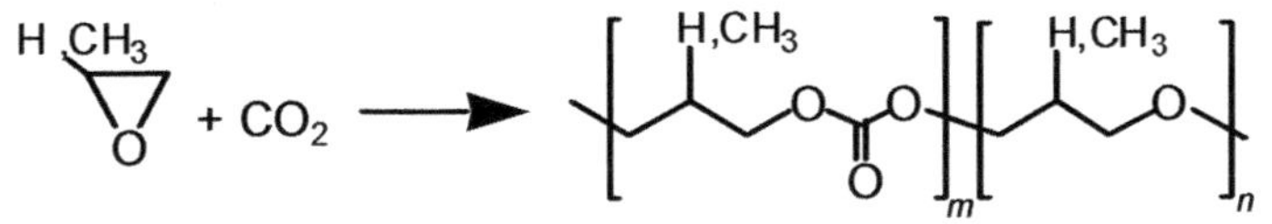

(8)

The working catalyst is most probably based on zinc derivatives with hydrogen donor cocatalyst (alcohol, water, or carboxylic acid).[24] Since the productivity is low, catalyst removal is a prerequisite for polymer commercialization. The production cost is therefore increased that is one reason for a restricted market.

5. INNOVATIVE TECHNOLOGIES

The on stream non-phosgene processes respond to the demand of safety and innovation in the chemical industry. However, safer technologies justify the assessment of chemical reactions based on carbon dioxide, especially carbonations. Carbon dioxide is non toxic, an advantage over phosgene and carbon monoxide, but less reactive.

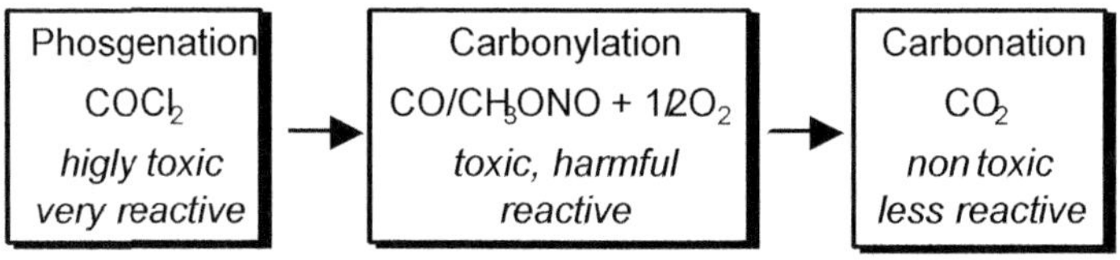

The development of new reaction methodologies for efficient and cost-effective synthesis of DMC, DPC, aromatic and aliphatic polycarbonates is a significant area of interest in both academic and industrial research groups. This paragraph outlines the most recent trends and features to addressing issues in CO_2 and non-CO_2 based technologies.

5.1. Oxidative carbonylation of alcohols

The catalytic oxidative carbonylation of methanol to DMC (Eq 3) is currently the state of the art commercial process for DMC production. The most recent published results point out that copper- and palladium-based catalysts are preferred, either under liquid or gas phase conditions. The reaction temperatures and pressures vary in the range 100-150°C and 0.1-8 MPa, respectively. Most of the described systems afford DMC selectivities higher than 80%, based on methanol. The addition of N-donor ligands to $CuCl_2$ increases the activity and selectivity for DMC, decreases its hydrolysis, and efficiently inhibits corrosion of the stainless steel batch reactor under liquid phase conditions.[25,26] Leaching of chloride ions preferentially occurs during the activation period of $CuCl_2$ supported on activated carbon. The formation of Cu(OH)Cl species and its relevance for DMC formation is underlined.[27] Not only catalyst modifications but also changes in conducting the reaction markedly improves DMC yield as shown by applying the pressure-swing technique for recycling and dewatering the reaction mixture containing a molten salt as catalyst.[28]

Although chloride is a source of corrosion, it has been recognized as an essential component for the copper-catalyzed systems to be efficient.[3] Its leaching from the reaction medium needs reajustement of its concentration through the addition of hydrogen chloride. This phenomenon occurs to a lesser extent with solid catalysts under gas phase conditions. Recent progress in catalyst design now shows that chloride is not required to catalyze the reaction at 130°C and atmopheric pressure with Cu(I)/Y-zeolite. Moreover a higher productivity and less deactivation are found compared to the known carbon-supported $CuCl_2$ precursor. It is suggested that the rate limiting step is CO insertion into the $(Cu\text{-}OCH_3)^+$ bond.[29] Chloride is also not needed with a Cu(I)/MCM-41 precursor. The system is totally selective in DMC, based on methanol, at 5% conversion and temperatures <130°C. The active center is described as $Si\text{-}O^-\text{-}Cu^+$.[30]

A bimetallic system, $PdCl_2\text{-}CuCl_2\text{-}CH_3COOK$ supported on activated carbon, exhibits a high productivity (240 mol-DMC/mol-Pd/h).[31] These types of palladium-based precursors are those which permit the direct synthesis of DPC from phenol according to Eq 9.

$$2C_6H_5OH + CO + 1/2O_2 \rightarrow (C_6H_5O)_2CO + H_2O \qquad (9)$$

The multicomponent composition of the catalyst system allow palladium (II), the active species for the stoechiometric formation of DPC, to be regenerated by oxidation of palladium (0) by organic or inorganic oxidants which are in turn reoxidized by dioxygen. Bases promote phenol activation, and among them ammonium halides may also act as surfactants.[32] Addition of dehydrating agents is optional to trapping water formed as a co-product. The major drawbacks of these systems are ligand oxidation and aggregation of palladium (0) which are the main causes for low turnover numbers (TON = DPC/Pd, molar ratio). Supported palladium is an option to limit palladium loss and facilitate its separation from the reaction medium.[33] Activated carbon is the preferred support due to its hydrophobic properties.[34] Side products from phenol, and CO_2 from CO, decrease DPC selectivity. Higher selectivities are better obtained at temperatures ≤130°C, but at the detriment of the reaction rate. For example, a selectivity of 99.8% based on GC analysis is reported with a DPC yield of 53% at 80°C under 0.8 MPa pressure. In this case, the catalyst has a TON value of 827 and is formed from a 6-component mixture: palladium acetate, manganese acetate, an heteropolyacid ($TBA_4SiWMo_{11}O_{40}$), TBABr (TBA = tetrabutylammonium), molecular sieves, and tetrahydrofuran. Molecular sieves act as a support for palladium and as a dehydrating agent.[35] Higher temperatures increase reaction rates up to 700 mol-DPC/mol-Pd/h at 135°C for a 4 h run, under 6 MPa pressure, when using a 5-component mixture composed of palladium acetate, cobalt acetylacetonate, benzoquinone, TBABr, and bidentate N-N ligands robust toward oxidation.[36]

The oxidative carbonylation of phenol is extended to bisphenol A for the direct synthesis of BPA-PC. The catalytic systems are as complex as in the phenol case. The nature of the terminal groups depends upon the catalyst precursor components.[37,38] Desirable high molecular weights are not yet obtained by this method.

5.2. Transesterification reaction

Transesterification is an equilibrium reaction and must be driven in the desired direction either by adding an excess of one reactant and/or by continuous removal of products. At the opposite of the oxidative carbonylation of alcohols, transesterification between alcohols and organic carbonates is catalyzed by a variety of acids and bases. Applied for the production of DPC,[3] this reaction is extended to the synthesis of DMC and PCs.

Transesterification of ethylene carbonate with methanol affords DMC, cogenerating ethylene glycol (Eq 10).

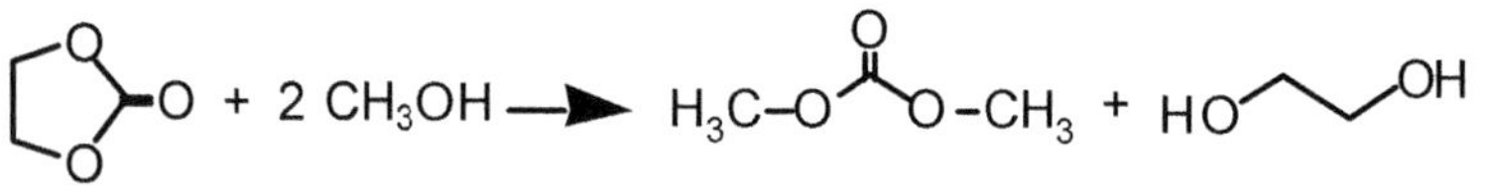

(10)

Both homogeneous and heterogeneous acid and base catalysts have been screened under moderate conditions (60-150°C). The base-catalyzed synthesis is faster and more selective (98% based on ethylene carbonate).[39] Inorganic solid base such as potassium-loaded titanosilicalite [40] hydrotalcite-type materials, [41] and cesium-loaded zeolites [42] are catalysts which can be easily separated from the reaction medium and recycled by calcination due their thermal stability. Correlations between activity/selectivity and basicity show that the mild basic oxide MgO gives the highest selectivity in DMC (100% based on ethylene or propylene carbonate) at 150°C, provided the reaction is conducted under 8 MPa of CO_2.[43] Turnover numbers are still low, and need to be improved for further development. The economic advantage of this method mainly relies on (*i*) reactants cost, (*ii*) adequation with the glycol demand [44] or its recycling to cyclic carbonate, (*iii*) enhancement of catalyst productivity, separation, recycling, and (*iv*) higher ceiling conversion by a drift in thermodynamic constraints.

Transesterification of urea with methanol to afford DMC is known since a long time to be catalyzed by tin complexes (Eq 11).[45]

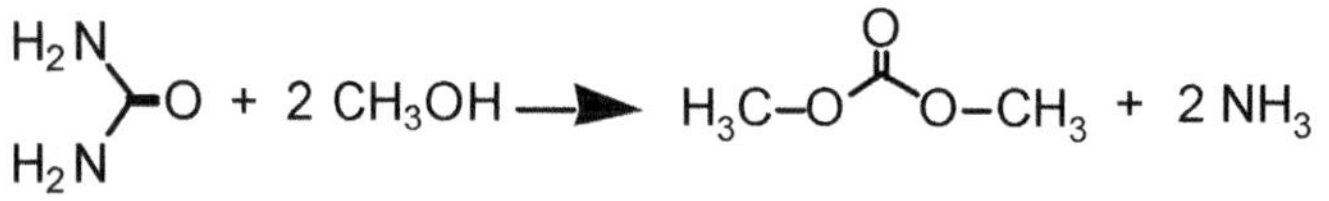

(11)

Yield and selectivity are drastically enhanced by running the reaction in a reactive distillation reactor with triethylene glycol dimethyl ether as high-boiling solvent. DMC is continuously eliminated from the boiler (180°C, ≈1-1.5 MPa) with a selectivity and conversion higher than 98%. This technology has the potential to lower significantly DMC cost production.[46] Furthermore, its integration with an urea facility could optimized ammonia recycling. The overall process would allow production of DMC from methanol and CO_2.

The melt transesterification method demonstrated for BPA-PC production is also feasible for the synthesis of poly(alkylene carbonate)s from a variety of aliphatic diols and DMC, with distannoxane catalyst at 220°C.[47] Another catalyst, $Ti(O^iPr)_4$, is active at 180-195°C for the formation of polycarbonate diols with molecular weight useful in polyurethane manufacture.[48] The utilization of DMC instead of DPC offers an economical advantage as DPC rely on DMC production with the EniChem process.[3] However, it is less reactive and escapes from the reaction medium while methanol is removed, by forming an azeotropic mixture.[49]

5.3. Carbonation reactions

The reaction of ethylene and propylene epoxides with CO_2 affords either the corresponding cyclic carbonate or poly(alkylene carbonate), according to the catalysts used (Scheme 6.1). Carbonation of these epoxides is one of the few successful industrialized process based on CO_2 chemical utilization. Many fundamental studies describe the behavior of new catalysts and approach the reaction mechanism. This domain is more mature than the other ones described in this review. Concerning copolymerization, numerous catalysts are reported mainly based on zinc and aluminium precursors.[50] They are active in the temperature range 20-100°C under moderated CO_2 pressure. The recent improvements in catalyst design increase the productivity by a factor of 100 and expand the range of epoxide monomers.[51] For cyclohexene oxide copolymerization, the most effective catalyst precursors are bulky β-diimine/Zn(II) complexes with TON values up to 494 based on converted epoxide in 2 h. Interestingly, the copolymer is monodispersed and predominates (95% carbonate linkage) with high molecular weight (M_n = 31000). An increase in CO_2 pressure to the supercritical region (Pc ≥7.4 MPa, Tc ≥ 31.1°C) also provides a high yield of copolymer, but the molecular weight is lower and molecular weight distribution broader.[52] Under such conditions, phase behavior and catalyst solubility strongly influence the kinetics and the copolymer characteristics.[53]

A much less documented reaction is the direct carbonation of alcohols (Eq 12).[54] The reaction occurs in the temperature range 140-190°C under CO_2 pressure.

$$2ROH + CO_2 \rightleftharpoons (RO)_2CO + H_2O \qquad (12)$$

The direct synthesis of DMC, according to Eq 12, is currently studied with catalysts based on either organometallic compounds or inorganic oxides. Dialkyldialkoxystannanes are selective towards DMC formation, but the major drawback is a very low activity; a maximum TON value of 3 is obtained, provided water traps are added.[55] Better yields are obtained after 50 h runs by applying a CO_2 pressure higher than 20 MPa.[56] Slow reaction is also found with a zirconia catalyst with an estimated TON value of 18.[57] Addition of phosphoric acid or ceria to zirconia enhances the activity.[58] To avoid the co-formation of water (Eq 12) and its removal during the reaction, trimethyl orthoesters or dimethyl acetals are used as dehydrated forms of methanol.[59] Although the economics of these two-step reaction is certainly not favorable, it demonstrates that TON can be increased up to a value of 340 with a DMC selectivity of 82% in the presence of ammonium or alkali metal iodide catalysts working at 180°C for 60 h.[60] Due to thermodynamic limitations, low reaction temperatures and non-conventional reaction conditions, like supercritical ones, should provide better yields.[61] However, the thermodynamic behavior of the reacting mixture under supercritical conditions needs to be investigated.[62]

6. PERSPECTIVES

The expanding market for open chain carbonates in a wide range of applications is unambiguously due to the commercialization of phosgene-free processes. These catalytic processes respond not only to the demand of greener chemistry but also to the demand of higher quality products, as exemplified by BPA-PC production. For DMC, the leading new technology is based on carbon monoxide with rises difficulties linked to corrosion, toxity, and potentially hazardous components. Transesterification reactions intervene for BPA-PC, DPC, and DMC syntheses with thermodynamic limitations that imply the design of reactors to drive the reaction in the proper direction. The diversity of catalysts makes this field very open to increase productivities.

One example of the few commercialized routes based on carbon dioxide is the carbonation of ethylene and propylene epoxides to poly(ethylene carbonate) and poly(propylene carbonate). The extension to other alkylene squeletons should widen the range of applications of these bio-compatible and degradable polymers. However, higher molecular weight polymers as well as enhanced catalyst activities should be obtained. DMC synthesis is still a challenge for large scale production. The CO_2-based technologies offer improvement of the selectivity but much more active catalysts have to be discovered. The constraints dictated by large-scale manufacture will imply the development of chemical engineering, particularly reactor design, to overcome the unfavourable thermodynamics and catalyst deactivation.

REFERENCES

1. Shaikh, A.-A.G., Sivaram, S. "Organic carbonates" Chem. Rev. 96: 951-976, 1996.
2. Tundo, P. "New developments in dimethyl carbonate chemistry" Pure Appl. Chem. 73: 1117-1124, 2001.
3. Rivetti, F. "The role of dimethyl carbonate in the replacement of hazardous chemicals" C R Acad. Sci. Paris Série II C Chimie 3: 497-503, 2000.
4. Pacheco, M.A., Marshall, C.L. "Review of dimethyl carbonate manufacture and its characteristics as a fuel additive" Energy Fuels, 11: 2-29, 1997.
5. (a) Thayer, A. "Polycarbonates: bright spot for a few producers" C&EN 76(51): 20, 1998; (b) Tullo, A.H. "New DVDs provide opportunities for polymers" C&EN 77(51): 14-15, 1999; "Polymer makers bridge the gap" C&EN 78(21): 21-24, 2000.
6. Sant'Angelo, J.G., Chen, X., McGinnis, V.D. (Empower Materials Inc.), "Functionalized poly(alkylene carbonate), and use thereof" US 6248860, 2001.
7. Tullo, A.H. "Fighting for position in polycarbonate" C&EN 79(15): 15-16, 2001.
8. (a) Inoue, S. "High polymers from carbon dioxide" CHEMTECH 6: 588-594, 1976; (b) Al-Azemi, T.F., Bisht, K.S. "Synthesis of novel bis- and tris-(cyclic carbonate)s and their use in preparation of polymer networks" Polymer 43: 2161-2167, 2002.
9. (a) Stinson, S.C. "Pharmaceutical fine chemicals" C&EN 78(28): 63-80, 2000; (b) Abrams, L., Cicha, W.V., Manzer, L.E., Subramoney, S. "A new catalyst for an old process driven by environmental issues" Stud. Surf. Sci. Catal. 130A: 455-460, 2000.
10. (a) Ryan, T.A., Ryan, C., Seddon E.A., Seddon K.R. "Phosgene and Related Carbonyl Halides" Amsterdam: Elsevier, 1996; (b) Senet, J.-P. "Phosgene chemistry and environment, recent advances clear the way to clean processes: a review" C R Acad Sci Paris Série II C Chimie 3: 505-516,2000.
11. Buysch, H.-J. "Carbonic Acid Esters." In Ullmann's Encyclopedia of Industrial Chemistry, Weinheim: VCH Publishers, A5: 197-201, 1992.
12. Société Nationale des Poudres et Explosifs. "Continuous manufacture of dialkyl carbonates". FR 2163884, 1973.
13. Volker, Serini. "Polycarbonates." In Ullmann's Encyclopedia of Industrial Chemistry, Weinheim: VCH Publishers, A21: 207-214, 1992.
14. Romano, U., Tesei, R., Massi Mauri, M., Rebora, P. "Synthesis of dimethyl carbonate from methanol, carbon monoxide, and oxygen catalyzed by copper compounds" Ind. Eng. Chem. Prod. Res. Dev. 19: 396-403, 1980.

15. Delledonne, D., Rivetti, F., Romano, U. "Developments in the production and application of dimethyl carbonate" Appl. Catal. A: General 221:241-251, 2001.
16. Curnutt, G.L. (Dow Chemicals) "Method for reactivating catalysts used in catalytic vapor phase process for producing dihydrocarbyl carbonates" US 5132259, 1992.
17. Uchiumi, S., Ataka, K., Matsuzaki, T. "Oxidative reactions by a palladium-alkyl nitrite system" J. Organomet. Chem. 576:279-289, 1999.
18. Brunelle, D.J. "Polycarbonates" In Kirk-Othmer Encyclopedia of Chemical Technology, New-York: Wiley-Interscience, 19: 584-608, 1996.
19. King, J.A., Jr "Synthesis of Polycarbonates." In Handbook of Polycarbonate Science and Technology, Donald G. LeGrand and John T. Bendler, eds. New-York: Marcel Dekker, 7-26, 2000.
20. Lemmon, J.P., McCloskey, P.J., Siclovan, O.P. (General Electric Comp) "Melt polycarbonate catalyst systems" US 6376640, 2002.
21. (a) Fukawa, I., Fukuoka, S., Komiya, K., Sasaki, Y. (Asahi Chemical Ind.) "Porous, crystallized, aromatic polycarbonate prepolymer, a porous, crystallized aromatic polycarbonate, and production methods" US 5214073, 1993; (b) Day, J., Varadarajan, G.S. (General Electric Comp.) "Method for conducting solid state polymerization of polycarbonates" US 6248858, 2001.
22. Inoue, S., Koinuma, H., Tsuruta, T. "Copolymerization of carbon dioxide and epoxide" J. Polym. Sci. Polym. Lett. B7:287-292, 1969.
23. (a) Dinjus, E., Fornika, R. "Carbon Dioxide as a C1-building Block" In Applied Homogeneous Catalysis with Organometallic Compounds, Boy Cornils, Wolfgang H., eds Weinheim: VCH Publishers, 2:1048-1072, 1996; (b) Aresta, M., Quaranta, E. "Carbon dioxide: a substitute for phosgene" CHEMTECH 27: 32-40, 1997.
24. Sun, H.-N. (Arco Chemical Comp.) "Soluble epoxide/carbon dioxide copolymerization catalysts" US 4783445, 1988.
25. Raab, V., Merz, M., Sundermeyer, J. "Ligand effects in the copper catalyzed aerobic oxidative carbonylation of methanol to dimethyl carbonate" J. Mol. Catal. A: Chemical 175:51-63, 2001.
26. Sato, Y., Yamamoto, T., Souma, Y. "Poly(pyridine-2,5-diyl)-$CuCl_2$ catalyst for synthesis of dimethyl carbonate by oxidative carbonylation of methanol: catalytic activity and corrosion influence" Catal. Lett. 65:123-126, 2000.
27. Tomishige, K., Sakaihori, T., Sakai, S., Fujimoto, K. "Dimethyl carbonate synthesis by oxidative carbonylation on activated carbon supported $CuCl_2$ catalysts: catalytic properties and structural change" Appl. Catal. A: General 181:95-102, 1999.
28. Kricsfalussy, Z., Waldmann, H., Traenckner, H.-J. "Oxycarbonylation of methanol to dimethyl carbonate using molten salts as catalyst" Ind. Eng. Chem. Res. 37:865-866, 1998.
29. King, S. T. "Mechanism of oxidative carbonylation of methanol to dimethyl carbonate in Cu-Y zeolite" J. Catal. 161:530-538, 1996.
30. Li, Z., Xie, K., Slade, R.C.T. "High selective catalyst CuCl/MCM-41 for oxidative carbonylation of methanol to dimethyl carbonate" Appl. Catal. A: General 205:85-92, 2001.
31. Yanji, W., Xinqiang, Z., Baoguo, Y., Bingchang, Z., Jinsheng, C. "Synthesis of dimethyl carbonate by gas-phase oxidative carbonylation of methanol on the supported solid

catalyst I. Catalyst preparation and catalytic properties" Appl. Catal. A: General 171:255-260, 1998.

32. (a) Hallgren, J.E., Lucas, G.M., Matthews, R.O. "The palladium-catalyzed synthesis of diphenyl carbonate from phenol, carbon monoxide, and oxygen" J. Organomet. Chem. 204:135-138, 1981; (b) Hallgren, J.E., Lucas, G.M. "The palladium-catalyzed synthesis of diphenyl carbonate from phenol, carbon monoxide, and oxygen II. aqueous sodium hydroxide as a base" J. Organomet. Chem. 212:135-139, 1981.
33. Ishii, H., Takeuchi, K., Asai, M., Ueda, M. "Oxidative carbonylation of phenol to diphenyl carbonate catalyzed by Pd-pyridyl complexes tethered on polymer support" Catal. Commun. 2: 145-150, 2001.
34. Song, H.Y., Park, E.D., Lee, J.S. "Oxidative carbonylation of phenol to diphenyl carbonate over supported palladium catalysts" J. Mol. Catal. A Chemical 154: 243-250, 2000.
35. Yin, G., Jia, C., Kitamura, T., Yamaji, T., Fujiwara, Y. "A new efficient Pd-catalyzed synthesis of diphenyl carbonate with heteropolyacid as a cocatalyst" J. Organomet. Chem. 630: 11-16, 2001.
36. Vavasori, A., Toniolo, L. "The promoting effect of chelating ligands in the oxidative carbonylation of phenol to diphenyl carbonate catalyzed by Pd-Co-benzoquinone system" J. Mol. Catal. A Chemical 151: 37-45, 2000.
37. (a) Goyal, M., Nagahata, R., Sugiyama, J., Asai, M., Ueda, M., Takeuchi, K. "Direct synthesis of aromatic polycarbonate from polymerization of bisphenol A with CO using a Pd-Cu catalyst system" Polymer 40: 3237-3241, 1999; (b) Goyal, M., Nagahata, R., Sugiyama, J., Asai, M., Ueda, M., Takeuchi, K. "Pd catalyzed polycarbonate synthesis from bisphenol A and CO: control of polymer chain-end structure" Polymer 41: 2289-2293, 2000.
38. Kim, W.B., Lee, J.S. "A one-step synthesis of reactive polycarbonate precursors by the coupled oxidative carbonylation of bisphenol-A and phenol with carbon monoxide" Chem. Lett. 1044-1045, 2001.
39. Knifton, J.F., Duranleau, R.G. "Ethylene glycol-dimethyl carbonate cogeneration" J. Mol. Catal. 67: 389-399, 1991.
40. Tatsumi, T., Watanabe, Y., Koyano, K. A. "Synthesis of dimethyl carbonate from ethylene carbonate and methanol using TS-1 as solid base catalyst" Chem. Commun. 2281-2282, 1996.
41. Watanabe, Y., Tatsumi, T. "Hydrotalcite-type materials as catalysts for the synthesis of dimethyl carbonate from ethylene carbonate and methanol" Microporous Mesoporous Mater 22: 399-407, 1998.
42. Chang, C.D., Jiang, Z., LaPierre, R.B., Schramm, S.E. (Exxon Mobil) "Process for co-production of dialkyl carbonate and alkanediol" US 6365767, 2002.
43. Bhanage, B.M., Fujita, S., Ikushima, Y., Arai, M. "Synthesis of dimethyl carbonate and glycols from carbon dioxide, epoxides, and methanol using heterogeneous basic metal oxide catalysts with high activity and selectivity" Appl. Catal. A: General 219: 259-266, 2001.
44. Kawabe, K. (Mitsubishi Chemical Corp) "Process for simultaneous production of ethylene glycol and carbonate ester" US 6380419, 2002.

45. Ball, P., Fullman, H., Schwalm, R., Heitz, W. "Synthesis of carbonates and polycarbonates by reaction of urea with hydroxy compounds" C1 Mol. Chem. 1: 95-108, 1984.
46. Ryu, J.Y. (Catalytic Distillation Technologies) "Process for making dialkyl carbonates" US 5902894, 1999.
47. Pokharkar, V., Sivaram, S. "Poly(alkylene carbonate)s by the carbonate interchange reaction of aliphatic diols with dimethyl carbonate: synthesis and characterization" Polymer 25:4851-4854, 1995.
48. Mizia, F., Rivetti, F. (EniChem) "Process for the preparation of polycarbonate diols with a high molecular weight" US 6384178, 2002.
49. Haba, O., Itakura, I., Ueda, M., Kuze, S. "Synthesis of polycarbonate from dimethyl carbonate and bisphenol-A through a non-phosgene process" J. Polym. Sci. Part A Polym. Chem. 37: 2087-2093, 1999.
50. (a) Kruper, Jr W.J., Swart, D.J. (Dow Chemical Comp.) "Carbon dioxide oxirane copolymers prepared using double metal cyanide complexes" US 4500704, 1985; (b) Aida, T., Inoue, S. "Metalloporphyrins as Initiators for Living and Immortal Polymerizations" Acc. Chem. Res. 29:39-48, 1996; (c) Darensbourg, D.J., Holtcamp, M. W. "Catalysts for the reaction of epoxides and carbon dioxide" Coord. Chem. Rev. 153:155-174, 1996.
51. Darensbourg, D.J., Holtcamp, M.W. "Catalytic activity of zinc(II) phenoxides which possess readily accessible coordination sites. Copolymerization and terpolymerization of epoxides and carbon dioxide" Macromol. 28:7577-7579, 1995.
52. Cheng, M., Lobkovsky, Coates G.W. "Catalytic reactions involving C1 feedstocks: new high-activity Zn(II)-based catalysts for the alternating copolymerization of carbon monoxide and epoxides" J. Am. Chem. Soc. 120: 11018-11019, 1998.
53. (a) Darensbourg, D.J., White Stafford, N. , Katsurao, T. "Supercritical carbon dioxide as solvent for the copolymerization of carbon dioxide and propylene oxide using a heterogeneous zinc carboxylate catalyst" J. Mol. Catal. A: Chemical 104:L1-L4, 1995; (b) Super, M., Berluche, E., Costello, C., Beckman, E. "Copolymerization of 1,2-epoxycyclohexane and carbon dioxide using carbon dioxide as both reactant and solvent" Macromol. 30:368-372, 1997; (c) Mang, S., Cooper, A.I., Colclough, M.E., Chauhan, N., Holmes, A.B. "Copolymerization of CO_2 and 1,2-cyclohexene oxide using a CO_2-soluble chromium porphyrin catalyst" Macromol. 33: 303-308, 2000.
54. (a) Genz, J., Heitz, W. (Bayer A.-G.) "Dialkyl and diaryl carbonates" DE 3203190, 1983; (b) Yamazaki, N., Nakahama, S. "Polymers derived from carbon dioxide and carbonates" Ind. Eng. Chem. Prod. Res. Dev. 18: 249-252, 1979; (c) Wagner, A., Löffler, W. (RWE) "Process for preparation of dialkyl carbonate" DE 4310109, 1994.
55. Kizlink, J., Pastucha, I. "Preparation of dimethyl carbonate from methanol and carbon dioxide in the presence of organotin compounds" Collect. Czech. Chem. Commun. 59: 2116-2118, 1994.
56. Ballivet-Tkatchenko, D., Douteau, O., Stutzmann, S. "Reactivity of carbon dioxide with n-butyl(phenoxy)-, (alkoxy-), and (oxo)stannanes: insight into dimethyl carbonate synthesis" Organometal. 19: 4563-4567, 2000.
57. Tomishige, K., Sakaihori, T., Ikeda ,Y., Fujimoto, K. "A novel method of direct synthesis of dimethyl carbonate from methanol and carbon dioxide catalyzed by zirconia" Catal. Lett. 58: 225-229, 1999.

58. (a) Ikeda, Y., Sakaihori, T., Tomishige, K., Fujimoto, K. "Promoting effect of phosphoric acid on zirconia catalysts in selective synthesis of dimethyl carbonate from methanol and carbon dioxide" Catal. Lett. 66: 59-62, 2000. (b) Tomishige, K., Furusawa, Y., Ikeda, Y., Asadullah, M., Fujimoto, K. "CeO_2-ZrO_2 solid solution catalyst for selective synthesis of dimethyl carbonate from methanol and carbon dioxide" Catal. Lett. 76: 71-74, 2001.
59. (a) Sakakura, T., Saito, Y., Okano, M., Choi, J.-C., Sako, T. "Selective conversion of carbon dioxide to dimethyl carbonate by molecular catalysis" J. Org. Chem. 63: 7095-7096, 1998; (b) Sakakura, T., Choi, J.-C., Saito, Y., Masuda, T., Sako, T., Oriyama, T. "Metal-catalyzed dimethyl carbonate synthesis from carbon dioxide and acetals" J. Org. Chem. 64:4506-4508, 1999; (c) Isaacs, N.S., O'Sullivan, B., Verhaelen, C. "High pressure routes to dimethyl carbonate from supercritical carbon dioxide" Tetrahedron 55: 11949-11956, 1999.
60. Cheong, M., Kim, S., Park, J. B. "Dimethyl carbonate synthesis via carbon dioxide activation in the presence of iodide catalysts" New J. Chem. 21: 1143-1145, 1997.
61. Zhao, T., Han, Y., Sun, Y. "Novel reaction route for dimethyl carbonate synthesis from CO_2 and methanol" Fuel Proc. Techn. 62: 187-194, 2000.
62. Camy, S., Pic, J.-S., Badens, E., Condoret, J.-S. "Fluid phase equilibria of the reacting mixture in the dimethyl carbonate synthesis from supercritical CO_2" J. Supercrit. Fluids, in press.

Chapter 11

ELECTROCHEMICAL CONVERSION OF CARBON DIOXIDE

Jan Augustynski[a], C. Jorand Sartoretti[a] and Piotr Kedzierzawski[b]
(a) Department of Chemistry, University of Geneva, 1211 Geneva 4, Switzerland
(b)Institute of Physical Chemistry of Polish Academy of Sciences, 01-224 Warsaw, Poland

1. INTRODUCTION

Electrochemical reduction of carbon dioxide had been explored for some time, this interest has largely been stimulated by analogies with the photosynthesis. More recently, the fact that carbon dioxide, being both available in nature and generated in huge amounts by human activities, constitutes an almost infinite source of carbon for the chemical industry, thus oriented numerous studies towards synthesis of fuels and intermediates. [1,2] In Table 11.1 are listed standard redox potentials corresponding formallèy to two, four, six and eight electron CO_2 reduction processes involving protons as the source of hydrogen. Also indicated are equilibrium potentials for these reactions in neutral solution, the medium most frequently used in actual electrolyses.

Table 11.1 - Standard and Equilibrium (at pH 7) Reduction Potentials of CO_2 in Aqueous Solutions (V versus SHE)

	E°	E_{eq} (pH 7)
$CO_2 + 2H^+ + 2e^- = HCOOH$	-0.199	-0.61
$CO_2 + 2H^+ + 2e^- = CO + H_2O$	-0.103	-0.52
$CO_2 + 4H^+ + 4e^- = CH_2O + H_2O$	-0.028	-0.44
$CO_2 + 6H^+ + 6e^- = CH_3OH + H_2O$	+0.031	-0.38
$CO_2 + 8H^+ + 8e^- = CH_4 + 2H_2O$	+0.169	-0.25
$2CO_2 + 2H^+ + 2e^- = H_2C_2O_4$	-0.49	-0.90

M. Aresta (ed.), Carbon Dioxide Recovery and Utilization, 279-292.

2. ELECTROLYSES OF AQUEOUS SOLUTIONS OF CO_2

2.1 Formation of formic acid

For a long time, formic acid (and formate ions) were considered as the only products of electrolytic reduction of carbon dioxide in aqueous solution (reported for the first time as early as in 1870). [3]

$$2\,CO_2 + H_2O + 2\,e^- \rightarrow HCOO^- + HCO_3^- \qquad (1)$$

The best faradaic efficiencies were obtained at metallic cathodes exhibiting high overvoltages for hydrogen evolution, the process competing with the reduction of CO_2. In this connection,

indium, tin, lead and mercury (as well as various amalgamated metals) appear as the most selective cathodes with the faradaic yields of formic acid exceeding in many cases 90%, confirmed also during electrolyses conducted under high pressure of CO_2. [4-8] However, the case of formic acid illustrates also the main drawbacks associated with any process of electrolytic reduction of CO_2, namely the low energy efficiency and the high dilution of the product(s). In fact, in order to attain any sizable current density at the cathode of the electrolysis cell (say on few mA/cm^2), overvoltages as large as 0.7-0.9 V are required. [5-6] However, under increased CO_2 pressure, the overvoltages in the range of 1 V allow much larger current densities to be reached, comparable with those at which electrolysis of water, to form hydrogen and oxygen, is industrially operated. A comparison with the latter process is certainly relevant here if one considers formic acid as a possible hydrogen storage agent (it is to be recalled that formic acid can be relatively easily decomposed into H_2 and CO_2 in the presence of a catalyst such as palladium). In fact, the production of formic acid by electrolysis of an aqueous solution of CO_2 would require at least a 50% larger expenditure of electrical energy than that associated with the electrolysis of water to form directly hydrogen (operated typically at 200 mA/cm^2 under a ca. 1.6 V bias). As already mentioned, a relatively low solubility of CO_2 in aqueous solutions (in the range of tens of mmol/L) restricts in practice the operation of the electrolysis cell to high CO_2 pressures. The corresponding increase of the CO_2 concentration (e.g., from 0.033 mol/L in water at 298 K under 1 atm to 1.17 mol/L under 60 atm) practically rules out any limitation connected with the mass transfer of the reactant to the electrode. The advantage in terms of productivity, associated with the high pressure electrolysis of CO_2, may be illustrated by the current densities exceeding 0.5 A/cm^2 attained at an indium cathode in 0.5 M $KHCO_3$ electrolyte under 60 atm of CO_2. [7]

Other means employed to increase the CO_2 concentration close to the electrode surface, namely alternative solvents exhibiting larger solubility of CO_2 and the use of gas diffusion electrodes will be discussed later.

Besides the above mentioned metallic cathodes, various transition metal complexes were investigated as potential electrocatalysts for the CO_2 reduction[9]. In most cases they are effective in non-aqueous-solvent-based electrolytes, leading to carbon monoxide as the main product. An important exception is provided by a class of $[Ru(L)(CO)_2]_n$ complexes (where L is a bidentate nitrogen ligand). These complexes, confined to the conducting electrode surface as polypyrrole-based polymeric films, yield formate ions as the principal CO_2 reduction product in aqueous electrolyte [10]. An interesting aspect of the latter study lies in the observed value of the onset potential associated with the formation of formate ions, only few tens of mV more negative than the equilibrium potential. Electrolyses conducted close to the onset potential, at low current densities, produced sodium formate with a faradaic yield of almost 100%. Sizable steady-state current densities of ca. 3 mA/cm^2 and 70% faradaic efficiency were attained under a 0.3 V overvoltage. Another significant feature of this work is the demonstrated ability to orientate the CO_2 reduction either towards the described formation of formate ions or that of carbon monoxide by the suitable choice of the ligand (L). Sodium formate was obtained selectively at the polymeric films bearing bpy substituted by ester groups as the ligand, suggesting a favourable formation of a formato bound structure with CO_2 .

A general problem, especially pertinent in the case of formic acid due to its large solubility in the reaction medium, regards the extraction of the final products from the solution. It is also to be mentioned, in this connection, that most of the published reports describe formation by the electrolysis of CO_2 of only highly diluted solutions of formic acid and that the feasibility of a continuous process leading to the product concentrations in the range of mol/L has not been conclusively demonstrated until now.

The perception of the electrolytic reduction of CO_2 markedly changed since the report in 1985 [5] of an unexpected activity of the copper electrode allowing, in particular, formation of hydrocarbons and alcohols, i.e., the $6e^-$ and $8e^-$ reduction products. Until then, copper had been considered as a low activity cathode for the CO_2 reduction and, in the absence of specific analyses, the gaseous reaction product was simply assigned to hydrogen. Systematic use of gas chromatography to monitor the CO_2 electrolyses products demonstrated that various metals exhibiting medium overvoltages for hydrogen evolution and, in particular, gold and silver, in addition to already mentioned copper, acted as efficient electrocatalysts for the CO_2 reduction.

At the same time, carbon monoxide, considered earlier as being formed exclusively in non-aqueous solutions, appeared as either the main product (at gold and silver cathodes) or the intermediate (on copper) for the CO_2

reduction in aqueous electrolyte. Such a particular behaviour of the group Ib metals may be associated with their low hydrogen coverage under cathodic polarization in aqueous solution (observed also in gas-phase experiments). The results of the electrolyses conducted using gold and silver cathodes also rise the question of the actual mechanism of CO_2 reduction. In fact, for these two metals, the onset potentials of CO formation are close to -1 V vs SHE, i.e., about 0.7 V more positive than the estimated standard reduction potential of the reaction. [11]

$$CO_{2(aq)} + e^- = CO_2^{-}{}_{(aq)} \qquad E° = -1.65 \text{ V vs SHE}$$

Generally considered as the first step in the electrochemical reduction of CO_2. Even if one takes into account a possible stabilization of the CO_2^- anion radical by adsorption at the electrode surface, the above mentioned potential difference still appears quite large. Moreover, formation of the CO_2^- radical has been suggested [12] to be followed by the reaction with water

$$CO_2^- + H_2O \rightarrow COOH_{(ads)} + OH^-$$

and the subsequent reduction to form carbon monoxide.

$$COOH_{(ads)} + e^- \rightarrow CO + OH^-$$

Such a reaction scheme appears in contradiction with the known $pK_a = -0.2$ corresponding to the $COOH/CO_2^-$ equilibrium [13], suggesting CO_2^- rather than COOH as being the dominant species in the 5-6 pH range in which carbon monoxide is formed with high faradaic efficiency.

An alternative mechanism has been proposed[14], involving two subsequent electron transfers to adsorbed hydrated CO_2 leading to the formation of CO via a dihydroxycarbene intermediate (Scheme 11.1).

The latter reaction mechanism may account for the observed behaviour of all the group Ib metals. In fact, while carbon monoxide is the final product of the CO_2 reduction at pure metallic gold and silver cathodes, it is formed only in the intermediate potential range on copper. At increasingly negative potentials, the latter cathode promotes formation of methane, ethylene and ethanol. Carbon monoxide appears as a possible intermediate of this process, being itself reduced to methane. [15]

Importantly, the electrochemical reduction of CO_2 at a copper cathode is normally accompanied by a severe poisoning of the electrode surface, causing a rapid decay of hydrocarbon production occurring within 0.5-1h of the electrolysis duration. [16] To counteract the poisoning process, an in situ activation method has been proposed. [16,17] This method, involving a rapid periodic anodic stripping of the nascent poisoning species from the electrode surface, enabled a high current efficiency of hydrocarbon formation (50-60%) to be maintained over prolonged (several-days-long) electrolysis runs. [18]

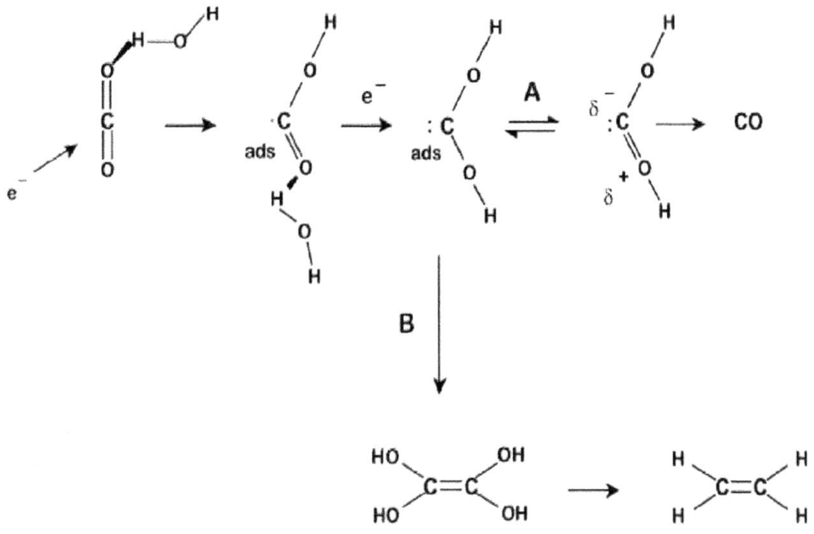

Scheme 11.1 - Electron transfer for the formation of CO

An additional effect of the periodic activation treatment is the increase of the effective surface area of the copper electrode, resulting in the current densities exceeding 100 mA/cm^2 reached at -1.7 V vs SHE in CO_2-saturated 0.5 M $KHCO_3$ solution. Although methane remained the main electrolysis product with 40-50% mean faradaic yields, the production of ethylene increased continuously to reach a yield of ca. 10% at the end of the run. Interestingly, modification of the copper surface with a porous layer of ZrO_2 or Cr_2O_3 directed the reduction of CO_2 towards the predominant formation of ethylene and ethanol (ca. 30% and 10% faradaic yields, respectively). [18]

It is to be recalled, in this connection, that copper is known as a component of Cu-ZnO/Al_2O_3 catalysts for the gas-phase synthesis of methanol. However, methanol is definitely absent among the products of the electrochemical CO_2 hydrogenation at the copper electrode.

To explain the formation of methane as the main product of the latter process, a general reaction scheme involving the species (such as surface formyl, $HCO_{(ads)}$, and carbene, $CH_{2(ads)}$), proposed earlier as intermediates in the Fischer-Tropsch synthesis, has been suggested (Scheme 11.2). [19] An alternative reaction route (indicated as route B in Scheme 11.1) may account for the preferential formation of ethylene and ethanol observed at the copper cathodes modified with oxides.

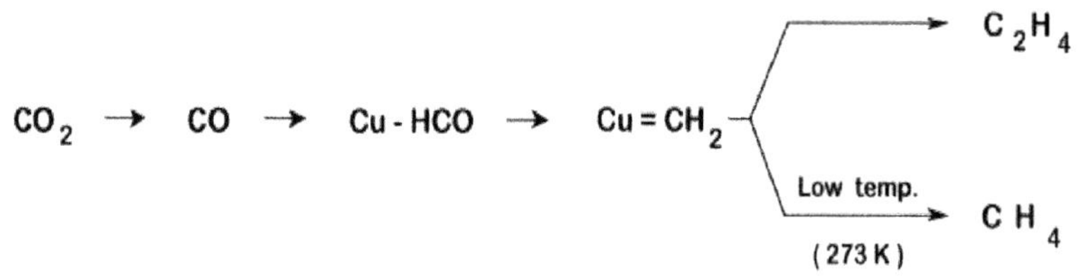

Scheme 11.2 - Formation of methane

The involvement of adsorbed CO and CH_x species as intermediates of the CO_2 reduction at the copper cathode has been confirmed by surface enhanced Raman scattering (SERS) measurements[20]. The observed time-dependent decay of the CO and CH_x SERS bands, reflecting the deactivation of the copper electrode, was accompanied by the increase of a new band attributed to the poisoning species and identified as a copper oxide patina.

2.2 Electrolyses under high CO_2 pressure

In an attempt to enhance the effective current densities at the copper cathodes, electrolyses were also conducted under increased CO_2 pressures. [21] High extent of CO_2 conversion was maintained up to very large current densities (700 mA/cm^2) under a pressure of 30 atm. Methane, the principal electrolysis product (obtained with maximum faradaic efficiency of 50%), was accompanied by smaller amounts of formate and ethylene.

The problem of deactivation of the copper electrode has not been addressed in that work, which report relatively short electrolysis runs. In fact, the poisoning of the copper surface is also observed under increased CO_2 pressure electrolysis conditions. However, a stable production of hydrocarbons during prolonged electrolysis runs could be maintained by applying the above mentioned in situ activation method to the copper electrode. [22]

2.3 CO_2 reduction using gas-diffusion electrodes

Besides employing increased CO_2 pressures, another possible means to overcome mass transfer limitations during the electrolysis process involves the use of gas-diffusion cathodes loaded with electrocatalytically active metals. The electrodes of this kind were developed for, and have been successfully employed for many years, in hydrogen-oxygen fuel cells confronted with a similar problem of low solubility of the gaseous reactants. A gas-diffusion electrode consists typically of a polytetrafluoroethylene-bonded porous carbon matrix impregnated with a suitable catalyst. The hydrophobic nature of such a structure favours formation of large three-phase (gas/electrolyte/electrode) contact areas inside the pores, enhancing the diffusion rate of the reactant towards the electrode. Application of electrodes of this kind, loaded with lead, indium or tin demonstrated their ability to produce formic acid at high rates (at current densities of ca. 100mA/cm^2) with faradaic yields ranging from 50% to almost 100%. [23] The best selectivity was obtained using lead-impregnated electrodes operated in acidic sulphate solutions (pH 2) at current densities exceeding 100mA/cm^2. However, the initial overvoltages (of the order of 1.3 V) tended to increase along the electrolysis run, indicating the build-up of the charge-transfer resistance. Even more spectacular results have been reported for the CO_2 reduction at copper-loaded gas-diffusion electrodes. [24] With faradaic yields close to 70%, ethylene was the main product of electrolyses conducted at very high current densities, 300-500 mA/cm^2, and largely negative cathodic potentials, -2.5 to -3.6 V (apparently uncorrected for ohmic drops). However, those experiments were performed using 1 M aqueous KOH as the electrolyte, known to exhibit a low surface tension but unstable in the presence of CO_2. The alkaline nature of the electrolyte might actually be the main reason for which, under such electrolysis conditions, ethylene becomes the dominant reduction product. In fact, similar electrolysis experiments, performed also using copper-loaded gas-diffusion electrodes in an almost neutral electrolyte (0.5 M $KHCO_3$) under increased CO_2 pressure (20 atm), yielded carbon monoxide and formate (both with faradaic efficiencies of ca. 30%) as the only CO_2 reduction products[25]. The latter electrolysis was conducted again at very high current density of 900 mA/cm^2 reached at a potential of only -1.05 V vs SHE (corrected for ohmic drop). Carbon monoxide was also produced at silver- and palladium-impregnated gas-diffusion electrodes with 86% and 58% faradaic yields, respectively. In all cases, only traces of hydrocarbons were detected among the electrolyses products. Unfortunately, the above data, obtained for particularly short electrolysis times (ranging from 3 to 8 minutes), do not allow any conclusions to be drawn with regard to the stability of the employed gas-diffusion electrodes.

With the aim of increasing the effective surface area of the electrocatalyst and improving the access of CO_2 to the active sites, high-surface area (ca. 1500 m^2/g) activated carbon fibers have been employed as the supporting material in gas-diffusion electrodes[26]. Interestingly, nickel-impregnated electrodes of this kind, employed in 0.5 M aq. $KHCO_3$ under 1 atm CO_2 pressure, exhibited the largest activity with ca. 65% faradaic efficiency for carbon monoxide formation attained at -1.35 V vs SHE and 80 mA/cm^2. It is to be recalled, in this connection, that the bulk nickel is practically inactive toward the CO_2 reduction in aqueous solutions due, apparently, to its low overvoltage for hydrogen evolution.

Although the carbon fiber support seems to promote electroreduction of CO_2 at the dispersed nickel, no such effect was observed for similar copper-impregnated electrodes at which only a limited CO_2 conversion into methane and carbon monoxide took place. A rapid poisoning of the copper surface may be an explanation for this differing behaviour.

3. ELECTROREDUCTION OF CO_2 IN NON-AQUEOUS SOLUTIONS

A number of studies have also been devoted to the electroreduction of CO_2 in non-aqueous electrolytes. Some organic solvents exhibit quite large CO_2 solubilities; this is, in particular, the case for dimethylformamide (DMF), propylene carbonate (PC) and methanol with, respectively, 20 times, 8 times and 5 times greater CO_2 solubilities than in water. On the other hand, much lower conductivities of the non-aqueous electrolytes imply larger ohmic drops and impose a limitation with regard to the current densities at which the electrolysis can be operated. The cathodic reduction of CO_2 in aprotic electrolytes leads to the formation of carbon monoxide and/or oxalate ions [27]:

$$2\,CO_2 + 2\,e^- \rightarrow CO + CO_3^{2-}$$

$$2\,CO_2 + 2\,e^- \rightarrow C_2O_4^{2-}$$

The synthesis of oxalate is the dominant reaction at lead, steel and mercury cathodes, in DMF, PC and acetonitrile-based electrolytes. There has been an interesting attempt to develop the latter process on a small pilot scale. [28] To avoid the unwanted oxidation of the electrolyte and also to reduce the cell voltage, a sacrificial zinc anode was employed. This not only enables the electrolysis to be operated in a single-compartment cell but, in addition, zinc cations generated at the anode form a precipitate with oxalate anions, allowing a relatively easy removal of the product by filtration :

$$Zn - 2e^{-} \rightarrow Zn^{2+} \xrightarrow{+\ C_2O_4^{--}} ZnC_2O_4 \downarrow$$

To minimize the ohmic losses, acetonitrile containing tetrabutylammonium perchlorate (forming a relatively conducting solution) has been chosen as the electrolyte. Faradaic yields of oxalate close to 90% were reached at stainless steel cathodes operated at current densities up to 80 mA/cm^2 during prolonged (up to 100h) electrolyses runs. Voltages not higher than 4 V could reasonably be expected for an optimized version of the electrolysis cell.

Importantly, the above work provides one of very few examples of an integrated electrochemical process involving utilization of CO_2 as a raw material. In this particular case, the use of a non-aqueous electrolyte is justified by the nature of the product and the overall efficiency of the process compensating for the increased cell voltage and operational difficulties.

Among various combinations of organic solvents and solutes employed as reaction media for the CO_2 reduction, the protic nature of methanol and its ability to form concentrated CO_2 solutions inspired numerous works. Electrolyses conducted at a copper cathode in benzalkonium chloride/methanol electrolyte at 258 K demonstrated, in fact, large extent of CO_2 conversion into methane, ethylene and carbon monoxide (with faradaic efficiencies of 39%, 4% and 24%, respectively). [29] The latter results were obtained at a cathodic potential equivalent to ca. -2 V vs SHE.

Solubility of CO_2 in methanol becomes very large under increased pressure, for example, at 40 atm and 25°C it reaches 8 mol/L. Electrolyses operated under such conditions with tetrabutylammonium tetrafluoroborate as a supporting electrolyte yielded carbon monoxide and methyl formate, stemmed from esterification of formic acid by the solvent, as the main products (with faradaic efficiencies of 46% and 34%, respectively). [30] The above yields refer to an extremely large current density of 500 mA/cm^2 at the copper cathode and to a very short electrolysis time (1 min). No information is given regarding longer-term activity of the copper electrode.

The latter example illustrates quite well the dilemma associated with the use of organic solvents in electrolysis processes. On the one hand, the increased reactant (i.e., CO_2) solubility enables one to operate at large current densities. Given low conductivity of the electrolyte, this leads, however, to prohibitively large ohmic losses (for the electrolysis conducted under above mentioned conditions, the contribution of the ohmic drop to the cell voltage amounted to 20 V). In addition, in order to avoid oxidation of the electrolyte and/or reaction products at the anode, an aqueous anolyte separated from the non-aqueous catholyte by a hypothetical membrane should be employed.

3.1 Electrochemical carboxylation

Reductive electrochemical activation enables incorporation of carbon dioxide into carbon-carbon double or triple bonds to form carboxylic acids. Such syntheses, requiring largely negative potentials, are frequently carried out in DMF using a sacrificial anode (in magnesium or zinc). The efficiency of electrocarboxylation reactions is enhanced in the presence of various catalysts, such as, for example, nickel (II) complexes (bipyridine, cyclam, phosphine) [31,32] acting as charge transfer mediators. The electrochemical method, involving operation under relatively mild conditions, opens an attractive route to the synthesis of a large number of industrially important intermediates.

4. CONCLUSIONS

The central question arising from the analysis of a large amount of the research work devoted to the electrochemical carbon dioxide reduction is about the possible future role of such a technique as a means for converting CO_2 into value added chemicals.
First, the choice of the electrochemical method can find a justification only when the electrical energy used in the process originates from a source other than fossil-fuels-fed power-stations. However, even if a cheap and abundant electrical energy, provided by nuclear or solar power stations, were available, the direct electrochemical CO_2 reduction process would face the comparison with an indirect route involving hydrogen production by water electrolysis to feed a conventional heterogeneous catalytic process.
It is to be recalled, in this connection, that the electrolysis of water is not only a well established industrial process operating at high current densities (> 200 mA/cm^2), but it is also less energetically demanding than the electroreduction of CO_2. Both processes having, in principle, a common anodic reaction (i.e., the oxygen evolution), the difference arises due to much larger overvoltages involved in the cathodic reduction of CO_2. In fact, as shown in Table 1, the standard redox potentials for the CO_2 reduction to various C_1 products are quite close to the standard (normal) hydrogen electrode potential. In addition, the gas-phase heterogeneous catalytic reactors use proven catalysts, such as e.g., Cu/ZnO, disposed in three dimensional structures (packed beds) which, from the point of view of the investment cost, offer an advantage in comparison with essentially two-dimensional electrodes.

However, unexpectedly enough, the electrochemical CO_2 reduction may reach, until now on the laboratory scale, much higher specific rates (expressed with respect to the electrode/catalyst area) than an equivalent conventional catalytic process. This confirms the fundamental superiority of the electric potential over high temperature and pressure conditions in enhancing the rates of reduction/oxidation reactions.

Taking into account all the above mentioned drawbacks of the direct electrochemical CO_2 reduction method, i.e., the fact of being an as yet unproven and energy demanding technology, involving rather high investment costs, one can try to identify prospects for its future applications. Storage of the electrical energy produced by solar or nuclear power stations appears to be the most promising option, provided the electroreduction of CO_2 will result in a selective formation of a liquid C_1 (or C_2) product, such as formic acid, methanol or ethanol. The latter compounds are, in fact, much easier to store than hydrogen formed through the electrolysis of water. In this connection, the two-electron CO_2 reduction product, formic acid, appears particularly attractive, due to its easy catalytic decomposition into hydrogen and CO_2. Formic acid may be formed selectively with high faradaic efficiency, exceeding 90%, when using lead, tin or indium as a cathode material. Reasonably high CO_2 reduction rates (above 100 mA/cm^2) can be attained using gas-diffusion electrodes, increased CO_2 pressures or the combination of both. However, the feasibility of obtaining relatively concentrated (in the range of mol/L) solutions of formic acid by electrolysis still remains to be proven.

Unlike in the case of formic acid, no efficient electrocatalyst allowing the CO_2 reduction to methanol has been found so far. On the other hand, ethanol is formed with faradaic yields of ca. 20%, among other products, in the course of the CO_2 electroreduction at a copper cathode. Apparently, the latter electrolysis process could be oriented towards selective formation of ethylene - an important feedstock product for the chemical industry. The latter option certainly warrants further study.

The use of non-aqueous electrolytes for the CO_2 reduction remains limited by their large resistivity. Obviously, this limitation becomes particularly severe when electrolyses are conducted at large current densities - made possible by the increased solubility of CO_2 in some organic solvents. An additional problem concerns the actual nature of the anodic reaction in organic electrolytes. This problem is avoided in aqueous electrolytes in which the proton-consuming CO_2 reduction at the cathode is actually combined with the oxygen evolution at the anode. The use of an aqueous electrolyte (anolyte) in the anodic compartment of the electrolysis cell associated with an organic catholyte requires inter-position of a sufficiently selective membrane and involves necessarily additional ohmic losses, i.e., a further decrease of energetic efficiency of the process. Accordingly, the choice of an organic electrolyte can only be justified by the specific nature of a high value product of the CO_2 reduction that cannot be synthesized in aqueous solution. The above described selective electrosynthesis of oxalic acid, performed in an acetonitrile-based electrolyte with a sacrificial zinc anode, is an example of such a rational use of a non-aqueous solvent. However, they are the electrocarboxylation reactions which constitute the most promising field of application for the non-aqueous CO_2 electrochemistry.

REFERENCES

1. Jitaru, M., Lowy, D.A., Toma, M., Toma, B.C., Oniciu, L. "Electrochemical reduction of carbon dioxide on flat metallic cathodes" J. Appl. Electrochem. 27, 875-89, 1997.
2. Arakawa, H., Aresta, M., Armor, J.N., Barteau, M.A., Beckman, E.J., Bell, A.T., Bercaw, J.E., Creutz, C., Dinjus, E., Dixon, D.A., Domen, K., DuBois, D. L., Eckert, J., Fujita, E., Gibson, D.H., Goddard, W.A., Goodman, D.W., Keller, J., Kubas, G.J., Kung, H.H., Lyons, J.E., Manzer, L.E., Marks, T.J., Morokuma, K., Nicholas, K.M., Periana, R., Que, L., Rostrup-Nielson, J., Sachtler, W.M.H., Schmidt, L.D., Sen, A., Somorjai, G.A., Stair, P.C., Stults, B.R., Tumas, W. "Catalysis Research of Relevance to Carbon Management: Progress, Challenges, and Opportunities" Chem. Rev. (Washington, D. C.) 101, 953-96, 2001.
3. Royer, M. E. "Réduction de l'acide carbonique en acide formique" Compt. Rend. 731-32, 1870.
4. Kapusta, S., Hackerman, N. "The electroreduction of carbon dioxide and formic acid on tin and indium electrodes" J. Electrochem. Soc. 130, 607-13, 1983.
5. Hori, Y., Kikuchi, K., Suzuki, S. "Production of carbon monoxide and methane in electrochemical reduction of carbon dioxide at metal electrodes in aqueous hydrogen carbonate solution" Chem. Lett., 1695-98, 1985.
6. Noda, H., Ikeda, S., Oda, Y., Imai, K., Maeda, M., Ito, K. "Electrochemical reduction of carbon dioxide at various metal electrodes in aqueous potassium hydrogen carbonate solution" Bull. Chem. Soc. Jpn. 63, 2459-62, 1990.
7. Todoroki, M., Hara, K., Kudo, A., Sakata, T. "Electrochemical reduction of high pressure CO_2 at Pb, Hg and In electrodes in an aqueous $KHCO_3$ solution" J. Electroanal. Chem. 394, 199-203, 1995.
8. Mizuno, T., Ohta, K., Sasaki, A., Akai, T., Hirano, M., Kawabe, A. "Effect of temperature on electrochemical reduction of high-pressure CO_2 with In, Sn, and Pb electrodes" Energy Sources 17, 503-08, 1995.
9. Leitner, W. "Carbon dioxide as a raw material: the synthesis of formic acid and its derivatives from CO_2" Angew. Chem., Int. Ed. Engl. 34, 2207-21, 1995.
10. Chardon-Noblat, S., Deronzier, A., Ziessel, R., Zsoldos, D. "Electroreduction of CO_2 catalyzed by polymeric [Ru(bpy)$(CO)_2$]n films in aqueous media: parameters influencing the reaction selectivity" J. Electroanal. Chem. 444, 253-60, 1998.
11. Frese, K.W.J. "Electrochemical reduction of CO_2 at solid electrodes" in: Electrochemical and Electrocatalytic Reactions of CO_2 (K. Christ, E. Guard, B.P. Sullivan Eds.), Elsevier: Amsterdam, p. 145-216, 1993.
12. Noda, H., Ikeda, S., Yamamoto, A., Einaga, H., Ito, K. "Kinetics of electrochemical reduction of carbon dioxide on a gold electrode in phosphate buffer solutions" Bull. Chem. Soc. Jpn. 68, 1889-95, 1995.

13. Jeevarajan, A.S., Carmichael, I., Fessenden, R.W. "ESR measurement of the pKa of carboxyl radical and ab initio calculation of the carbon-13 hyperfine constant" J. Phys. Chem. 94, 1372-76, 1990.
14. Kostecki, R., Augustynski, J. "Electrochemical reduction of CO_2 at an activated silver electrode" Ber. Bunsen-Ges. Phys. Chem. 98, 1510-15, 1994.
15. Hori, Y., Murata, A., Takahashi, R., Suzuki, S. "Electroreduction of carbon monoxide to methane and ethylene at a copper electrode in aqueous solutions at ambient temperature and pressure" J. Am. Chem. Soc. 109, 5022-23, 1987.
16. Jermann B., Augustynski, J. "Long-term activation of the copper cathode in the course of CO_2 reduction" Electrochim. Acta 39, 1891-96, 1994.
17. Augustynski, J., Carroy, A., Feiner, A.S., Jermann, B., Link, J., Kedzierzawski, P. Kostecki, R. "Enhanced electrochemical CO_2 reduction at copper, silver and gold electrodes" Proc. of the Intern. Conf. on Carbon Dioxide Utilisation, Bari, 26-30 Sept 1993, p. 331-38.
18. Augustynski, J., Kedzierzawski, P., Jermann, B. "Electrochemical reduction of CO_2 at metallic electrodes" in: Advances in Chemical Conversions for Mitigating Carbon Dioxide, (T. Inui, M. Anpo, K. Izui, S. Yanagida, T. Yamaguchi Eds.), Elsevier Science B.V.: Amsterdam 1998, p. 107-16.
19. DeWulf, D.W. Jin, T., Bard, A. J. "Electrochemical and surface studies of carbon dioxide reduction to methane and ethylene at copper electrodes in aqueous solutions" J. Electrochem. Soc. 136, 1686-91, 1989.
20. Smith, B.D., Irish, D.E., Kedzierzawski, P., Augustynski, J. "A surface enhanced Raman scattering study of the intermediate and poisoning species formed during the electrochemical reduction of CO_2 on copper" J. Electrochem. Soc. 144, 4288-96, 1997.
21. Hara, K., Tsuneto, A., Kudo, A., Sakata, T. "Electrochemical reduction of CO_2 on a Cu electrode under high pressure. Factors that determine the product selectivity" J. Electrochem. Soc. 141, 2097-103, 1994.
22. Kedzierzawski, P., Augustynski, J., Silvestri, G. "Electroreduction of carbon dioxide on copper at elevated pressures", presented at the Joint Meeting of the Electrochem. Soc. / ISE, Paris, Aug. 31-Sept. 5 1997, p. 726.(Ext. Abstr. No 613)
23. Mahmood, M.N., Masheder, D., Harty, C.J. "Use of gas-diffusion electrodes for high-rate electrochemical reduction of carbon dioxide I. Reduction at lead, indium- and tin-impregnated electrodes" J. Appl. Electrochem. 17, 1159-70, 1987.
24. Cook, R.L., MacDuff, R.C., Sammells, A.F. "High rate gas phase carbon dioxide reduction to ethylene and methane using gas diffusion electrodes" J. Electrochem. Soc. 137, 607-08, 1990.
25. Hara, K., Sakata, T. "Large current density CO_2 reduction under high pressure using gas diffusion electrodes" Bull. Chem. Soc. Jpn. 70, 571-76, 1997.
26. Tryk, D.A., Yamamoto, T., Kokubun, M., Hirota, K., Hashimoto, K., Okawa, M., Fujishima, A. "Recent developments in electrochemical and photoelectrochemical CO_2 reduction: involvement of the $(CO_2)_2$.cntdot.- dimer radical anion" Appl. Organomet. Chem. 15, 113-20, 2001.
27. Gressin, J.C., Michelet, D., Nadjo, L., Saveant, J.M. "Electrochemical reduction of carbon dioxide in weakly protic medium" Nouv. J. Chim. 3, 545-54, 1979.
28. Fischer, J., Lehmann, T., Heitz, E. "The production of oxalic acid from carbon dioxide and water" J. Appl. Electrochem. 11, 743-50, 1981.
29. Naitoh, A., Ohta, K., Mizuno, T., Yoshida, H., Sakai, M., Noda, H. "Electrochemical reduction of carbon dioxide in methanol at low temperature" Electrochim. Acta 38, 2177-79, 1993.

30. Saeki, T., Hashimoto, K., Fujishima, A., Kimura, N., Omata, K. "Electrochemical Reduction of CO_2 with High Current Density in a CO_2-Methanol Medium" J. Phys. Chem. 99, 8440-46, 1995.
31. Bringmann, J., Dinjus, E. "Electrochemical synthesis of carboxylic acids from alkenes using various nickel-organic mediators: CO_2 as C_1-synthon" Appl. Organomet. Chem. 15, 135-40, 2001.
32. Tascedda, P., Weidmann, M., Dinjus, E., Dunach, E. "Nickel-catalyzed electrochemical carboxylation of epoxides: mechanistic aspects" Appl. Organomet. Chem. 15, 141-44, 2001.

Chapter 12

CARBON DIOXIDE REDUCTION TO C_1 OR C_n MOLECULES

Michele Aresta
METEA Research Centre and Department of Chemistry, University of Bari, Campus Universitario, 70126, Bari – Italy
aresta@metea.uniba.it

1. INTRODUCTION

In Chapter 9 the chemical utilization of carbon dioxide as a whole molecule has been considered. Alternatively, carbon dioxide can be reduced to other C_1 or C_n molecules richer in energy that could be used as energy vectors. Such conversion is endoergonic, as can be seen in Table 12.1, and has high kinetic barriers. Therefore, it requires ad hoc engineerized catalysts and energy, that may be supplied in various forms, e.g. radiations, heat, electrons, dihydrogen, or hydrocarbons.

Table 12.1 - Free energy of formation of various C_1 molecules

Species	$\Delta G°f$ (kJ/mole)
CH_4(g)	-51
HCHO(l)	-102
CO(g)	-137
CH_3OH(l)	-166
HCOOH(l)	-361
CO_2(l)	-386
CO_2(g)	-394

Whatever would be the energy source, if the reduction of carbon dioxide is intended for recycling carbon, the chemical energy stored in the final product should be higher than the energy supplied for the conversion of CO_2 to occur. If this were not the case, the conversion of carbon dioxide would

M. Aresta (ed.), Carbon Dioxide Recovery and Utilization, 293-312.

generate more carbon dioxide than is converted, making the process a non-sense. Therefore, the right combination of reactions must be selected that brings to a favorable thermodynamics of the reduction process. Also, finding catalysts, that may drive the reduction reaction with low kinetic barriers, is a major need. The moment being, considered that the reaction conditions for carbon dioxide conversion into energy richer products are still quite hard, the use of carbon dioxide as source of carbon for energy products does not represent a short term option, but it may be better considered as a medium term option.

In the following paragraphs a number of reactions will be discussed that are today considered the most promising approaches to carbon recycling, namely:

- Carbon dioxide reduction via methane dry reforming
- Tri-reforming process
- Methanol from carbon dioxide

The plasma technique will be also considered besides thermal processes. As the carbon dioxide dissociation to CO and oxygen is a step that may play a key role in several reduction processes, it will be considered first.

1.1 Carbon dioxide dissociation to CO mediated by metal centers

Carbon dioxide can interact with a metal center, either metal surfaces or metal atoms, with formation of a variable number of bonds that imply either only one atom (η^1-C or O-mode), or two atoms (η^2-C,O or –O,O mode) or else all atoms of the trinuclear molecule. The modes of bonding are shown in Scheme 12.1. They have been determined by XRD for a number of metal complexes, or by spectroscopic techniques at low temperature for metal atoms or surfaces. It is a general feature that the η^1-C or η^2-C,O or η-μ modes cause an elongation of the C-O bond and, eventually, its cleavage. Such event is promoted by a back-donation of electrons from the metal center to the antibonding LUMO of carbon dioxide that allows the dissociation to occur under milder conditions than observed for the gas phase dissociation. In fact, the latter can be performed by irradiating with high-energy UV radiations having $\lambda < 160$ nm [1]. Also, carbon dioxide can be vibrationally excited to reach a dissociative state. [1] As the co-ordination of CO_2 to a Ni-centre resulted to favor the oxygen transfer to an oxophile [2], the interaction of carbon dioxide with either metal units [3], surfaces [4] over a wide range of temperature [5] or with metal atoms at low temperature has been intensively investigated. Various modes of bonding (Scheme 12.1) of CO_2 have been ascertained [6], which imply a different perturbation of the molecule.

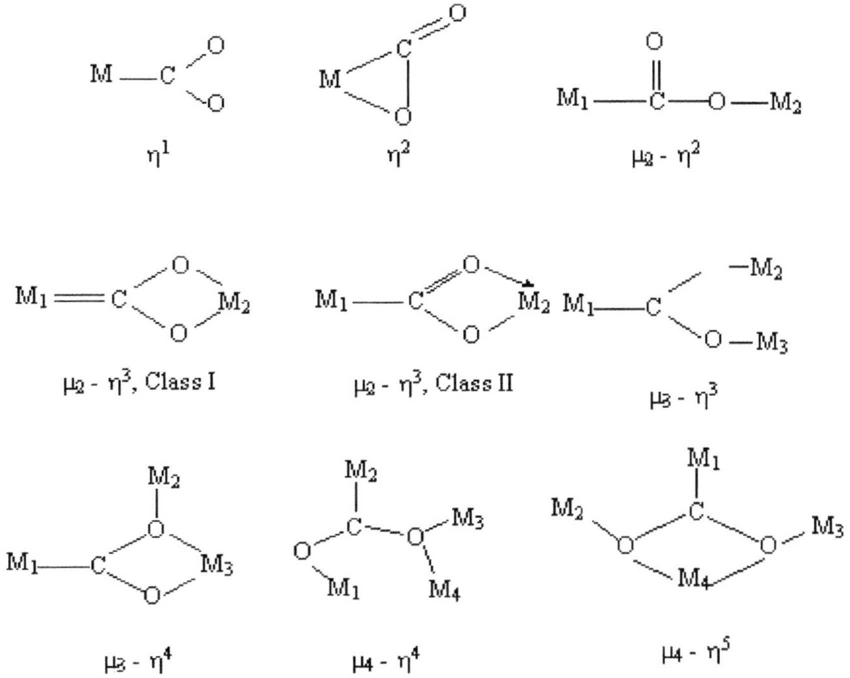

Scheme 12.1 - Modes of bonding of carbon dioxide

1.1.1 Dissociation on metals surfaces

Carbon dioxide has been shown to interact with several transition-metal-clean-surfaces under various temperature and pressure conditions and its molecular perturbation may be markedly crystal face-specific. The interaction can result to be either non-dissociative or dissociative to afford carbon monoxide.

Polycrystalline Rh-foil or Rh(111) single crystal surfaces interact with carbon dioxide under UHV conditions [7] with a dissociation probability ranging from 10^{-5} at 750 K to 10^{-8} at 470 K at 1 torr. The heterocumulene is supposed to bind via one of the oxygen lone pairs. The calculated heat of adsorption is 5.4 kcal/mol, a value that is common to most transition metals with a variation of only a few tenths of a kcal/mol. Surface techniques (TDS and EELS) confirmed the non-dissociative adsorbtion as they could not detect CO, which, if formed, should be strongly bonded to Rh and, thus, easily detectable by both methods. Calculations [8], as well as experimental data [9] show that the rate of dissociation of CO_2 becomes detectable only at surface temperatures above 300 K and at pressures of the order of one atmosphere. However, the adsorption can be dissociative in presence of traces of boron, that is a frequent impurity for Rh. Also hydrogen pre-adsorption or co-adsorption assists CO_2 dissociation. Additionally, a prominent difference between the behaviour of smoothed and stepped surfaces has been found [10].

Pd single crystal faces adsorb carbon dioxide in a non-dissociative mode [11]. On a Pt-film CO_2 is only physisorbed, with a heat of adsorption equal 10 kcal/mol [12]. Silver has been carefully investigated, due to its industrial application for ethene epoxidation and shown to adsorb very weekly CO_2 on a clean surface. CO_2 was not adsorbed on clean Cu (110), Cu (111) or Cu(100) up to 400 K and several kPa of pressure [13,] while dissociation to carbon monoxide and oxygen was observed on stepped surfaces, as well as on polycrystalline Cu under UHV conditions and at temperatures below 150 K.

As reported above, the C-O bond elongation occurs when there is a charge transfer from the metal system to CO_2. Therefore, the work function of the metal surface may play an important role in CO_2 dissociation. It seems, indeed, that a difference of 0.55 eV, typical of smooth and stepped surfaces, can be enough to promote the dissociation. This has been reported above to occur with copper, and is also true for iron: while the smooth Fe(110) surface is inactive towards the adsorption of CO_2, the Fe(111) surface and vicinal faces of the Fe(110), which have a smaller work function than the Fe(110) surface, are more reactive [14]. The interaction of CO_2 with Al-foil is pressure dependent and results in the formation of metastable adsorbates at about 100 K that dissociate readily at higher temperatures to give carbon in the form of carbide $C_{(a)}^{\delta-}$ and graphite $C_{(a)}^{0}$ [15]. CO_2 undergoes a dissociative adsorption on Ni even at 150 K: an electron transfer from the metal generates the bent CO_2^- species with C_{2V} symmetry that is co-ordinated to the metal through the oxygen atoms. This pre-dissociative state has been confirmed by diffuse LEED structure analysis [16]. Clean Mg (001) exhibits a high reactivity towards CO_2, that dissociates even at 100 K with formation of MgO. The latter interacts with carbon dioxide and produces a carbonate, that remains confined to the topmost layer, with the oxide forming a thick, uniform layer beneath [17].

It is a general feature that the presence of oxygen adatoms on the metal surface facilitates the interaction with carbon dioxide that is converted into a carbonate moiety that can be desorbed without decomposition, as in the case of silver [17]. An ad hoc study using carbon dioxide and ^{18}O has shown that the three oxygen atoms of the carbonate moiety are chemically equivalent, pointing at a fluxional system that may equalize the three oxygen atoms through the interconversion of mono- and bi-dentate carbonate forms.

Group 1 and 2 metals are known to react with carbon dioxide [18] to afford the radical anion $CO_2^{\cdot-}$ that can either interact with another carbon dioxide molecule or dimerize to oxalate. (Scheme 9.7) If added to a transition metal surface, Group 1 and 2 metals can significantly modify the surface adsorptive properties, by increasing the carbon dioxide binding coefficient (at a moderate coverage rate), or generate new adsorption sites, where carbon dioxide may strongly bind. Moreover, Group 1 and 2 metals promote the dissociative reactivity as they may increase the negative charge donated from the surface to carbon dioxide, favoring the η^2-C,O co-ordination mode that is a good pre-dissociation state to afford CO.

Metal atoms have been reported to react with carbon dioxide at low temperature (14 K) in a gas matrix to afford adducts in which carbon dioxide may have a different co-ordination mode [19]. Rising the temperature to 40 K may often cause dissociation of CO_2.

1.1.2 CO_2 adsorption on oxides

The physi- and chemi-sorption of CO_2 in its distorted configuration with a OCO angle close to 130° has been modeled for regular and defect sites of MgO(100) using the ab initio cluster calculations [20]. The chemisorption of the CO_3^{2-} moiety, takes place at defect sites (step sites), through a charge flow from the HOMO ($2p_z$) of O^{2-} to the LUMO ($2\pi^*_u$) of CO_2. This trend is common to most oxides capable to form carbonates (TiO_2 and Al_2O_3 among others). Symmetric, monodentate, bidentate and bridged carbonates can be observed. Depending on the nature of the surface [21], various modes of co-ordination of the carbonate can be observed: monodentate or bidentate, either chelating or bridging, that easily react with protons.

CO_2 upon reaction with an oxide matrix, may generate either surface- or a bulk-vacancy that may support an efficient redox couple. Doping a support with small amounts of Cr^{3+} or La^{3+}, leads to the formation of a host-dopant-type solid solution with a considerable amount of Schottky defects as cationic vacancies [22], that increases the mobility of lattice oxide. Carbon dioxide can fill-up the oxygen vacancies in the reduced support (1) through its reduction to CO, and facilitates the reoxidation:

$$CeO_{2-x} + xCO_2 \rightarrow CeO_2 + xCO \qquad (1)$$

The addition of a basic oxide to a support may influence the interaction of the host with carbon dioxide. In fact, the adsorption of CO_2 is driven by the negative charge of the host oxide and this may be influenced, in turn, by the nature of the added oxide: lower is the electronegativity of the added metal, higher will be the charge on the oxide and more ready the interaction with carbon dioxide. This effect has been encountered when alumina was doped with oxides of either rare earth metals (La, Ce, Nd, Pr), or Group 2 metals (Mg, Ca, Sr, Ba) or Group 1 metals (Na, K, Rb, Cs) [23]. Equation (2) has been found to correlate the formation of carbonate to the negative charge on oxide and to the electronegativity of the basic metal-atom:

$$\delta_O = [(S_{int} - S_O) / 2.08 * S_O]^{1/2} \quad (2)$$

where δ_O is partial charge on oxygen, S_{int} is the geometric mean of electronegativities and S_O is the electronegativity of oxygen. The added metals facilitate the charge flow into the $2\pi^*_u$ orbital of CO_2, with stabilisation of the carbonate complex, that on clean surfaces is usually unstable towards dissociation, except at sites with coordinatively unsaturated oxides [20].

Also, the formation of surface color centers (F^+) generated by the added metal atom may bring to the formation of (Me-F^+) complexes that could modify the Madelung potential at the sites of nearby O^{2-} anions in a way that carbonate formation becomes facile at such sites.

2. METHANE REFORMING WITH CO_2

The direct reaction of methane with carbon dioxide (known as “dry reforming” of methane), has a great interest from several points of view [24]. In fact, it allows to convert two undesired “greenhouse” gases into Syngas (H_2+CO), (Eq. 3) that may be used for making methanol (Eq. 4) or Cn hydrocarbons (Eq. 5), meeting thus environmental and industrial targets simultaneously.

$$CH_4 + CO_2 \rightarrow 2H_2 + 2CO, \qquad \Delta H^0_{298K} = 247 \text{ kJ/mol} \quad (3)$$

$$CO + 2\,H_2 \rightarrow CH_3OH \qquad \Delta H^0_{298K} = -128.6 \text{ kJ/mol} \quad (4)$$

$$nCO + 2nH_2 \rightarrow [CH_2]n + [H_2O]n \quad (5)$$

Additionally, the exploitation of dry-reforming of methane would make possible the direct conversion of LNG, that is composed by 65-70% of methane and 35-30% of carbon dioxide, into liquid fuels at the LNG extraction site. This would result in a further benefit as the distribution of a liquid fuel is less expensive that the transport of the gas. Despite its great potential, this technology has not been implemented so far, as to its full exploitation a few barriers exist, that must be removed.

As Eqns. 4 and 5 show, the ratio H_2/CO produced in dry reforming is lower than required by either the synthesis of methanol (Eq. 4) or Fischer-Tropsch (Eq. 5). This problem can be easily solved by coupling the dry- with wet-reforming (or steam-reforming) (Eq.6) of methane so to produce the correct $H_2/CO>1.66$ ratio.

$$CH_4 + H_2O \rightarrow 3H_2 + CO \qquad \Delta H^0_{298K} = 206 \text{ kJ/mol} \qquad (6)$$

An issue of great relevance is that both reactions 3 and 6 are strongly endothermic. The energetics of Syngas production can be made less negative by coupling the two previous processes with the partial oxidation of methane Eq. (7) that is an exothermic process.

$$CH_4 + 1/2O_2 \rightarrow 2H_2 + CO \qquad \Delta H^0_{298K} = -38 \text{ kJ/mol} \qquad (7)$$

Coupling Eqns. 3, 6, and 7 is known as the "tri-reforming" process that minimizes the negative aspects of each of the three reactions [24].

However, the week point of dry-reforming is the catalyst deactivation due to the carbon deposition. Coke is readily formed either via methane dehydrogenation (Eq. 8) or in the Baudoir equilibrium (CO disproportionation) (Eq.9).

$$CH_4 \rightarrow C + 2H_2 \qquad \Delta H^0_{298K} = 206 \text{ kJ/mol} \qquad (8)$$

$$2CO \rightarrow C + CO_2 \qquad \Delta H^0_{298K} = -172 \text{ kJ/mol} \qquad (9)$$

If the "tri-reforming" approach is used, the coke formation is minimized [24]. Alternatively catalysts should be designed that minimize the coke formation. It is worth to note that the thermodynamics of the Baudoir equilibrium (Eq. 9) says that it is possible to shift equilibrium to the left working at a temperature above 1230 K. Analogously, working at a similar temperature the coke formation in methane dehydrogenation is suppressed [25]. The industrial application of dry-reforming requires that the reaction is performed at a much lower temperature than 1200 K. Therefore, the catalyst design is a key point for running the dry reforming at industrial level. It has been reported [26] that the rate of coke formation depends on the metal used. For example it has been found that it decreases in the order Ni >> Pt > Ru. Due to the cost of Pt and Ru, it would be interesting to develop catalysts based on Ni. This has pushed in the direction of searching new Ni-based catalysts that may reduce the coke formation and allow to run the reaction at lower temperatures (below 900 K). When Ni supported on SiO_2 [27] is used under the steady state reaction conditions (T=870-970 K, CH_4/CO_2=1), carbon adspecies are originated from both methane and carbon dioxide and are deposited on the catalyst. Carbide-like species (Ni_3C and Ni_2C, a single monolayer, roughly) are continuously formed on the surface of the catalyst (site S1) by the dissociative reaction of methane with gas-H_2 evolution. (Eq. 10) Such carbon particles have been proposed to undergo oxidation to CO (Eq. 11) upon reaction with oxygen atoms formed

$$CH_4 + S_1 \rightarrow C_{ads} + 2H_2 \qquad (10)$$

$$CO_2 + S_2 \rightarrow CO + O_{ads} \qquad (11)$$

$$C_{ads} + O_{ads} \rightarrow CO + S_1 + S_2 \qquad (12)$$

upon dissociation of CO_2 on a different site S_2 (Eq. 12). In order to explain the topotactic growth of the nickel particles, the formation of a thin interface has been proposed, with the presumable structure $Ni_3[SiO_5](OH)_4$ on Ni/SiO_2 [28]. The catalyst deactivation occurs when carbon filaments (pol ymorphic graphite) accumulate at such amount that Ni particles are completely encapsulated. The rate of carbon accumulation and catalysts deactivation can be controlled by using several techniques, among which sulphur-passivation. Adsorbed O-atoms, generated in the decomposition of CO_2, increase the dissociative chemisorption of methane on Ni (100). They differ from O-atoms at hollow centers because the latter have a higher activation energy and are more endothermic.

Noble metal, such as Pt, Rh, Ir, are active catalysts for dry reforming. They differ from Ni because are less active towards CH_4 dissociative adsorption. Such difference could arise from the redox state of the metal particles: the rather high oxygen coverage found for Ni could provide an additional pathway for methane activation. The oxidation of methane takes place over rhodium oxide even in absence of gaseous oxygen [29]. The size of Rh clusters increases with Rh loading and the relative amount of on-top oxygen species over the catalyst surface decreases as does the contribution of oxygen assisted dissociation of methane. Consequently, CO formation is almost unaffected by rhodium loading above 0.05%. The reaction mechanism for Rh supported over Al_2O_3 , as well as for supported-Ru, under normal pressure at 970 K can be described as a step-wise process: the dissociative adsorption of CH_4 to afford active C and H_2 is the rate determining step, followed by the direct fast reaction of CO_2 with adsorbed carbon [30]. Such hypothesis is supported by the fact that a deuterium isotope effect has been ascertained in methane conversion [31]. Whether CO_2 dissociates on noble metals as it does on Ni, is still an open question. For the latter, besides the activation on the metal surface, the interaction with the interface support-metal has been proposed [32] that should be more effective for the inhibition of the formation of carbon.

The support has been proposed to play a key role not only in the adsorption and subsequent dissociation of carbon dioxide, but also in the formation of CO and H_2 and adsorbed oxygen atoms transport [33]. It has also been shown that addition of basic oxides like MgO, Al_2O_3 and TiO_2, to a noble metal like Rh supported over an acid oxide such as SiO_2 influences the dry reforming reaction of methane, leading to the conclusion that carbonate species may play a key role in the dry-reforming process [34]. γ-Alumina seems to play a peculiar role due to the OH mobility. The latter may be involved in the conversion of either metal carbides (Eq. 13) or CHx (Eq. 14) species into CO [35].

$$M\text{-}OH + M\text{-}C \rightarrow 2\,M + \tfrac{1}{2}H_2 + CO \qquad (13)$$

$$C\text{-}H + M\text{-}OH \rightarrow M + CO + H_2 \qquad (14)$$

However, literature data are some controversial and several issues still need deeper investigation. The role of supports like Al_2O_3, SiO_2, TiO_2, and ZrO_2 on noble metals and Ni still requires investigation [36]. Recently, ZrO_2-supported platinum has been shown to be both more active and stable than Pt/Al_2O_3 or Pt/TiO_2 [37, 38]. A great attention has been paid to understanding the mode of lay-down of surface species. The use of a sequential pulsing technique [39] has shown that the reacting surface of Pt, Rh, Ru, or Ir supported on silica is covered by a pool of intermediates and reactants and not just a ill-defined carbonaceous deposit. The accumulated C-species drive the dry-reforming activity and determine the H_2:CO ratio.

Metal carbides have been questioned for their role in dry reforming [40-42]. Both WC and Mo_2C have been shown to be active and their activity depends on the CO pressure. In fact, they are easily deactivated by carbon dioxide as are converted into the respective dioxides (Eqns. 15 and 16).

$$WC + 3\,CO_2 \rightarrow WO_2 + 4\,CO \qquad (15)$$

$$MoO_2 + 5\,CO_2 \rightarrow 2\,MoO_2 + 6\,CO \qquad (16)$$

Keeping a high CO pressure by recycling the product gas stabilizes the catalyst.

Lantanium-oxide derived catalysts of formula $LaNi_xFe_{1-x}O_3$ (where $0.3 < x < 0.8$) have been shown to perform well in dry reforming. The H_2 and CO yield depends on the temperature and the value of x [43]. The following mechanism (Eqs. 17-20) on Ni/La_2O_3 has been proposed:

$$CH_4 + Ni \rightarrow NiC + 2H_2 \tag{17}$$

$$La_2O_2CO_3 + Ni \rightarrow CO + NiO + La_2O_3 \tag{18}$$

$$NiC + NiO \rightarrow CO + 2Ni \tag{19}$$

$$CO_2 + La_2O_3 \rightarrow La_2O_2CO_3 \tag{20}$$

Also, the amount of deposited carbon depends on x, and reaches the maximum at x=1.

Catalysts of formula $LaCo_xFe_{1-x}O_3$ have also been investigated. For various values of x (0.3, 0.5, and 0.7) the perovskite structure is preserved. The activity increases with x and the reaction temperature. Ni supported on olivine has been shown to be able to afford a yield of methane conversion in dry reforming as high as 95% with hydrogen yield of 80% at 1030 K. This catalyst has been used in catalytic biomass gasification in a fluidized bed pilot plant, with a net control of tar formation. The use of combinatorial catalysis may play an interesting role in new catalyst development and testing in this area [44, 45].

The CO_2 reforming of methane has only one commercial application in the MIDREX process [46] in the steel industry, that operates at 1300 K. The mixture of gases produced (hydrogen and carbon monoxide) is used directly in the shaft furnace for reducing the iron ores. The CO_2 reformer consists essentially of two main sections: a bottom fired furnace with several parallel rows of tubes charged with catalyst, and a convection section, where heat is recovered from the flue gas by the duties of process. The thermal efficiency of the reformer is highly improved by the heat recovery system. Sensible heat is recovered from the reformer flue gas and used to preheat both the combustion air used in the reformer burners, as well as the mixture of top gas and natural gas fed to the reformer tubes.

The carbon dioxide reforming of methane may have a relevance to developing a technology for producing carbon monoxide-free hydrogen that is necessary for use in fuel cells that accept only a low concentration of CO. In this case, the decarbonization of methane produces pure dihydrogen (Eq. 21) and graphitic carbon. The latter is then converted into carbon dioxide by reaction with dioxygen. (Eq. 22)

$$CH_4 \rightarrow C_{graph} + 2\,H_2 \tag{21}$$

$$C_{graph} + O_2 \rightarrow CO_2 \tag{22}$$

The catalyst used in reaction 21 is the Ni-active phase used for the dry reforming of methane.

3. SYNTHESIS OF METHANOL

Carbon dioxide is since long added to Syngas ($CO + H_2$) for the industrial production of methanol. As reported in Chapter 9, some 2 Mt/y of carbon dioxide are used to this end. Various copper-catalysts have been developed to drive the reaction. More recently, pure carbon dioxide has been converted into methanol by using Cu-Zn catalysts added with some promoters like $La_2Zr_2O_7$, or supported on zirconia [47-50]. The direct conversion of carbon dioxide with dihydrogen has been developed at the pilot plant scale in Japan at RITE [49]. This reaction has been for long questioned as, compared to the conversion of carbon monoxide, it uses one mol of hydrogen in excess (Eqns. 23 and 24)

$$CO + 2\,H_2 \rightarrow CH_3OH \tag{23}$$

$$CO_2 + 3\,H_2 \rightarrow CH_3OH + H_2O \tag{24}$$

Unless dihydrogen would be produced via non fossil-fuel technology, route 24 would not greatly contribute to carbon recycling as carbon dioxide would be evolved in the production of dihydrogen. Nevertheless, reaction 24 could be meaningful if the reduction of carbon dioxide to methanol would proceed via a path not implying carbon monoxide, with better yield and selectivity, and lower energy barriers. Under more performing conditions, the better use of dihydrogen would compensate the higher stoichiometric amount required by reaction 24 with respect to reaction 23. As a matter of fact, two different reaction paths have been proposed, based on the:

(i) direct hydrogenation of carbon dioxide via formate (Scheme 12.2 i), or the
(ii) reduction of carbon dioxide to carbon monoxide (Scheme 12.2 ii) with subsequent hydrogenation to methanol.

Route (ii) would be started by a Reverse Gas Shift Reaction (RWGS) [51] that would consume one mol of dihydrogen per mol of carbon monoxide (Eq. 25)

$$CO_2 + H_2 \rightarrow CO + H_2O \tag{25}$$

The two alternative pathways are presented in Scheme 12.2

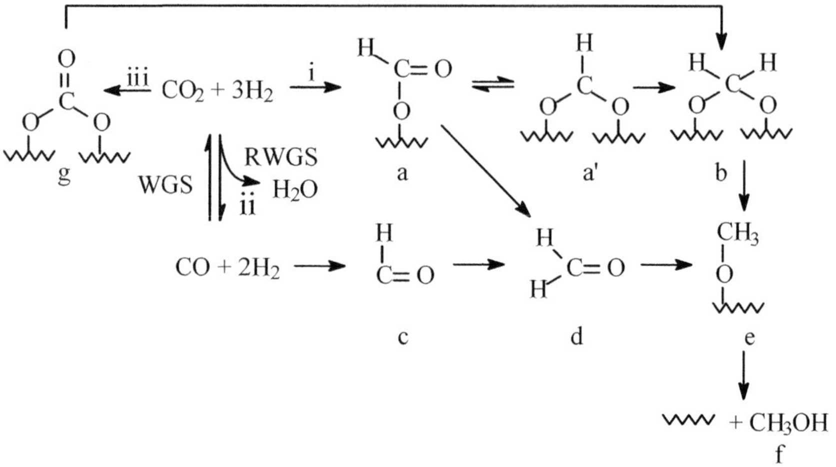

Scheme 12.2 -The two alternative pathways for CO_2 reduction to methanol

The direct reduction of carbon dioxide [52] goes through the formation of formates (a) that can be then converted into the geminal glicolate (b) that originates the methoxo group (a) from which methanol (f) is obtained. Alternatively, the formate could be step by step converted into formaldehyde (d), that affords the methoxo group (e), and finally methanol. The preliminary conversion of CO_2 into CO through the RWGS, continues with the formation of the formyl moiety (c) that originates formaldehyde. The two processes have some common intermediates. A third way could be the interaction of carbon dioxide with the oxide surface of the catalysts to afford the carbonate anion species (g in Scheme 12.2 iii) that can be converted into the geminal glycolate (b), [53] following then the route of carbon dioxide reduction.

The energetic analysis of the synthesis of methanol from carbon dioxide or carbon monoxide demonstrates that the recovery of carbon dioxide is a process demanding less energy than Syngas production [54]. Therefore, supposed that dihydrogen were used that is obtained via a non fossil fuel route, as specified above, the synthesis of methanol from recovered carbon dioxide is convenient with respect to the synthesis based on the formation of Syngas.

The economics of the formation of methanol from carbon dioxide or carbon monoxide also shows the conditions for which it is convenient to use recycled carbon dioxide [55].

4. CONVERSION OF CARBON DIOXIDE INTO Cn ALCOHOLS OR HYDROCARBONS

Differently from the use of carbon monoxide, for which the synthesis of Cn alcohols has been demonstrated to occur and for which a plausible mechanism has been proposed [56, 57], for carbon dioxide the mechanism is still uncertain, although the production of high alcohols is documented [58, 59], with yields of the order of 30-35 % in C_2-C_5 [59]. Several catalysts have been developed based on Co or Mo or Co-Mo, that may also be combined with Cu-$La_2Zr_2O_7$.

The conversion of carbon dioxide into hydrocarbons is documented in the literature for a direct process [60]. Alternatively, a two step process can be envisaged in which methanol is first formed and then converted into gasoline [61]. Zeolite supported Fe-ZnO, [62] or Cu-Zn-Cr, [63] or Cu-La-Zr [64] have been used that have shown a good resistance to the temperature and an interesting working ability.

5. CONVERSION OF CARBON DIOXIDE UNDER PLASMA CONDITIONS

The plasma technique, either thermal or non-thermal, has been used for the conversion of methane [24]. The thermal plasmas produce unsaturated hydrocarbons (acetylene) or carbon black. Low temperature (300 K) plasmas have been more recently used for the simultaneous conversion of methane and carbon dioxide. Several techniques have been used, like corona, microwave, radio-frequency discharge that produce ethylene, carbon monoxide, dihydrogen. The use of the dielectric-barrier discharge has opened a way to more complex products like oxygenates. Using carbon dioxide as oxidant for methane [65] has allowed the synthesis of formic acid or methanol. The material of the metal-electrode influences the nature and distribution of the products [66]. Higher hydrocarbons can be also formed [67]. Very interestingly, acetic acid is formed through the direct coupling (Eq. 26) of methane with CO_2 [68].

$$CH_4 + CO_2 \rightarrow CH_3COOH \qquad (26)$$

The observed selectivity (5.2% in the best case with a conversion of methane and carbon dioxide of 64.3 and 43.1 %, respectively) is a promising result that can be ameliorated with the continuous elimination of the oxygenated product formed or the use of ad hoc heterogeneous catalysts under plasma conditions.

This area appears to be quite innovative and interesting as it implies the simultaneous conversion of two greenhouse gases into useful products. Recently, their selective conversion into Syngas has been achieved under non-equilibrium pulsed discharge conditions at room temperature [69]. The use of Ni-MgO catalysts has allowed to reach a significant conversion yield of the equimolar CH_4-CO_2 feed mixture, with formation of acetylene and CO-H_2 with a selectivity in the range 15-44 for the C_2 molecule and 84-56 for CO. In all cases the ratio CO/H_2 was 1.1.

6. BIOTECHNOLOGICAL PROCESS FOR CARBON DIOXIDE CONVERSION INTO METHANOL

The conversion of carbon dioxide by means of enzymes (either in solution or supported) is an outstanding way to recycling carbon. This approach, that would be mimetic of natural facts, might bring unforeseeable results and benefits.

To date, only two reports are in the literature that concern either the fixation of the entire CO_2 molecule onto an organic substrate or the reduction of carbon dioxide to C_1 molecules mediated by enzymes. The biotechnological conversion of phenol into 4-OH-benzoic acid [70] has been described in Chapter 9. Here we report on the conversion of carbon dioxide into methanol. The process consists of several steps, as depicted in Eqns. (27)-(31), each driven by a specific enzyme.

$$CO_2 + 2\,H^+ + 2e^- \rightarrow CO_{(bound)} + H_2O \quad (27)$$

$$CO_2 + 2H^+ + 2e^- \rightarrow HCOOH \quad (28)$$

$$CO + 2H^+ + 2e^- \rightarrow CH_2O \quad (29)$$

$$HCOOH + 2e^- + 2H^+ \rightarrow CH_2O + H_2O \quad (30)$$

$$CH_2O + 2H^+ + 2e^- \rightarrow CH_3OH \quad (31)$$

Reaction (27) is catalyzed by the enzyme Carbon-monoxide dehydrogenase (CODH), a Ni-enzyme that converts carbon dioxide into bound carbon monoxide [71]. The reaction has been fully mimicked using Ni-complexes [72]. Reaction (28) is driven by the enzyme formate dehydrogenase, a tungsten enzyme. The insertion of CO_2 into a M-H bond to afford a formate complex has been documented for several metal systems [73]. Moreover, the synthesis of formic acid from carbon dioxide and dihydrogen has been fully performed using metal systems like Ru or Rh complexes [74]. The formic acid reduction step (Eq. 30) involves the enzyme Formaldehyde dehydrogenase (FDH). Also this reaction can be fully mimicked by metal complexes [75] and is considered to occur on heterogeneous catalysts in the formation of either methanol or hydrocarbons from Syngas [52]. The final step (Eq. 31) is the conversion of formaldehyde into methanol and requires the intervention of the enzyme Alcohol dehydrogenase (ADH). The three steps (Eqs. 28, 30 and 31) have been realized by supporting the relevant enzymes on a silica sol-gel. Using NADH as reducing agent, methanol was obtained from carbon dioxide [76]. Although the turnover number of the reaction is very low and the cost quite high, nevertheless this example coupled with the biotechnological conversion of phenol into 4-OH-benzoic acid mentioned above, shows that the biotechnological way has a good potential for carbon dioxide conversion into useful chemicals.

REFERENCES

1. Aresta, M "Utilization of carbon dioxide as a source of carbon: a challenge of great intellectual significance for scientists" in NATO ASI Series, Series C: Mathematical and Physical Sciences Vol. 314: Enzymatic and Model Carboxylation and Reduction Reactions for Carbon Dioxide Utilization, 1-18, 1990.
2. Aresta, M., Nobile, C.F., Albano, V.G., Forni, E., Manassero, M. "New nickel-carbon dioxide complex. Synthesis, properties, and crystallographic characterization of (carbon dioxide)bis(tricyclohexylphosphine)nickel" J. Chem. Soc., Chem. Commun., 15, 636-7, 1975.
3. Aresta, M., Quaranta, E., Tommasi, I., Giannoccaro, P., Ciccarese, A. "Enzymic versus chemical carbon dioxide utilization. Part I. The role of metal centers in carboxylation reactions" Gazz. Chim. Ital., 125(11), 509-38, 1995.
4. Ozin, G.A., Huber, H., McIntosh, D. "Metal atom chemistry and surface chemistry: Carbon dioxide silver, $Ag(CO_2)$. A localized bonding model for weakly chemisorbed CO_2" Inorg. Chem., 17, 1472, 1978.
5. a) Mascetti, J., Galan, F., Papai, I. "Carbon dioxide interaction with metal atoms: matrix isolation spectroscopic study and DFT calculations" Coord. Chem. Rev., 190-192, 557-576, 1999; b) Galan, F., Fouassier, M., Mascetti, J., Papai, I. "CO_2 coordination and cooperative effects in ligand binding to transition metal atoms: matrix isolation and density functional studies" International Conference on Low Temperature Chemistry, 2nd, Kansas City, Mo., Aug. 4-9, 59-62, 1996.
6. Gibson, D.H.,"The organometallic chemistry of carbon dioxide" Chem. Rev. 96, 6, 2063-2095, 1996.
7. a) Solymosi, F., Bugyi, L. "Adsorption and dissociation of carbon dioxide on a potassium-promoted rhodium(111) surface" J. Chem. Soc., Far. Trans. 1: Phys. Chem. Cond. Phas., 83(7), 2015-33, 1987; b) Solymosi, F., Klivenyi, G. "HREELS study on the

formation of CO-2 on a K-promoted Rh(111) surface" Surf. Sci. 315(3), 255-68, 1994; Goodman, D.W., Peables, D.E., White, J.M. "Carbon dioxide dissociation on rhodium: measurement of the specific rates on rhodium(111)" Surf. Sci. 140(1), L239-L243, 1984; c) Solymosi, F., Kiss, "The effect of boron impurity on the adsorption and dissociation of carbon dioxide on rhodium surfaces" Chem. Phys. Lett. 110(6), 639-42, 1984.

8. Weinberg, W.H. "Why carbon dioxide does not dissociate on rhodium at low temperature" Surf. Sci., 128 (2-3), L224-L230, 1983.
9. a) Campbell, C.T., White, J.M. "The adsorption, desorption, and reactions of carbon monoxide and oxygen on rhodium" J. Catal., 54(3), 289-302, 1978; b) Thiel, P.A., Williams, E.D., Yates, J.T., Weinberg, W.H. "The chemisorption of carbon monoxide on rhodium (111)" Surf. Sci., 84(1), 54-64, 1979.
10. Ledoux, M.J., Meunier, F., Heinrich, B., Pham-Huu, C., Harlin, M.E., Outi A., Krausel. "n-Butane dehydrogenation on unsupported carbon modified MoO3 (MoOxCy). Part I. Effect of steam on the catalyst stability" Appl. Catal. A: General 181(1), 157-170, 1999.
11. a) Engel, T., Ertl, G. "Elementary steps in the catalytic oxidation of carbon monoxide on platinum metals" Adv. Catal., 28, 1-78, 1979; b) Engel, T., Ertl, G. "A molecular beam investigation of the catalytic oxidation of carbon monoxide on palladium (111)" J. Chem. Phys., 69(3), 1267-81, 1978.
12. Norton, P.R., Richards, P.J. "Photoelectron spectroscopic study of the adsorption of carbon monoxide and carbon dioxide on platinum" Surf. Sci., 49(2), 567-76, 1975.
13. a) Clendening, W. D.; Rodriguez, J. A.; Campbell, J. M.; Campbell, C.T. "The chemisorption and coadsorption of water and oxygen on cesium-dosed copper(110)" Surf. Sci., 216(3), 429-61, 1989; b) Krause, J., Borgmann, D., Wedler, G. "Photoelectron spectroscopic study of the adsorption of carbon dioxide on Cu(110) and Cu(110)/K - as compared with the systems Fe(110)/CO_2 and Fe(110)/K+CO_2" Surf. Sci., 347(1/2), 1-10, 1996; c) Pohl, M., Otto, A. "Adsorption and reaction of carbon dioxide on pure and alkali-metal promoted cold-deposited copper films" Surf. Sci., 406(1-3), 125-137, 1998.
14. Behner, H., Spiess, W., Wedler, G., Borgmann, D. "Interaction of carbon dioxide with iron(110), stepped iron(110) and iron(111)" Surf. Sci., 175(2), 276-86, 1986.
15. Carley, A.F., Gallagher, B.E., Roberts, M.W. "Activation of carbon dioxide at low temperatures at aluminum surfaces" Surf. Sci., 183(1-2), L263-L268, 1987.
16. Heinz, K., Saldin, D.K., Pendry, J.B. "Diffuse LEED and surface crystallography" Phys. Rev. Lett., 55(21), 2312-15, 1985.
17. Ochs, D., Brause, M., Maus-Friedrichs, W., Kempter, V. "CO_2 chemisorption at Mg and MgO surfaces: a study with MIES and UPS(He I)" Surf. Sci., 397(1-3), 101-107, 1998.
18. a) Aresta, M., Schloss, J.V.; Editors. NATO ASI Series, Ser. C: Mathematical and Physical Sciences, Vol. 314: Enzymatic and Model Carboxylation and Reduction Reactions for Carbon Dioxide Utilization,451 pp., 1990; b) Aresta, M., Quaranta, E., Tommasi, I., "The role of metal centers in reduction and carboxylation reactions utilizing carbon dioxide" New J. Chem., 18(1), 133-42, 1994.
19. Mascetti, J., Tranquille, M. "Fourier transform infrared studies of atomic titanium, vanadium, chromium, iron, cobalt, nickel and copper reactions with carbon dioxide in low-temperature matrices" J. Phys. Chem., 92(8), 2177-84, 1988.
20. a) Pacchioni, G., Ricart, J.M., Illas, F. "Ab Initio Cluster Model Calculations on the Chemisorption of CO_2 and SO_2 Probe Molecules on MgO and CaO (100) Surfaces. A Theoretical Measure of Oxide Basicity" J. Am. Chem. Soc., 116(22), 10152-8, 1994; b) Pacchioni, G. "Physisorbed and chemisorbed carbon dioxide at surface and step sites of the magnesia(100) surface" Surf. Sci., 281(1-2), 207-19, 1993.

21. Kung, H.H., "Transition Metal Oxides, Surface Chemistry and Catalysis", Elsevier Science Publ., Amsterdam, 1989.
22. Zhang, H.-B., Chen, P., Lin, G.-D., Tsai, K.-R. "Development of coking-resistant Ni-based catalyst for partial oxidation and CO_2-reforming of methane to syngas" Applied Catalysis A: 166, 2, 343-350, 1998.
23. Horiuchi, T., Hikada, H., Fukui, T., Kubo, Y., Horio, M., Suzuki, K., Mori, T. "Effect of added basic metal oxides on CO_2 adsorption on alumina at elevated temperatures" Appl. Catal. A: General, 167, 195-202, 1998.
24. Song, C., Gaffney, A.M. and Fujimoto K. Eds, "CO_2 conversion and utilization", ACS Symp Series 809, 2002
25. Gadalla, A.M., Bower, B. "The role of catalyst support on the activity of nickel for reforming methane with carbon dioxide" Chem. Eng. Sci., 43(11), 3049-62, 1988.
26. Rostrup - Nielsen, J. R. "Syngas for C_1-chemistry. Limits of the steam reforming process" Stud. Surf. Sci. Catal, 36 (Methane Convers.), 73-8, 1988.
27. a) Swaan, H.M., Kroll, V.C.H., Martin, G.A., Mirodatos, C. "Deactivation of supported nickel catalysts during the reforming of methane by carbon dioxide", Catal. Today, 21(2-3), 571-8, 1994; b) Kroll, V.C., Swaan, H.M., Lacombe, S., Mirodatos, C. "Methane reforming reaction with carbon dioxide over Ni/SiO_2 catalyst. II. A mechanistic study", J. Catal., 164(2), 387-398, 1996; c) Kroll, V.C., Swaan, H.M., Mirodatos, C. "Methane reforming reaction with carbon dioxide over Ni/SiO_2 catalysts I. Deactivation studies" J. Catal., 161(1), 409-422, 1996.
28. Quinlan, M.A., Wood, B.J., Wise, H. "Detection of surface intermediates in the oxidation of methane on nickel(100) by electron energy loss spectroscopy", Chem. Phys. Lett., 118(5), 478-80, 1985.
29. Mark, M.F., Maier, W.F. "CO_2-reforming of methane on supported Rh and Ir catalysts", J. Catal., 164(1), 122-130, 1996.
30. Wang, H.Y., Au, C.T. "Carbon dioxide reforming of methane to syngas over SiO_2-supported rhodium catalysts", Appl. Catal. A: General 155(2), 239-252, 1997.
31. Tsipouriari, V.A., Efstathiou, A.M., Verykios, X.E. "Transient kinetic study of the oxidation and hydrogenation of carbon species formed during CH_4/He, CO_2/He, and CH_4/CO_2 reactions over Rh/Al_2O_3 catalyst" J. Catal., 161(1), 31-42, 1996.
32. Huinink, J.P. Doctorate Thesis, Eindhoven University of Technology, The Nehterland, 1995.
33. a) Wang, D., Dewaele, O., de Groote, A.M., Froment, G. "Reaction mechanism and role of the support in the partial oxidation of methane on Rh/Al_2O_3" J. Catal., 159(2), 418-426, 1996; b) Buyevskaya, O.V., Wolf D., Baerns, M. "Rhodium-catalyzed partial oxidation of methane to CO and H_2. Transient studies on its mechanism" Catal. Lett., 29(1,2), 249-60, 1994.
34. a) Tsipouriari, V.A., Efstathiou, A.M., Verykios, X.E., Zhang, Z. L. "Reforming of methane with carbon dioxide to synthesis gas over supported Rh catalysts" Catal. Today, 21(2-3), 579-88, 1994; b) Ergohelyi, A., Cserenyi, J., Solymosi, F. "Activation of methane and its reaction with carbon dioxide over supported rhodium catalysts" J. Catal., 141(1), 287-99, 1993.
35. a) Scuurman, Y., Marquez-Alvarez, C., Kroll, V.C.H., Mirodatos, C. "Unraveling mechanistic features for the methane reforming by carbon dioxide over different metals and supports by TAP experiments" Catal. Today, 46(2-3), 185-192, 1998; b) Ruckenstein, E., Hu, Y.H. "Interactions between Ni and La_2O_3 in Ni/La_2O_3 catalysts prepared using different Ni precursors"J. Catal., 161(1), 55-61, 1996.
36. a) Seshan, K.; ten Barge, H.W.; Hally, W.; van Keulen, A.N.J.; Ross, J.R.H. "Carbon dioxide reforming of methane in the presence of nickel and platinum catalysts

supported on ZrO_2" Stud. Surf. Sci. Catal., 81(Natural Gas Conversion II), 285-90; 1994; b) Bitter, J.H., Hally, W., Seshan, K., van Ommenn, J.G., Lercher, J.A. "The role of the oxidic support on the deactivation of Pt catalysts during the CO_2 reforming of methane" Catal. Today, 29(1-4), 349-353, 1996.

37. Bitter, J.H., Hally, W., Seshan, K., Lercher, J.A. "The state of zirconia supported platinum catalysts for CO_2/CH_4 reforming" J. Catal., 171(1), 279-286, 1997.
38. Tournayan, L., Auroux, A., Charcosset, H., Szymanski, R. Ads. Sci. Tech., 2, 55, 1985.
39. RUCADI Report – RUCADI Project BRRT-CT98-5089.
40. Lofberg, A., Frennet, A., Leclercq, G., Lerclercq, L., Girudon, J.M. "Mechanism of WO_3 Reduction and Carburization in CH_4/H_2 Mixtures Leading to Bulk Tungsten Carbide Powder Catalysts" J. Catal., 189(1), 170-183, 2000.
41. Satterfield C.N. "Heterogeneous Catalysis in Practice" Chem. Eng. Series, McGraw-Hill Book Company (1980).
42. Sehested, J., Jacobsen, C.J.H., Rokni, S., Rostrupt-Nielsen, J.R. "Activity and Stability of Molybdenum Carbide as a Catalyst for CO_2 Reforming", J. Catal., 201(2), 206-212, 2001.
43. European Programme JORCT 970196 "Hydrogen rich gas from biomass gasification"
44. Holzwarth, A., Denton, P., Zanthoff, H., Mirodatos, C. "Combinatorial approaches to heterogeneous catalysis: strategies and perspectives for academic research", Catal. Today, 67(4), 309-318, 2001.
45. Farrusseng, D., Baumes, L., Hayaud, C., Vauthey, I., Denton, P., Mirodatos, C. "The combinatorial approach for heterogeneous catalysis: a challenge for academic research. NATO Science Series, II: Mathematics, Physics and Chemistry, 69 (Principles and Methods for Accelerated Catalyst Design and Testing), 101-124, 2002.
46. Xu, G.L., Shi, K.Y., Gao, Y., Wei, Y.D. "Studies of reforming natural gas with carbon dioxide to produce synthesis gas X. The role of CeO_2 and MgO promoters", J. Mol. Catal. A: Chem., 147(1-2), 47-54, 1999.
47. Klier, K., Chatikavanij, V., Herman, R. G., Simmons, G. W. "Catalytic synthesis of methanol from carbon monoxide/hydrogen. IV. The effects of carbon dioxide", J. Catal., 74(2), 343-60, 1982.
48. Kieffer, R., Fujiwara, M., Udron, L., Souma, Y. "Hydrogenation of CO and CO_2 toward methanol, alcohols and hydrocarbons on promoted copper-rare earth oxide catalysts" Catal. Today, 36(1), 15-24, 1997.
49. Saito, M., Takeuchi, M., Fujitani, T., Toyir, J., Luo, S., Wu, J., Ushikoshi, K., Mori, K., Watanabe, T. "Advances in the research between NIRE and RITE for developing a novel technology for methanol synthesis from CO_2 and H_2" ICCDU V Karlsruhe September 5-10, 1999 .
50. Sahibzada, M., Metcalfe, I.S., Chadwick, D. "Methanol synthesis from CO/CO_2-H_2 over $Cu/ZnO/Al_2O_3$ at differential and finite conversions", J. Catal., 174(2), 111-118, 1998.
51. Klier, K. "Methanol synthesis", Adv. Catal., 31 243-313, 1982.
52. Deluzarche, A., Hindermann, J.P., Kieffer, R., Kiennemann, A. "On the role of formate species in the reactions of carbon oxides (carbon monoxide-carbon dioxide) with hydrogen or water", Rev. Chem. Interm., 6(3), 255-74 1985.
53. Chaumette, P.; Courty, Ph.; Barbier, J.; Fortin, T.; Lavalley, J. C.; Chauvin, C.; Kiennemann, A.; Idriss, H.; Sneeden, R. P. A.; Denise, B. "Carbonates, a key for selecting methanol synthesis catalysts" Proc. - Int. Congr. Catal., 9th, 2, 585-93, 1988.
54. Aresta, M., Caroppo, A., Dibenedetto, A., Narracci M."Life cycle assessment (LCA) applied to the synthesis of methanol. Comparison of the use of syngas with the use of CO_2 and dihydrogen produced from renewables". Book on "Environmental Challenges and Greenhouse Gas Control for Fossil Fuel Utilization in the 21st Century" edited by Maroto Valer et al., Kluwer Academic/Plenum Publisher, New York, 331, 2002.

55. Mignard, D., Sahibzada, M., Duthie, J.M., Whittington, H.W. "Methanol synthesis from flue-gas CO_2 and renewable electricity: a feasibility study" Intern. J. Hydrogen Energy, 28(4), 455-464, 2003.
56. Chaumette, P., Coutry, P., Durand, D., Grandvallet, P., Travers, C. "Synthesis of a mixture of primary alcohols from synthesis gas in the presence of a catalyst containing copper, cobalt, zinc and aluminium" Ger. Offen., DE 3516928,1985.
57. Kiennemann, A.; Diagne, C.; Hindermann, J. P.; Chaumette, P.; Courty, P. "Higher alcohols synthesis from carbon monoxide + hydrogen on cobalt-copper catalyst: use of probe molecules and chemical trapping in the study of the reaction mechanism", Appl. Catal., 53(2-3), 197-216, 1989.
58. Chu, W., Kieffer, R., Kiennemann, A., Hindermann, J.P. "Conversion of syngas to C1-C6 alcohol mixtures on promoted $CuLa_2Zr_2O_7$ catalysts" Appl. Catal. A: General, 121(1), 95-111, 1995.
59. Chu, W., Kieffer, R.; Kiennemann, A.K.; Xiong, G.X. "Influence of molybdenum additive onto the Co/CuLaZr catalysts" Chinese Chem. Lett. 5(5), 425-6, 1994.
60. Jun, K. W., Shen, W. J., Rama Rao, K.S., Lee, K.W. "Residual sodium effect on the catalytic activity of $Cu/ZnO/Al_2O_3$ in methanol synthesis from CO_2 hydrogenation" Appl. Catal.A: Gen. 174, 231-38, 1998.
61. a) Chang, C.D.; Chu, C. T. W. "Carbene intermediates in methanol conversion to hydrocarbons. Reply to van Hooff" J. Catal., 79(1), 244-5, 1983; b) Chang, C.D.; Lang, W.H.; Smith, R.L. "The conversion of methanol and other O-compounds to hydrocarbons over zeolite catalysts" J. Catal., 56(2), 169-73, 1979.
62. Fijuwara, M., Kieffer, R., Ando, H., Souma Y. "Development of composite catalysts made of Cu-Zn-Cr oxide/zeolite for the hydrogenation of carbon dioxide" Appl. Catal. A: General 121, 113-124, 1995.
63. Fujiwara, M., Kieffer, R., Udron, L., Ando, H., Souma, Y. "Hydrogenation of carbon dioxide over copper-pyrochlore/zeolite composite catalysts" Catal. Today 29, 343-348, 1996.
64. Fujimoto, K., Shikada, T., Fujimoto, K., Shikada, T. "Selective synthesis of C2-C5 hydrocarbons from carbon dioxide utilizing a hybrid catalyst composed of a methanol synthesis catalyst and zeolite" Appl. Catal., 31(1), 13-23, 1987.
65. (a) Larkin, D.W., Lobban, L.L., Mallinson, R. "Production of organic oxygenates in the partial oxidation of methane in a silent electric discharge reactor" Ind. & Eng. Chem. Res., 40(7), 1594-1601, 2001; (b) Matsumoto, H., Tanabe, S., Okitsu, K., Hayashi, Y., Suib, S.L. "Selective oxidation of methane to methanol and formaldehyde with nitrous oxide in a dielectric-barrier discharge-plasma reactor" J. Phys. Chem. A 105, 5304-5308, 2001; Okumoto, M., Su, Z., Katsura, S., Mizuno, A. "Dilution effect with inert gases in direct synthesis of methanol from methane using nonthermal plasma" IEEE Trans. Ind. Appl., 35(5), 1205-1210, 1999.
66. Li, Y., Xu, G.-H., Liu, C.-J., Eliasson, B., Xue, B. "Co-generation of syngas and higher hydrocarbons from CO_2 and CH_4 using dielectric-barrier discharge: effect of electrode materials" Energy&Fuels 15(2); 299-302, 2001.
67. Liu,C.-J., Xue, B., Eliasson, B., He, F., Li,Y., Xu, G.-H., "Methane conversion to higher hydrocarbons in the presence of carbon dioxide using dielectric-barrier discharge plasmas" Plasma Chem. and Plasma Process., 21(3), 301-310, 2001.
68. He, M.-X., Zou, J.-J., Li, Y., Eliasson, B., Liu, C.-J. "Oxygenate synthesis from methane and carbon dioxide using dielectric-barrier discharges" Preprints of Symposia - ACS, Division of Fuel Chemistry , 47(1), 328-329, 2002.

69. Kado, S., Urasaki, K., Sekine, Y. "Methane coupling and reforming using non-equilibrium pulsed discharge at room temperature" 223th ACS National Meeting, Orlando, FL, Fuel Chemistry Division Preprints 47(1), 326-327, 2002.
70. Aresta, M., Quaranta, E., Liberio, R., Dileo, C., Tommasi, I. "Enzymic synthesis of 4-hydroxybenzoic acid from phenol and carbon dioxide: the first example of a biotechnological application of a carboxylase enzyme" Tetrahedron, 54(30), 8841-8846, 1998.
71. Aresta, M., Narracci, M., Tommasi, I. "Influence of iron, nickel and cobalt on biogas production during the anaerobic fermentation of FVG residual biomass" Chem. and Ecology 2003, in press.
72. a) Tommasi, I., Aresta, M., Giannoccaro, P., Quaranta, E., Fragale, C. "Bioinorganic chemistry of nickel and carbon dioxide: an Ni complex behaving as a model system for Carbon Monoxide Dehydrogenase enzyme" Inorg. Chim. Acta, 272(1,2), 38-42, 1998; b) Aresta, M., Quaranta, E., Tommasi, I., "Reduction of coordinated carbon dioxide to carbon monoxide via protonation by thiols and other Broensted acids promoted by nickel systems: a contribution to the understanding of the mode of action of the enzyme carbon monoxide dehydrogenase" J. Chem. Soc., Chem. Commun., (7), 450-2, 1998.
73. a) Darensbourg, D.J., Wiegreffe, P., Riordan, C.G. "Mechanistic aspects of decarboxylation reactions of Group 10 metal formato hydrido tricyclohexylphosphine $[(PCy_3)_2M(H)O_2CH]$ derivatives" J. Am. Chem. Soc., 112(15), 5759-62, 1990; b) Frost, B.J.; Darensbourg, D.J. "Investigation of CO_2 insertion into the tungsten-nitrogen bond of the $[W(CO)_4(NHC_5H_4N)]$- anion" Book of Abstracts, 216th ACS National Meeting, Boston, August 23-27, Inorg.-507, 1998; c) Aresta, M., Dibenedetto, A., Amodio, E., Pápai, I., Schubert G. "Synthesis, characterization and reactivity of cationic hydride $[HPd(diphosphine)_2]^+$ $CF_3SO_3^-$, the missing member of the family $[HM(dppe)_2]^+X^-$ (M = Ni, Pd, Pt). DFT QM/MM structural predictions for the $[HPd(dppe)_2]^+$ moiety" Inorg. Chem., 41(25), 6550-6552, 2002.
74. Jessop, P.G., Linehan, J.C., Main, A.D., Pitts, J.M. "Recent development in the hydrogenation of CO_2 catalyzed by ruthenium phosphine complexes" VI International Conference on Carbon Dioxide Utilization, Breckenridge – Colorado, p. 31, September 9-14, 2000.
75. Aresta, M., Quaranta, E., Tommasi, I., "Spectroscopic characterization of (diphos)Rh(η^6-$PhBPh_3$) in solution and study on its reactivity as a total phenylating agent. Formation of C-C and C-C-C bonds by reaction with aldehydes" Gazz. Chim. Ital., 123(5), 271-8, 1993.
76. Jiang, Z., Wu, H., Xu, S., Huang, S. "Enzymatic conversion of carbon dioxide to methanol by dehydrogenases encapsulated in sol-gel matrix", 223th ACS National Meeting, Orlando, FL, Fuel Chemistry Division Preprints 47(1), 306, 2002.

SECTION V

BIOLOGICAL UTILIZATION

This Section summarizes the activity performed within WP5B of the BRRT-CT98-5089 RUCADI Project, “Recovery andUtilization of Carbon Dioxide” by the following Partners:

Angela Dibenedetto, *University of Bari*, Italy
James J, Leahy, *University of Limerick*, Ireland
Immacolata Tommasi, *Nowatech Ass.*, Italy
Piergiorgio Zappelli, *Enitecnologie*, Italy

Chapter 13

BIOLOGICAL UTILIZATION OF CARBON DIOXIDE: THE MARINE BIOMASS OPTION

Angela Dibenedetto and Immacolata Tommasi
Department of Chemistry and NOWATECH Ass., University of Bari, Campus Universitario, 70126, Bari – Italy
a.dibenedetto@chimica.uniba.it

1. INTRODUCTION

The carbon based fuels produce 80% of the energy used worldwide and emit 22 billion t (22 E∗9 t) per year of CO_2, or 6 E∗9 t of carbon. Despite the natural carbon cycle uses a quantity of carbon close to 200 E∗9 t per year, it cannot recycle all the carbon emitted by anthropic activities that represents only 3% of the total amount. As a matter of fact, the rate of fossil carbon formation is over 100 000 times slower than the rate of fossil carbon consumption and this makes the fossil fuels a "non renewable" source of energy. Carbon dioxide, a green-house gas, is steadily increasing its concentration into the atmosphere since 200 years, rising serious concerns about the effects on global warming and climate change, should higher limits be reached in next two decades. The utilization of biomass represents an environmentally and economically feasible alternative to fossil fuels, moving a step towards the "zero emission" option. As a matter of fact, the forecast is that biomass may contribute to the global energy balance with a share of more than 10% by 2050, [1] with a fivefold increase with respect to the actual 2%. Such expansion of the market would be possible if biomass for energy were specifically grown and used in addition to the limited amount of terrestrial or residual biomass used today as energy source. However, aquatic biomass may represent a convenient solution, because it has a higher (from three to four times) growth-rate with respect to terrestrial

M. Aresta (ed.), Carbon Dioxide Recovery and Utilization, 315-324.

plants, duen tomthe higher photosynthetic efficiency. Micro-algae have been extensively studied so far[2]. More recently, marine macro-algae have been considered with increasing attention. [3, 4] In both cases carbon dioxide recovered from either power plants or industrial flue gases could be bubbled in the culture medium implementing a recycle of carbon.

2. MICRO-ALGAE AS ENERGY SOURCE

Micro-algae can be grown in open ponds or in bioreactors. The culture in open ponds is more economically favorable, but rises the issues of land cost and water availability, appropriate climatic conditions, nutrients cost and production. In bioreactors, micro-algae can grow under light-irradiation and temperature controlled conditions, with an enhanced fixation of carbon dioxide that is bubbled through the culture medium. This technology is still quite expensive as requires solar light harvesting and distribution. Although technical solutions have been developed [5] their cost remains still too high for a practical application. Therefore, the open pond culture remains the preferred option, to date.

The production of micro-algae in open ponds depends on the climatic conditions. The solar irradiation and temperature are the most important factors affecting the farming process and its productivity. These two parameters drive the growing period and, thus, the economics of the process. The availability of land and water is a key factor for developing open ponds cultures. So far, semi-desert flat lands non suitable for tourism, industry, agriculture, municipal development were selected. In these conditions, water becomes a limiting factor. Either ground saline-water, or effluents from various anthropic activities, or draining from agricultural areas have been considered for use. None of these is the ideal solution, as the operational costs could result quite high. Recently, coastal areas have been considered for algal growth, that make possible the use of waste-water, rich in nutrients, from activities like fishpond, agriculture, municipal water treatment plant, and others. The site-location of algae-ponds requires the consensus of population and authorities so that lagoon areas, or sites where relevant productive activities (fishpond) already exist, represent the low risk option. Also, open-sea farms would in principle be a practicable solution if the algae-pond could be protected from predators, that increases the cost. Noteworthy, the depth of the pond is not a demanding requisite. In fact, ponds deep from 0.6 to 1.3 m are usually used for such farming activities, in which the limiting factor is the light penetration and the density of cultures. Nutrients (N- and P-compounds, micronutrients) represent one of the major costs for algal growth. The use of waste water (sewage, fisheries, some industrial waters) rich in N-, and P-nutrients is an economic option with a double benefit represented by recovery and utilization of useful inorganic compounds, and the production of clean water that, finally, can be reused or discharged into natural basins. Another major issue with this technology, that aims at carbon recycling, is to enrich the algal cultures with inorganic carbon by direct sparging of flue gases produced from power plants. This requires that algae should be resistant to the pollutants that are usually present in the flue-gas stream, namely nitrogen- and sulphur-oxides. Studies have shown that 150 ppm of NO_2 and 200 ppm of SO_2 do not affect the growth of some algal species [6]. Anyway, it must be noted that the resistance to NOx and SOy is not a common feature of all algal species, and this may represent a limitation to the direct use of flue gases. Another point that demands clarification is the optimal concentration of carbon dioxide in the culture, as CO_2 addition lowers the pH of the medium, also by 2.4 units. Although the response to the concentration of carbon dioxide may be different for the various algal species, operating at pH close to 6 may in general strongly affect the algal growth.

The issues discussed above suggest that R&D is still necessary for the optimization of the conditions for algal growth in open ponds. Studies on confined systems have allowed an economic assessment of the potential of the technology for carbon dioxide capture and energy production.

The characteristic parameters for the production of energy from microalgae cultivated in open ponds, using either ground-water added with nutrients or waste-water that already contains N- and P-nutrients, show a low degree of homogeneity. The growing season may vary from 200 to 280 d, with a perspective of 300 d. The productivity obviously depends on the species and growing conditions. For a productivity of 33 t ha^{-1} a cost of the algae of 403 US$ t^{-1} has been calculated, with a perspective of 300 US$ t^{-1} under optimized conditions and scale economy. The lipid content of microalgae may largely vary with the species, typical values ranging around 20-40% of the dry weight for different ponds. The microalgae have not been fully characterized for their composition so far. An average heat content of the dry biomass equal to 21GJ t^{-1} has been estimated.

3. MACRO-ALGAE FOR ENERGY PRODUCTION

Despite the fact that macroalgae are extensively grown since long in China, Korea, Philippines, and Japan, for purposes like food production, or chemicals extraction, their use for energy production has received less attention with respect to microalgae, so far. The productivity of natural basins is in the range 1-20 kg m^{-2} y^{-1} dry weight (10-150 t_{dw} ha^{-1} y^{-1}) for a 7-8 month culture. Either brown algae (*Laminaria, Sargassum*) or red algae have been used. Interestingly, macro-algae are very effective in nutrients (N, P) uptake from sewage and industrial waste water. The estimated recovery capacity is 16 kg ha^{-1} d^{-1} [7]. To this end macroalgae have been used in Europe for cleaning municipal waste-water [8] and in Europe and Japan for the treatment of fishery effluents [9, 10] and for recycling nutrients. The use of macroalgae for cleaning up effluents from fisheries has an economic value as macroalgae can reduce the concentration of nitrogen derivatives like urea, amines, ammonia, nitrite or nitrate to a level that is not toxic for fishes allowing the reuse of water, reducing, thus, the cost of their growth. Also, improving the quality of the effluent water, penalties are avoided should it be discharged into natural basins. In Europe macroalgae are grown in experimental fields, and natural basins. The capacity of macroalgae as biofilters or nutrient uptake has been tested in the north-western Mediterranean Sea, along French coasts [11] using *Ulva lactuca* or *Enteromorpha intestinalis* that adapted to non-natural basins. Also in a colder climate, macro-algae grow at an interesting rate. For example, in Denmark the Odjense Fiord produces *ca.* 10 kt per day of dry-weight-biomass equivalent to *ca.* 10 t per year per ha.

Although macroalgae can grow in both hemispheres, the climatic factors may affect the productivity by reducing either the rate of growth or the growing-season. The Mediterranean Sea has ideal climatic conditions for a long growing season, with good solar irradiation intensity and duration, and with a correct temperature. Moreover, along the coasts of several EU Countries (Italy, Spain, France, Greece) fish-ponds exist that may be the ideal localization for algae-ponds.

As reported above, a point that requires a careful evaluation is the influence of the carbon dioxide concentration on the algae growth. In fact, macroalgae may either use CO_2 or HCO_3^- as source of carbon, making less strict the pH requirements. In fact, algae that use hydrogencarbonate would prefer basic water, while algae that use carbon dioxide would better grow in more acidic media, where the free carbon dioxide concentration is higher. However, the pH of the growing medium is not a strict prerequisite for those algae that contain the Carbonic anhydrase enzyme, that drives the kinetics of hydration of carbon dioxide to afford hydrogencarbonate.

The photosynthesis of macroalgae is saturated at different levels of carbon dioxide, ranging from 500 to 2000 ppm, [12, 13] that means that with carbon dioxide concentration up to five times the atmospheric concentration, under the correct light conditions and nutrient supply, macroalgae may grow with the same or better performance they show in natural environments [14].

4. ECONOMICS OF ALGAE GROWING

Microalgae are currently produced for fishpond feeding in several countries all around the world. They have also been used for municipal water treatment where microalgae provide oxygen to bacteria that metabolize organic compounds. Microalgae have found an application in nutrients capture in marine basins receiving water from rivers draining large areas of agricultural land. More recently, microalgae have been grown with the specific task of assessing their capacity of capturing carbon dioxide and producing biofuel. Several assessment studies are available on this specific topic [6, 15]. Microalgae have also been grown in bioreactors with solar light capture and distribution using optical fibers [16-18]. The growth of microalgae is often combined with their fermentation to produce hydrogen [19]. The anaerobic fermentation to afford biogas is also extensively practized.

The use of microalgae for nutrients abatement has been tested on a large scale in natural basins in the USA and an economic estimate has been done [20]. *Spirulina* is already grown in ponds of 0.5 ha, and ponds of 8 ha are designed for microlagae to be used for biofuel production. Major requisites are the mixing of the culture and nutrients supply. The latter problem should be solved with the use of wastewater or drainage waters, otherwise the operational cost of the plant and the cost of fuel would be prohibitive. Stirring requires paddlewheels, flue gases should be bubbled through the pond, better if with no further treatment of the gas produced by the power plant in order to reduce costs. The sparging technique must be reliable and cheap. Under most favorable conditions, assuming the lowest possible cost for harvesting, avoiding centrifuging, with a limited amount of chemicals added for flocculation a cost of 40-50 kUS$ per ha has been calculated, with a productivity per ha of 120-140 t per year. The cost of biogas produced in this way would range around 0.56 US$ per MJ, still too high for competing with existing fuels. It must be emphasized that the costs reported above represent the most optimistic estimate for a growing system for which all the external costs and the energy input were minimized. As a matter of fact, the current price for the production of *Spirulina* in non optimized conditions is ca. 5000 US$ per t_{dw}, that is fifty times higher than the target of 100 US$ per t_{dw}-microalgae, estimated for the production of methane at the cost given above. Much R&D is still necessary for an economic exploitation of this option.

Macraolgae have been produced or grown for years in China, Japan, Korea, and Philippines. Japan is the leading Country with more than 500 000 t per year produced in the nineties. The added value of algae depends on their use. The utilization as food is the most remunerative. The production cost depends on the algal species and a value ranging around 0.10 US$ per kg_{dw} has been established by FAO in 1990 [21]. The cost of energy production from macroalgae depends on their biofuel content and on the extraction technology. The estimate cost of feedstock may vary over a large interval, depending on many factors such as: the algal species, the cultivation technique, the need of nutrients, the collection technique, others. Typical values of 5 to 60 US$ per GJ of produced energy are commonly encounterd. Such values are higher than those of coal, oil, and LNG. An update of the cost assessment is required as most recent revision of the cost of energy derived from macroalgae is as old as 1987 [22].

The productivity of macroalgae under most performant conditions ranges from 150 to 600 t_{fw} per hectar per year, that must be compared with the typical value for sugarcane that ranges from 70 to 170 t_{fw} ha^{-1} y^{-1}. Also considering the different water content, it comes out that under the most appropriate operative conditions macroalgae would be a better energy source than terrestrial biomass. The minimum target value for a seaweed farm to be economically valuable is around 200 t per ha per year.

Interestingly, because of their edible use, several macroalgae have been fully characterized for their lipid content and nutritional properties. Consequently, a number of macroalgae are better characterized for their composition with respect to micro-algae. This also has contributed to gather useful information from the perspective of the evaluation of the biofuel production. Table 6.1 gives the classes of major compounds isolated from the algae so far treated. [23, 4] They are either hydrocarbons or fatty acids. Both such classes of compounds may be directly used as biofuel, that makes interesting the biofuel extraction with respect to the combustion or gasification. Noteworthy, depending on the algal species, a different composition of the extract is observed, that gives a different heat content to the fuel. It must be emphasized that also methyl esters of acids may be used as fuels without any major problem.

Table 13.1 - Composition of the fatty acid fraction extracted from some algae

Fatty acid	Species and relative percentage of organic compounds			
number of carbon atoms/number of unsaturated bonds	*Ulva lactuca*	*Enteromorpha compressa*	*Padina pavonica*	*Laurencia obtusa*
Saturated $C_{12} \rightarrow C_{20}$	15%	19.6%	23.4%	30.1%
Monounsaturated $C_{14} \rightarrow C_{20}$	18.7%	12.3%	25.8%	9%
Poly-unsaturated $C_{16/2} \rightarrow C_{16/4}$ $C_{18/2} \rightarrow C_{18/4}$, $C_{20/2}$	66.3%	68.1%	50.8%	60.9%

The energetic value of the algae depends on the amount of compounds extracted that may be used as fuels. The content of fuels is very variable with the species considered. (Table 13.2).

Table 13.2 - Lipid content in some marine macroalgae

Species	Lipid Content
Palmaria palmata	0.3 g/100g wet wt
Fucus serratus	2.1% dry wt
Codium harveyi	8.8-12.1% dry wt
Codium duthiae	12.2-20.7% dry wt
Codium fragile	21.1% dry wt
Hypnea charoides	18- 41.7% dry wt

It is quite evident that the lipid content may vary over a wide range, with the algae species. A content of 10-20 % dw is quite usual, with a peak value of 40%.

The net energy (NE) obtainable from either micro- or macro-algae, once it is known the energy needed for their growth (Eg), harvesting (Eh) and processing (Ep), is given by Eq. 1, where ΔH is the heat content of the extracted fuel.

$$\text{Net Energy} = \Delta H - (Eg + Eh + Ep) \tag{1}$$

For some selected species an average heat content of the extracted biofuel equal to 6.2 GJ t^{-1} has been calculated [4], that should be compared with the heat content of coal (10-28 GJ t^{-1}), oil (40-46 GJ t^{-1}) or LNG (30-60 MJ m^{-3}). In order to have the total heat content of dry algae, the energy of the residual dry mass should also be considered.

5. TREATMENT OF THE ALGAE AND TECHNOLOGIES FOR BIOFUEL EXTRACTION

Several techniques have been used for the use of algae as source of energy. Besides the direct combustion or co-combustion of the dry material, the extraction of products that may be used as biofuels and the conversion of algae into Syngas or biogas are taking a predominant role. The anaerobic methanation is a particularly suitable technology when fresh biomass is used. For the extraction of biofuels, several technologies can be considered, as the extraction with supercritical-carbon dioxide ($scCO_2$) [4, 24], or with $scCO_2$ and methanol as co-solvent [4], or else the extraction with organic solvents. Conversely, the pyrolysis can be used for the production of Syngas.

The use of $scCO_2$ for biofuels extraction is quite interesting as this technology would use recovered carbon dioxide as extraction fluid and for fixation in algae, implementing a cycle based on recovered carbon dioxide. Because of the low permeability and resistance of the cell membrane of the algae, the efficacy of the $scCO_2$ extraction depends on the pre-treatment of the algae that must be milled for an efficient extraction of chemicals, biofuel, or lipids. The extraction conditions depend on whether only $scCO_2$ [24] is used (30 MPa, 350 K) or a co-solvent (methanol [4]) is added (373 K and 8 MPa). In the former case, a dry residual mass is obtained that can be used for other purposes.

Processing algae at high temperature, as required by pyrolysis or liquefaction, produces tars that need further work-up or must be disposed. Methanation affords biogas and a sludge that can be composted and used as soil additive.

In conclusion, while algae appear a promising source of renewable energy, R&D is necessary for establishing the optimal growing condition, with full optimization of the yield and minimization of costs, so to make the fuel production an economic option. Table 13.3 comparates the properties of macro- and micro-algae.

Table 13.3 – Comparison of the properties of macroalgae and microalgae

Parameters	Microalgae	Macroalgae
Growing season, d	250-280	210-240
Productivity, t ha^{-1}, dw	33	10-150
Lipid content, % dw	20-40	0.3-42
Production cost, \$ t^{-1}, dw	*100*/5000	100
Heat value, GJt^{-1}, dw	21	12.2–25
Energy cost, \$ MJ^{-1}	0.56	0.05-0.6

A five- to ten-fold reduction of the actual best production cost is necessary for algae could be considered as energy feedstock. Large ponds (10-20 hectare per pond, 1000 hectare surface) may produce a scale economy. The use of coastal areas for growing macro-algae with nutrients recovery from waste-water, or the seaweed co-production with fish-pond may represent an economic option. The most convenient algal mass treatment or biofuel extraction technology must be selected for minimizing costs of fuel production and for developing the best use of the entire algal mass.

REFERENCES

1. Aresta, M., Dibenedetto A, Tommasi I."Developing Innovative Synthetic Technologies of Industrial Relevance based on Carbon Dioxide as Raw material" Energy&Fuels, 15, 2, 269-273, 2001.
2. (a) Wilde, E.W., Benemann, J.R. "Bioremoval of heavy metals by the use of microalgae", Biotechnology Advances, 11, 781-81, 1993; (b) Kadam, K.L. "Power plant flue gas as a source of CO_2 for microalgae cultivation: economic impact of different process options" Energy Convers. Mgmt., 38, PS 505-510, 1997.
3. Gao, K., McKinley, K.R. "Use of macroalgae for marine biomass production and CO_2 remediation: a review" J. Appl. Phycol., 6, 45-60, 1994.
4. Aresta, M., Dibenedetto, A., Tommasi, I., Cecere, E., Narracci, M., Petrocelli, A., Perrone, C. "The use of marine biomass as renewable energy source for reducing CO_2 emissions", Elsevier 2002, Special Issue Dedicated to GHGT-6, Kyoto, October 2002.
5. Ono, E., Cuello, J.L. "Design Parameters of Solar Concentrating Systems for Algae-Based CO_2-Emission Mitigation", Elsevier 2002, Special Issue Dedicated to GHGT-6, Kyoto, October 2002.
6. Zeiler, K.G., Heacox, D.A., Toon, S.T., Kadam, K.L., Brown, L.M. "The use of microalgae for assimilation and utilization of carbon dioxide from fossil fuel-fired power lant flue gas" Energy Convers. Mgmt., 36, 707-712, 1995.
7. Ryther, J.H., DeBoer, J.A.,. Lapointe, B.E "Cultivation of seaweeds for hydrocolloids, waste treatment and biomass for energy conversion" Proceedings International Seaweed Symposium, 9, 1-16, 1979.
8. Schramm, W. "Seaweed for waste water treatment and recycling of nutrients"in Guiry M.D., and Blunden, G. Eds "Seawed Resources in Europe: Uses and Potential",.John Wiley & Sons, Chichester, p. 149-168, 1991.
9. Cohen, I., Neori, A. "Ulva lactuca biofilters for marine fishpond effluents I. Ammonia uptake kinetics and nitrogen content" Bot. Mar., 34, 977-984, 1991.
10. Hirata, H., Xu, B. "Effects of feed addictive Ulva produced in feedback culture system on the growth and color of Red Sea Bream. Pagure major" SUISANZOSHOKU, 38, 177-182, 1990.
11. Sauze, F. "Increasing the productivity of macroalgae by the action of a variety of factors. in Stub, A., Chartier, A., Schleser, P., Schleser, G. Eds, "Energy from Biomass", Elsevier Applied Science, London, 324-328, 1983.
12. Brown, D.L., Tregunna, E.B. "Inhibition of respiration during photosynthesis by some algae" Can. J. Bot., 45, 1135-1143, 1967.
13. Smith. R.G., Bidwell, R.G.S. "Carbonic anhydrase-dependent inorganic carbon uptake by the red macroalga Chondru crispus" Plant Physiol., 83, 735-738, 1987.
14. Gao, K., Aruga, Y., Asada, K., Kiyohara, M. "Influence of enhanced CO_2 on grand photosynthesis of he red algae Gracilaria sp. And G. chilensis" J. Appl. Phycol., 5, 563-571, 1993.
15. Benemann, J.R., Oswald, W.J. "Systems and Economic Analysis of Microalgae Ponds for Conversion of CO_2 to Biomass", Final Report, Pittsburg Energy Technology Center, 1996, p. 260.
16. Hase, R., Oikawa, H., Sasao, C., Morita, M., Watanabe, Y. J. "Photosynthetic production of microalgal biomass in a raceway system under greenhouse condition in Sendai City" Bioscience & Bioengineering 89(2), 157-163, 2000.

17. Nagase, H., Yoshihara, K., Okamoto, Y., Murasaki, S., Yamashita, R., Hirata, K., Miyamoto, K. "Uptake pathway and continuous removal of nitric oxide from flue gas using microalgae" Biochem. Eng. J., 7, 241-246, 2001.
18. Nakajima, Y., Tsuzuki, M., Ueda, R. "Improved productivity by reduction of the content of light-harvesting pigment in Chlamydomonas perigranulata" J. Appl. Phycol., 13, 95-101, 2001.
19. Ikuta, Y., Akano, T., Shioji, N., Maeda, I., in "Biohydrogen", Zaborsky O. Ed., Plenum Press., New York, 1998, p. 319-328.
20. Weissman, J.C., Goebel, R.P. "Design and Analysis of Pond Systems for the Purpose of Liquid Fules", Solar Energy Res. Inst., Golden CO, SERI/STR-231-2840, Final Report, 1997.
21. FAO "Training manual on Gracilaria Culture and Seawood Processing in China", in Training manual, 6, 2-46, 1990.
22. Bird, K.T. "Cost analysis of energy from marine biomass" in Bird, K.T., Benson, P.H. Eds., "Seaweed Cultivation for Renewable Resources", Elsevier, Amsterdam, 327-350, 1987.
23. Cheung, P.C.K. "Temperature and pressure effects on supercritical carbon dioxide extraction of n-3 fatty acids from red seaweed" Food Chemistry, 65, 399-403, 1999.
24. Wahbeh, M.I. "Amino acid and fatty acid profiles of four species of macroalgae from Aqaba and their suitability for use in fish diets" Aquaculture, 159, 101-109, 1997.

Chapter 14

ENERGY FROM BIOMASS

Piergiorgio Zappelli* and James J. Leahy[§]
*EniTecnologie, Monterotondo (Italy); pzappelli@mail.enitecnologie.eni.it
[§]University of Limerick, Limerick (Ireland)

1. INTRODUCTION

Renewable energy refers to a wide range of naturally occurring and replenishable energy sources that can give a contribution to the needs of the five basic energetic markets: centralized electricity supply, decentralized electricity supply, heat supply services and liquid fuels for transport.

Renewables contributed in 1995 71.5 Mtoe to EU energy markets, corresponding to 5.1% of energy demand. This was mostly provided by large-scale hydro plants and by biomass, mainly in the form of wood-fired heating. The proportion of EU energy supplies derived from renewables has grown in recent years, particularly due to increased uptake of newer technologies, like wind power, landfill gas photovoltaics and solar heating.

By 2010, with the present policies and trends the contribution of renewables in the EU will increase, but the EU increasing energy demand will limit the renewables energy contribution to 5.9% of EU demand [1].

In terms of technological and market status the technologies for renewable energy production vary very widely. Some technologies, like hydropower and conventional domestic combustion of wood are fully mature. Others, like wind power, photovoltaics and solar water heating, are becoming established in limited niche markets, but with the potential to move into larger markets. Tidal and wave power, and pyrolysis of biomass are at the R&D stage.

Renewable energy technologies typically have relatively high capital costs and very low running costs, and this has a strong influence on how they are perceived in the market. Although the cost of several of the

M. Aresta (ed.), Carbon Dioxide Recovery and Utilization, 325-376.

renewables has fallen dramatically in recent years, many, when compared with the conventional market alternatives, still appear uncompetitive.

Renewable energy can play a major role in furthering a wide range of EU policy goals. Development of a successful EU renewables sector would make a useful long-term contribution to diversity, security and self-sufficiency of energy supply. In addition, as low environmental impact is one of their main benefits, renewables could play a leading role in mitigating the environmental effects of energy use.

Recent developments in market deployment and technological innovation have generated confidence that the world is beginning the transition to a sustainable energy system with a significant contribution from renewable resources.

Under the "Factor 4 Strategy" [2] which addresses the question of the time scale to reduce CO_2 by 50% while phasing out nuclear energy, the scenario outlined in Table 1 is obtainable. Many developing countries with emerging economies lack sufficient energy supplies and are deficient in conventional energy resources. This scenario envisages that they would "leapfrog" to sustainable energy by adopting the most energy efficient principles and conversion technologies while introducing advanced renewable energy options using local resources. A key finding of the analysis was that total energy consumption would stay at about 10 Gtoe through improved energy efficiency particularly in the transport and building sectors.

Table 14.1 - Primary Energy under a factor 4 Scenario [2]

Gtoe	1995	2020	2050
Coal	2.5	2.3	0.3
Oil	3.0	3.0	2.5
Gas	1.7	1.8	1.2
Renewables	1.8	2.6	6.3
Nuclear	0.5	0.2	0.0
Total	9.5	9.9	10.3

With the exception of hydro and bioenergy many renewable energy options are limited in market deployment to niche applications. To broaden application within energy markets renewable energy technologies must have improved efficiency and reliability to the point where they are comparable with conventional energy.

Drivers for renewable energy are:

- energy security: the dependency on oil imports for the IEA region is currently 54% of total oil requirements and is forecast to grow significantly as with increasing demand in non-OECD countries.
- concerns about environmental degradation and the Kyoto protocol [3] employment: there are significant job creation benefits from a

strategy for greater promotion of renewable energy with forecasts of up to 500,000 jobs in the EU region by 2010, (TERES II).

2. THE BIOMASS

Biomass represents the renewable raw material in which large amounts of carbon dioxide can be fixed by the photosynthetic process. The energetic conversion of biomass allows the production of energetic vectors for the substitution of fossil fuels, finalized to the reduction of CO_2 emissions.

Biomass, as renewable raw material, is the subject of many studies, principally devoted to point out:

- cultivation procedures, harvesting, transport and storage of the raw material
- transformation technologies in different types of products (solid, liquid or gaseous fuels, chemicals etc.)
- alternative use of products (for example ethanol "100" or in mixture with gasoline)

The above activities have been joined to economic studies and to general evaluations finalized to the choice of the most favorable local situation for the biomass exploitation, and for the definition of the strategies for overcoming the barriers hampering their wide utilization.

Recently, with the availability of new tools for the evaluation of the environmental impact (LCA, ecobalance, etc.) and of sophisticated procedures of environmental-economic analyses, the amount of available literature has grown further.

After the Kyoto agreement (8.0% CO_2 emission reduction, referred to 1990, at European level, and 6.5% reduction at Italian level) the debate on the use of biomass and, in general on renewable energies, has moreover increased.

At present, however, the use of biomass is still limited, and only few of the potential technologies are utilized: direct combustion and alchoolic fermentation, on large scale; biomethanation and biodiesel on limited amount.

The topic, in spite of the wide mass of studies, still presents improvement margins.

3. REFERENCE SCENARIO

The 1996 energetic needs, at global level, have been evaluated amounting to 8380 Mtoe (value estimated only on the base of the commercially exchanged fuels). To the above figure, energy coming from the traditional use of biomass (930 Mtoe) could be added.

The renewable energies as a whole, amount to 1660 Mtep. A synthetic picture on the future global energetic sources is found in two scenarios published by Eurostat 1996 (4).

- The first scenario, called *"Business as usual"* shows an increase in the innovative use of biomass from 172 to 270 Mtoe in the period 2000-2020. Moreover, the overall energetic demand will grow, for the same period, from 9,000 to 15,000 Mtoe, and the contribution from all (innovative and traditional) the renewable sources, would reach 2,800 Mtoe, corresponding to about 21% of the global energetic needs.
- In the second scenario, called *"Environmentally driven"* the use of biomass would grow, during the above interval, from 170 to 670 Mtoe. The overall renewable would amount to 3,300 Mtoe, corresponding to 30% of the forecast global energy needs.

We observe that in both scenarios, biomass is considered to play an important role as energetic source. The positive trend to increase the energetic use of biomass in Europe as outlined by EU legislation (5,6) is influenced by:

- increase in population sensitivity towards environmental problems;
- progressive reduction of nuclear power generation;
- unbalance of the EU energetic sources (56% of the needs).

In this purpose EU has promoted the financing of studies on renewable energies. Both the EU Commission and The World Energy Council forecast for the renewables by 2020 are quite similar, ranging around 14% of the total (7,8).

4. BIOMASS FEEDSTOCKS

Biomass refers to all forms of plant-based material that can be converted into usable energy including short rotation wood crops, herbaceous energy crops, halophytes and some annual crops and oilseeds as well as agricultural and forestry residues. The use of biomass based energy systems as a substitute for fossil fuel offer potential as a means of reducing CO_2 emissions. Such systems are carbon neutral, since when combusted, biomass returns to the atmosphere CO_2 captured by the feedstock. There are however CO_2 costs embedded in the inputs required for biomass production, fertilizer and diesel for example, but there are also savings in sequestration of carbon in the soil.

World wide biomass ranks fourth as a energy source, providing 14% of the world's energy needs; biomass is the most important source of energy in developing countries providing ~ 35% of their energy [9]. Projections of potential biomass contribution to global energy requirements vary depending on which models are used to assess future global energy needs but estimates of 17-38% by 2050 have been proposed [10]. This estimate forecasts that by 2050 biomass could provide nearly 38% of the world's direct energy fuel use and 17% of the world's electricity or a total contribution of 206 EJ 1EJ = 10^{18}J (see Table 14.2).

When estimating biomass feedstock and the potential of biomass to meet future demand several questions need to be addressed including land availability, (short and long term), productivity's, species and mixtures, geographical location/environmental sustainability. Biomass production on a scale that will allow a significant substitution of fossil fuels cannot be accomplished solely on marginal lands, but will require large areas of prime agricultural land and substitution of food crops and industrial crops (e.g. cotton, tobacco) for biomass crops.

The EU White Paper on renewable energy sources estimates that 12% of total energy requirements (190 Mtoe) will be supplied from renewable sources by 2010 (Mtoe = million tons of oil equivalent) of which bioenergy will represent 90 Mtoe. [12] This 90 Mtoe comprises 15 Mtoe from biogas, 30 Mtoe from agriculture and forestry wastes with dedicated energy crops supplying 45 Mtoe. The contribution of dedicated energy crops can be further subdivided into 18 Mtoe from liquid biofuels, with ligno-cellulosic materials including short rotation forestry supplying 27 Mtoe. Kuiper et al. [13] have analyzed the total supply of roundwood, industrial wood residues and waste wood in the EU, (see Table 14.2).

Table 14.2- Biomass supply for renewables intensive global energy scenario (EJ $year^{-1}$) [11]

	Forests	Residues		Plantations		Total	
Year	2025/2050	2025	2050	2025	2050	2025	2050
Africa	2.43	6.81	9.38	18.94	31.81	28.18	43.6
Latin America	1.59	10.92	13.59	32.3	49.6	44.81	64.78
S.& E Asia	3.13	13.61	20.42	-	-	16.74	23.55
C.P Asia	1.21	3.85	4.16	5	15	10.06	20.37
Japan		0.89	0.95		-	0.89	0.95
Australia/NZ	0.02	1.14	1.39	--	-	1.16	1.41
USA	0.61	5.86	1.42	9.6	9.6	16.07	15.89
Canada	0.04	1.43	4.86	1.2	1.2	2.67	2.66
OECD Europe	0.31	4.85	0.23	9.0	9.0	14.16	14.17
Middle East	0.02	0.18		-	-	0.02	0.25
Total	9.36	49.54	56.04	76.04	116.21	134.76	187.63

Table 14.3 - Amounts of biomass required by 2010 as proposed by EU White Paper on Renewable energy [13]

	Mtoe	Amounts proposed by EU White Paper	
		10^6 m^3 roundwood equivalents	million tons
Woody biomass			
Forestry residues	30	176	
SRF	27	159	
Non-wood biomass			
Agricultural residues	30		85
Energy crops	27		76

The development of large scale energy production from biomass in Europe will probably rely in the future on specifically grown energy crops such as tree plantations sugar cane, perennial herbaceous plants and oil plants. The types of crops can be divided into the three categories

- conventional agricultural crops e.g. wheat, maize potato, sugar beet sunflower and rape
- new industrial crops such as sweet sorghum and miscanthus
- agro-forestry crops i.e. eucalyptus and willow

4.1 Agro-forestry products

The projected demand from the agro-forestry industry is shown in Table 14.4.

Table 14.4 - Total supply and demand for roundwood, wood residues and woody biomass from energy crops in 2010 in million m^3 roundwood equivalents and (Mtoe)

	Roundwood	Residues/ waste wood	Energy crops	Total
Supply	356 (60.4)	47 (8)		403
Demand	298 (50.6)	177 (30)	159 (27)	634 (107.6)
Total	58 (9.8)	-130 (-22)	-159 (-27)	-231 (-39.8)

The projected maximum supply of roundwood in 2010 is 356×10^6 m^3 and that of non used wood residues 47×10^6 m^3, while the projected demand by forest industries is 298×10^6 m^3 of roundwood, 177×10^6 m^3 residues and waste wood and 159×10^6 m^3 of energy crops. This leaves a surplus of 58×10^6 m^3 of roundwood with a 130×10^6 m^3 shortfall of wood residue, which will have to make up from dedicated energy crops or agricultural residue. Non-wood fibers such as cereal straw, rapeseed straw and linseed straw with a total production output within the EU of approximately $120x10^6$ tons per annum have energy content corresponding to 43 Mtoe.

4.2 Agricultural crops

Within the EU there are considerable difference in agricultural productivity's e.g. the average yields of wheat and potato are 6.4 and 6.8 dry t ha^{-1} in countries such as France, Germany, UK and Netherlands (middle Europe) compared with 2.0 and 3.9 dry t ha^{-1} for Portugal, Spain, Italy and Greece (Southern Europe), as reported by Diamantidis et al., [14]. They assessed the crops cultivated on agricultural land and their residues (Table 14.5) [14]. They found that wheat is the dominant crop both in terms of area 9.9 Mha and in terms of tonnage on a dry basis followed by maize and sugar beet. This ranking remained the same when crop residues were taken into account.

Diamantidis et al., also assessed feedstocks of conventional crop residues (Table 14.6). They divided Europe into three regions North, Middle and South on the basis of crop types, biomass growth rates and labor costs.

Table 14.6 - Availability of crop residues as bioenergy feedstock [14]

Crop	Residue	Residue yield (dry tons/ha)	Potential production (M dry tons/y
Wheat	Straw	3.5	34.2
Maize	Cobs corn stover	9.4	20.7
Rape	Straw	3.0	7.4
Sunflower	Straw, heads	6.0	6.2
Potato	Tops & leaves	2.1	2.1
Sugar beet (M)	Tops & leaves	4.1	6.3
Sugar beet (S)	Tops & leaves	3.7	1.9
Total			78.6

The estimated available quantity of residues, based on both the cultivated area of each crop and the respective product/residue ratio is 78.6 M dry tons per year. Panoutsou [15] estimated the total amount of agricultural residues in the EU at 153 M dry tons.

Table 14.5 Productivity of cultivated crops and residues in Europe

Crop/tree	Area (Mha)	Production (M fresh t)	Yield (dry tons/ha)	N inputs (kg N/ha)	Energy inputs (GJ/ha)	Labor (man-hours /ha)	Energy content (GJ/dry ton)	By products
Middle Europe								
Wheat	9.9	68.2	4.6-7.7 (6.4)	100-180	120	17.2	16.5	Straw
Maize	2.2	17.1	6.3-8.0 (7.2)	110	160	17	16.5	Cobs and stovers
Sunflower	1.0	2.2	1.8-2.0 (1.9)		50-75	16	16	Straw, heads
Rape	2.5	7.5	2.7-4.5 (3.0)		50-100	13.9	18.2	Straw
Sugar Beet	1.5	82.8	6.5-10 (8.2)		55-110	41.4	15	Tops & leaves
Potato	1.0	34.6	5.4-8.2 (6.9)		75-110	45.8	15	Tops & leaves
Miscanthus			20.0-21.0 (20.3)		75	15	17	
Poplar			9.0-12.0 (10)		50	24	17	Tops & leaves
Willow	0.018		5.0-20 (10)		70	24	17	
South Europe								
Sugar beet	0.5	23.3	6.5-9.8 (7.4)		55-110	41.4	15	
Miscanthus			22-26 (24)		75	15	17	
Sweet Sorghum			22-45 (35))		75-125	13.4	17	
Poplar			6-10 (8.0)		75	24	17	
Eucalyptus	0.5		10-25 (16.0)		75	24	18	

The dominant use of agricultural biomass is food production. About 55 M fresh tons p.a. are directly used for human consumption with 51 M fresh tons being used as animal feed to support meat production for human consumption. The reform of the Common Agricultural Policy (CAP) in 1992 has encouraged the use of agricultural land for non-food production. The selection of future crops will be based on criteria such as high yields, efficient water use high dry matter content at time of harvest, high energy density and lowest possible environmental impact [16]. Thus industrial crops (e.g. miscanthus and sweet sorghum) offer significant potential which may be further enhanced with genetic engineering technologies. Plant breeding and husbandry for conventional crops is such that nowadays the total biomass from these crops that can be utilized (main-product + residues) for food and energy is of the same order of magnitude as that of agro-forestry crops. These yields are expected to improve further with selective breeding programs. Diamantidis et al. have projected yields from selected crops as well as costs (Table 14.7).

In order for biomass cultivation as an energy resource to increase, feedstock prices must be attractive in order for farmers to replace food production with non-food crops and the costs of conversion of biomass to energy must be competitive with fossil fuels. Production costs vary widely with the single greatest contribution to the variation arises from location which determine the variable cash expenses (seeds, fertilizer, chemicals, fuel, labor) fixed costs (taxes, overheads, depreciation). Land availability is also an issue affecting the biomass cultivation. According to scenario studies for Western Europe the available land for industrial crops will be about 20-25 Mha [17]. This land will support the production of about 90 Mtoe of energy, (see Table 14.8) [14], which agrees with the amount of biomass required to fulfil the goal of 12% share of renewables by 2010.

Table 14.7- Yield (Dry t / ha) and costs (1995) (ECU per dry ton) [14]

	1990	2000	2020	2050	cost
Wheat	5.9	6.7	7.4	7.7	110-211
Maize	7.1	7.4	8.2	8.6	135-211
Rape	2.7	2.8	3.5	3.5	305-488
Sunflower	2.3	2.5	2.8	2.8	176-245
Potato	6.4	7.1	7.9	7.9	390-595
Sugarbeet	8.4	8.8	9.2	9.2	203-316
Miscanthus		23.6	27.4	29.5	76.0-83.3
Poplar	10	11.1	12.2	12.2	101.1-143.4
Sweet Sorghum		35	40.6	43.8	40.0-63.0
Willow	10	11.1	12.2	12.2	69.4-75.4
Eucalyptus	16	17.7	19.5	19.5	63.0-74.8

Table 14.8 - Estimated amount of available bioenergy in Europe 2010 [14]

	North	Middle	South	West.Europe
Crop cultivated (Mha)				
Willow	4	4		8
Cereals		4		4
Rape		4		4
Sorghum			3	3
Eucalyptus			3	3
Total	**4**	**12**	**6**	**22**
Product (Mtoe)				
Ethanol		6	15	21
Diesel		4		4
Heat & electricity	15	30	20	65
Total	**15**	**40**	**35**	**90**

4.3 Waste

Waste production is originated by the following main categories, reported in decreasing order of importance: manufacturing industry, mining, energetic industry, agriculture, municipalities (Fig.14.1).

With reference to energetic industry, at the end of the 80s, the European refineries were producing about 800.000 tons of sludge, corresponding approximately to 2 kg/t of processed oil. In the effort to reduce sludge production, this industry introduced a lot of minimization and recycling procedures, and incineration of oily wastes, with an ultimate waste production of 150.000 t, corresponding to 0.2 kg/t of processed oil, with a total recovery/recycle of 80-90%.

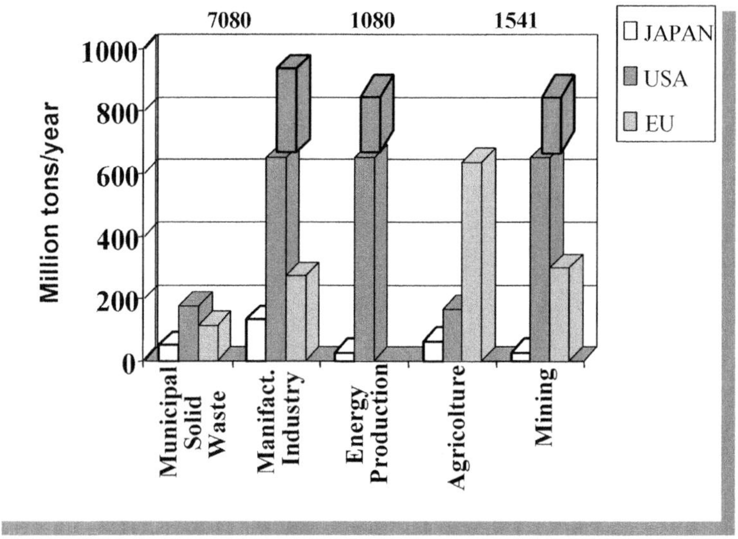

Figure 14.1- Main categories of waste in developed countries

Waste originated by forestry and agriculture has high potential for energetic valorization. The energetic value of forestry and agricultural residues in Europe is 4.5 x 10^{12} MJ/y. Forestry residues are currently utilized as feedstock for integrated gasification processes, whereas there is considerable under-utilization of agricultural residues.
While waste production in the industrial sectors is generally decreasing, the trend of waste production from municipalities seems to be opposite. Waste production in developed countries ranges from 0.8 to 2 kg per capita/day, and it tends to remain almost stable, whereas in developing countries said production, presently at level of 0.2 kg/capita/day is continuously increasing. Municipal solid waste (MSW) remains one of the most urgent environmental issues to be tackled, as in addition to waste production it must be considered the increasing environmental concern and the shortage of landfill sites. Energetic exploitation of wastes is considered one of the most promising options [18]. Municipal solid waste could give a useful contribution to the energy budget of many nations. For instance, if part of waste produced in Italy were recycled, according to the present Italian legislation, and the remaining were incinerated in conventional industrial plants, more than 3% of the Italian electrical consumption could be obtained. Presently incineration is performed at large scale only in a few countries: Japan (75%), Denmark (65%), Sweden (55%), Germany (40%), France (42%), whereas in the other developed countries said technology has limited application (in US only 15% of the total MSW is incinerated) as reported in Fig. 14.2 [19].

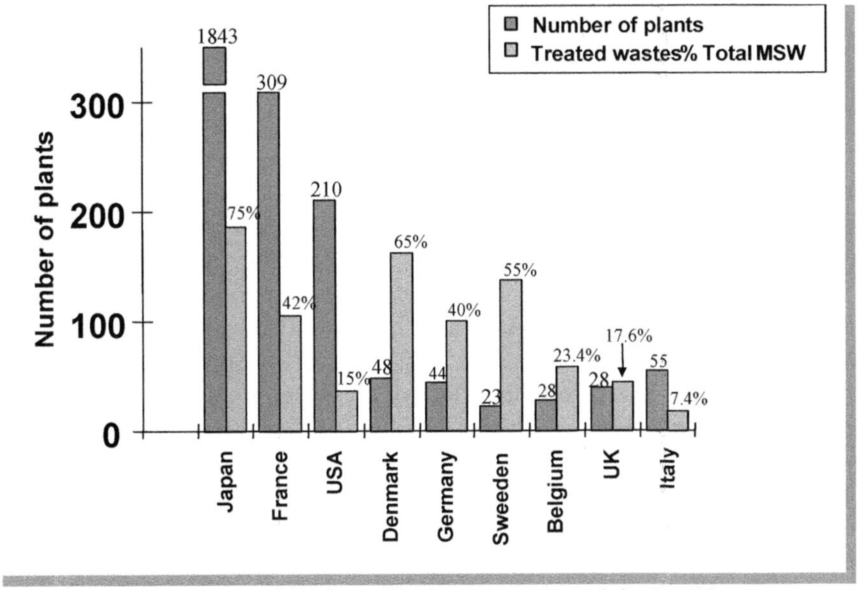

Figure 14.2 - The diffusion of incineration of municipal solid wastes

4.4 Landfill gases

Landfill gases, consisting mainly of methane and carbon dioxide, is produced when organic wastes decay in landfill sites. The methane content can be utilized as fuel for electricity and heat generation.

The technology is well established and consists of perforated plastic pipes drilled into the waste that are joined together by a plastic pipe-work and connected to a suction pump to extract the gas. The gas is cleaned and burned off in a flare stack or used as fuel.

Landfill gas has been widely adopted in the EU since the early '80s and currently provides 553 MW through heat and electricity generation, figure that is expected to increase to 1366 MW in 2010. The world market for landfill gas use is estimated to be about 4500 MW by 2010. Over the next 10-20 years landfill gas use will increase rapidly in the EU due to improved waste management and new environmental measures, but in the subsequent period there will be a decline of utilization, due to the EC Landfill Directive aimed to reduce the biodegradable material entering landfill sites, thereby reducing the generation of landfill gas. The largest area of growth is in the world's emerging economies.

The sector is developed by small-medium enterprises with technical know-how and financial skills to put landfill gas use packages together.

Landfill gas technology needs little in the way of further development.

5. EXPLOITATION OF BIOMASS POTENTIAL IN EUROPE BY 2010

The potential for exploitation of biomass in Europe by 2010 has been estimated in a White Paper by the European Commission. [12] The indicative objective estimated by the plan by 2010 is a 12% share of renewables of the overall energetic consumption. Currently biomass accounts for about 3% of total inland energy consumption (EU15). However in some member states like Austria, Finland and Sweden, this renewable source already accounts for 12%, 23% and 18% respectively of primary energy supply. On condition that effective measures are adopted to reach the objective, a three times increase of the present amount of produced bioenergy (44.8 Mtoe) is considered to be possible. This would mean additional biomass amounting to 90 Mtoe, equivalent to 8.5% of the projected total energy consumption by 2010, with an overall bioenergy amount of 135 Mtoe. The present biomass utilization includes, in addition to woody biomass and the residues of wood working industry, energy crops, agricultural residues and agro-food effluents, manure as well as the organic fraction of municipal solid waste.

The additional estimated bioenergy use of 90 Mtoe would be derived from agricultural, forest and forest industry residues, waste streams as well as new energy crops.

Biomass exploitation has the double benefit of exploiting an important renewable energy resource and also of improving the environment and the climate.

The advantages of exploiting biomass, based on new technologies, can be clearly seen in the case of biogas exploitation. This consists largely of methane, a gas with a large greenhouse impact. It has been estimated that the total energy content of landfill gas and digestible agricultural wastes in the EU exceeds 80 Mtoe. The contribution that could be made by biogas exploitation from livestock production, agro-industrial effluents, sewage treatment and landfill by 2010 is estimated at 15 Mtoe.

As far as solid residues are concerned, there is a huge potential so far unexploited in the form of wood and agricultural residues, which exceeds 150 Mtoe/y. It is estimated that 30 Mtoe can be mobilized annually by 2010 for the power and the heating and industrial process heat market.

Energy crops also need to be considered with the aim of reaching the objective. 45 Mtoe is the amount estimated, of this 18 Mtoe could be in the form of liquid biofuels and 27 Mtoe as biomass for heat and/or power.

In substantial agreement with the above figures are the evaluations made by IEA Bioenergy (20) (see. Fig. 14.3).

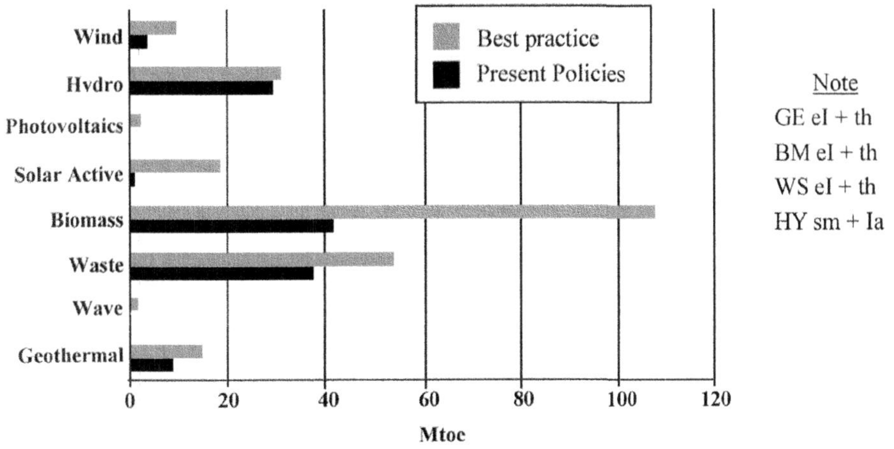

Figure 14.3 - Renewable energy penetration by Technology in EU by 2010 [20]

Bioenergy currently contributes 12-14 % or about 50 EJ to global energy supply and has the ability to increase its share of energy supply to200-300Ej by 2010. (IEA Bioenergy Agreement). Biomass has a dual role in CO_2 mitigation since it can act as a source of sustainable energy to substitute for fossil fuels as well as acting as a carbon store. The IEA estimate that for every MWh of renewable energy generated there may be as much as 1MT of CO_2 eliminated.

Main R&D priorities identified for Biomass, by the IEA are

- improved conversion technologies in combustion systems, co-firing of biomass and biomass wastes in coal fired generating plants,
- process modelling.

6. BIOMASS ENERGETIC CONVERSION OPTIONS

A schematic representation of the principal options for the biomass conversion into energetic products is reported in Fig. 14.4.

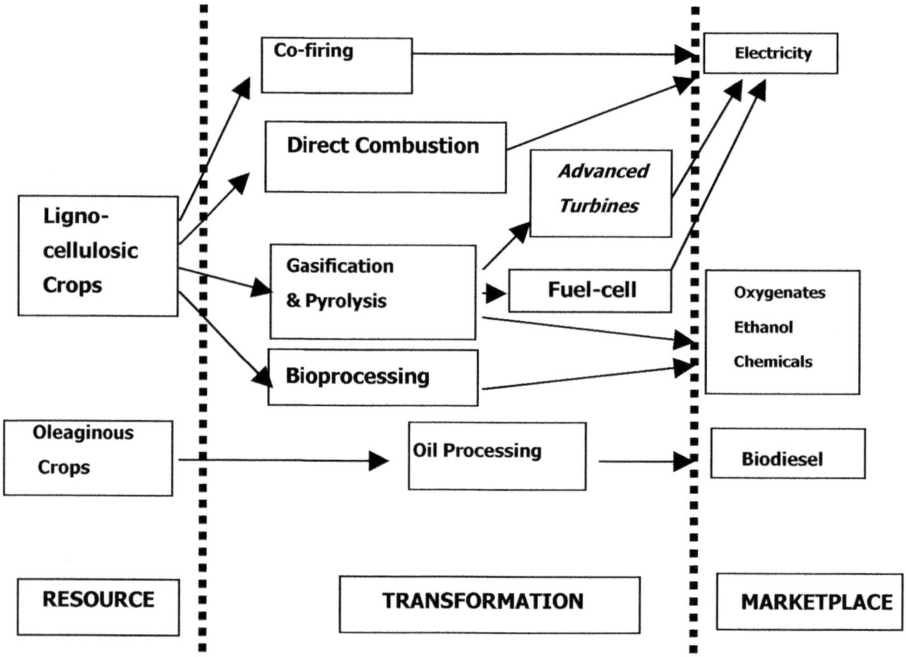

Figure 14.4- Biomass conversion to energetic products

The most appealing opportunities for the biomass energetic exploitation are bioelectricity and bio fuels (biodiesel and ethanol). At present, the costs of the above products are more expensive of their counterparts obtained from fossil energetic sources. The production cost of a power plant based on biomass gasification reaches 8.8 US cents/kWh, as compared to the corresponding cost of 3.7 US cents/kWh, obtained by a combined cycle gas turbine power plant. [21] In the same way, the costs of biofuels are generally higher than the corresponding traditional fuels.[12] The principal reasons are: low prices of fossil sources, high costs of biomass, and, at lesser extent, the more complex technologies needed for biomass utilization. This situation could remain unchanged for a long period unless new threat/opportunity happens. The best way to get the maximum advantage from biomass is to convert it into an energy vector such as fuel (liquid or solid) or into electric power. In Table 14.9 the most common biomass conversion processes are reported.

Table 14.9 - Bioenergy conversion options

Process	Energetic vector
Biological Processes	
Anaerobic Digestion	Biogas (mixture methane-CO_2 60/40)
Fermentation - Bioconversion	Ethanol; Hydrogen; Volatile fatty acids
Chemical Process	
Transesterification	Biodiesel
Thermochemical Processes	
Combustion	Production of electric power by steam cycles
Gasification	Syn-gas and fuel gas for fueling gas turbines
Pyrolysis	Liquid hydrocarbons and syn-gas

7. CONSOLIDATED POWER TECHNOLOGIES FROM BIOMASS

7.1 General considerations

As discussed before, biomass is obtained from numerous sources, including by-products from the timber industry, agricultural crops, raw material from the forest, major part of household waste, and demolition wood. Several factors make it an attractive option for power generation. Biomass, if properly grown and managed, is a renewable resource. It does not add carbon dioxide to the atmosphere because it absorbs the same amount of carbon in growing as it releases when consumed as fuel. Its low sulfur content means biomass combustion is much less acidifying than coal, for example.

The use of biomass for power generation has increased over the last decade. In the United States, electricity generation from biomass grew by 7% each year between 1990 and 1994, reaching 59,000 GWh in 1994. Such growth could result in an industry with a capacity of approximately 30-40 GW, producing 150,000-200,000 GWh of electricity by 2020.

In Europe biomass energy currently accounts for about 2% of total consumption, and the European Commission predicts that figure will reach 15% in the European Union over next 15 years.

At Italian level, the trend for the production of bioelectricity has been pointed out in the National Program for Renewable Energies from Biomass. [22] The figures reported are very exciting, indicating a fast growing of the installed power: from 100 MW by 1999, to 4000 MW by 2012.

Unlike other renewable-based systems that require costly advanced technology (such as solar photovoltaics), electricity can be generated from the same type of equipment and power plants used for burning fossil fuels. Many innovations in power generation with other fossil fuels my also be adaptable to the use of biomass fuels.

From the development of new technologies, like the combined cycle pressure gasification, and co-firing (biomass plus coal) a substantial reduction of power costs is expected, approaching the one obtained from fossil sources.

However, various factors have hindered the growth of the renewable energy resource.

Most power plants operating today are characterized by low boiler and thermal-plant efficiencies; both the fuels characteristics ant the small size of most facilities contributes to these efficiencies. In addition, such plants are costly to build. Today's best biomass-based gasification power plants require investments amounting to approximately $2,000 per We, with a thermal efficiency of about 40%, while large coal-fired stations cost about $1,500 per We, with a thermal efficiency of approximately 45%.

The main challenge to using biomass for power generation, therefore, is to develop more efficient, lower cost systems.

The description of the principal thermal technologies for the energetic conversion of biomass is reported.

Thermal processes include combustion, which is the most established technology, as well as advanced technologies, such as gasification and pyrolysis, which are developing the potential to compete with combustion. With the development of new combustion technologies such as co-firing (biomass plus coal) a substantial reduction in power costs is expected, approaching that obtained for fossil fuels.

A promising alternative to combustion is the gas turbine fueled by gas produced from biomass by means of thermochemical decomposition with a restricted air supply. Gas turbines have comparable to lower capital costs than their combustion counter-parts, can be considerably more efficient and have good prospects for improvements of both parameters.

Pyrolysis is also becoming a feasible and viable route to renewable liquid fuels, chemicals and derived products. Pyrolysis process results in a 21% increase in electricity output per ton biomass, as compared to combustion processes. The liquid fuel obtained (bio-oil) offers significant advantages in terms of storage and transport, as compared to gasification.

For dealing with moist biomass, a promising approach, although at exploratory level, is represented by the hot compressed water biomass gasification, whose product is a hydrogen rich fuel gas.[23]

Advanced biomass-based systems for power generation require fuel upgrading, combustion and cycle improvement, and better flue gas treatment. Future biomass-based power generation technologies have to provide superior environmental protection at lower cost by combining sophisticated biomass preparation, combustion, and conversion processes with post combustion cleanup. Such systems include fluidized bed combustion, biomass integrated gasification, and biomass externally fired gas turbine.

Being power generation the most attractive option in the short term, our attention has been addressed particularly to this task, describing in more detail the principal aspects of the related technologies.

7.2 Thermal conversion: electricity and heat

Biomass fuels are derived from four sources: forestry residues, agricultural residues, agro-processing residues, and energy crops specifically grown for use as fuel.

The conventional combustion technology is expensive and has limited development potential for biomass electricity. Advanced technologies that convert the biomass to gas or liquid before combustion show the promise of lower overall costs. Co-firing with fossil fuels in an existing boiler is potentially the lowest cost option, but is limited to use in areas with existing coal plants. In EU Nordic countries (Sweden, Finland, Denmark) and Austria, district heating using wood chips and straw is diffused practice. Agricultural residues are a difficult fuel to burn efficiently but a technology is being developed in Denmark. Agro-processing residue are not an important resource in EU, but represent an export opportunity for the power engineering industry. Energy crops (mostly willow and poplar) are important for the EU log term energetic strategy, because they are the only biomass fuel that can be expanded sufficiently to significantly shift the pattern of EU energy supply.

The electricity produced from biomass in the EU in 1994 was 11.9 TWh, that is projected to increase to 27 TWh in 2010. At world level the today power production from biomass is 127 TWh/y that is set to rise to 291 TWh/y in 2010. [1] The industry consists of a wide range of enterprises ranging from multinational power engineering companies in the Nordic countries to smaller companies involved in the development of fuel supply systems.

EU, principally thanks to Finland and Sweden is the market leader in electricity generation using biomass in conventional steam cycle power plant, and is moving towards the same position in advanced biomass technologies. The projected increase in biomass electricity generation (2400 MWe by 2010). If more favorable policies to renewables are adopted, the already strong market position of EU industry could be further improved.

Many are the barriers to the further development of the biomass electricity. Among them we consider: lack of understanding of the benefits of biomass for power; lack of integrated policies for achieving specific targets at EU and member states level; high capital and fuel biomass costs, relative to fossil fuels, lack of farmer willing to plant energy crops, due to costs and uncertainty of policies, uncertainty about future European standards for atmospheric emissions.

7.3 Municipal solid waste combustion

Municipal solid waste combustion (MSWC) reduces waste to a sterile ash, thereby reducing transport cost and landfill requirements, and saving on the consumption of fossil fuels. It also eliminates the production of methane, a potent greenhouse gas, associated with the degradation of waste that has been sent to landfill. The approach is widely practiced in the EU and the energy may be recovered for power production alone or for CHP (combined heat and power). The technology is well established and fully commercial although plant capital and operating costs are high. The competitiveness of MSWC is expected to improve due to the rising cost of alternative disposal routes.

There are currently over 250 MSWC facilities in the EU, mostly in Northern Europe, with typical waste throughputs of 100,000-200,000 tons/y. The dominant technology is direct conversion to power, providing a total of about 1500 MW, that is about a half the world production of 3200 MW. If the EU achieves the EC landfill directive, diverting 25-30% of waste to combustion, by 2010 the potential power output is estimated to be in the range 2000-2400 MW. The world market will also increase to 5500 MW, whose value is estimated to be 1000-2000 MEuro/y of which 500-900 MEuro/y would be in the EU. [1]

EU companies dominate the world market in MSWC. The technology was developed within the EU and has been steadily improved over the last decade to meet the increasingly demanding standards for environmental performance and power output.

MSWC is highly developed and commercialized, but, in spite of improvements in environmental performance, public perception of the technology is still poor and the companies face many political, legislative and financial barriers to its deployment.

7.4 Direct combustion

This technology transforms directly the chemical energy of biomass into thermal energy through a sequence of operations such as drying, pyrolysis, gasification, combustion, and flue gas treatment. It is a mature approach and the marketed technologies differ for the combustor type (fixed grid, mobile grid, rotating drum, fluidized bed (ebullated or recirculated).

The energetic vector of all these technologies is steam, which is utilized in thermal cycles for power production and for co-generation purposes.

In general, fueling of such plants with lignocellulosic material, involves the cultivation of large surfaces (500-600 ha/MWe) in the case of species with carbon uptake capacity of 5tC/ha/y.

The cost of the produced power is in the range 8-12 cents/kWh (US$ 1997) with plants in the range 10-50 MW. Said cost is chiefly influenced by the biomass cost itself, by the efficiency and by the investment cost.

The cost of biomass is related to many variables (cultivated/refuse biomass, type of plant, site of production, water availability, insolation etc.) and can't be generalized. The efficiency, that at present is around 23%, is forecast (2020) to grow substantially till 34%. The associated increase of power plant size from 50MW (today) to 184 MW (2020) would allow to reduce the operating costs from 5.5 to 3.87 $/kWh and the investment costs from 1965 to 1115 $/kW.

7.5 Fluidized bed combustion

The development of fluidized bed combustion technology over the past 20 years has significantly increased the use of various biomass and waste products in power and heat generation. [24] In Finland, for example, the installed capacity of fluidized bed boilers has already reached 1,200 MW.

The technology has been applied to a very wide range of fuels including wood-based fuels and residues such as wood chips, sludge from paper mills and deinking plants, and Refuse Derived Fuels (RDF). Two different combustion methods can be used: the bubbling fluidized bed (BFB), in which all the solid material is stationary in the bed; and the circulating fluidized bed (CFB), in which the solid material circulated through the bed to a cyclone, than back to the bed. Fluidized bed boilers are now available with capacities up to 100 MW with BFB and 250 MW with CFB.

The choice between BFB and CFB technology is largely linked to the choice of fuels. BFB, the simpler and cheaper technology, has been favored for plants exclusively fueled with biomass or similar low grade fuels containing highly volatile substances. Enhanced CFB designs, on the other hand, may be competitive even in smaller biomass fired plants. In either case the low operating temperature of fluidized bed boilers means that effectively no thermal nitrogen oxides are formed. Also, because of the low sulfur content of biomass, sulfur emission control is not required.

The benefits from the fluidized bed combustion technology have attracted considerable interest in biomass heat and power generation. In January, Brista Kraft, a new 122 MW biomass CFB plant (44 Mwe and 75 MWth output) was activated in the area around Stockholm, Sweden. The boiler is a compact second generation CFB from Foster Wheeler Energia in Oy Finland, and the steam turbine is from ABB in Stal, Sweden. Wood chips are used as the main fuel, and peat is the support fuel. The plant is designed to burn 250,000-300,000 t of wood waste annually.

7.6 Co-combustion

The direct co-combustion can be used to co-combust supplementary fuels by mixing and blending them with the coal and pretreating them in the same manner as conventional coal using existing facilities (Fig. 14.5). Because they are using existing equipment with limited capacity for renewable fuels the usage of these energy feedstocks is limited in these utilities.

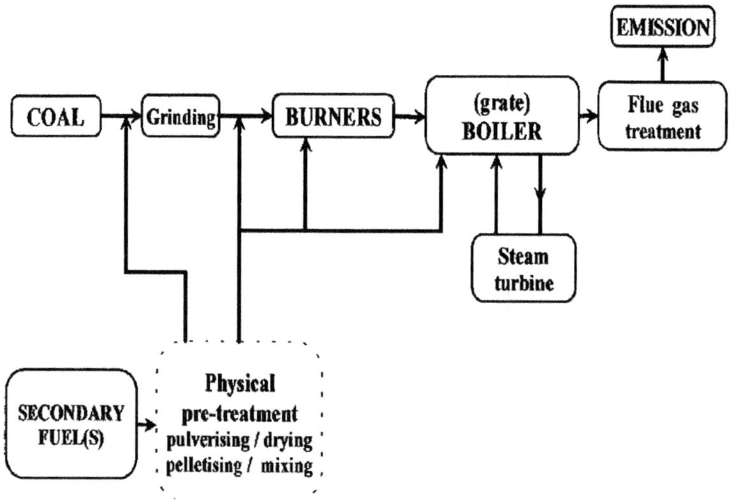

Figure 14.5 - Direct co-combustion process

This barrier can be overcome by indirect co-combustion where separate pre-treatment streams exist for the two feedstocks and they are mixed and blended in the boiler. Dutch experience on energy recovery from biomass and wastes is primarily based on direct co-combustion in coal-fired stations but some in-direct co-combustion trials have been undertaken (Table 14.10). [25]

Table 14.10 - Overview of direct co-combustion power plants in Holland [25]

Power plant	Plant Capacity [MW_e]	Secondary fuel	Co-combustion Mass %	Comments
Amer-8	645	Paper sludge	3	Commercial operation max 10% tested
Amer-9	600	Petroleum cokes	5 and 10	4 week test program, slight increase in carbon in fly ash
Borssele-12	403	Phosphor oven gas Paper sludge Sewage sludge Industrial biomass wastes		
Gelderland-13	600	Demolition wood	5	Commercial operation
Hemweg-8	630	Sewage sludge	3 & 6	Large scale tests
Massvlakte	2x518	Biomass pellets	8	Commercial operation

Both the Amer units have been in operation since 1996 and are suitable for direct co-combustion of paper sludge. The phosphor production plant at Borssele which previously flared its oven gas now feeds it to a nearby coal fired power plant. The plant at Gelderland co-combusts waste wood with coal and has been in operation since 1995. In 1997 the Hemweg plant was licensed to co-combust 75 Kt/year of dried sewage sludge. The plant at Massvlakte uses either biomass pellets or sewage sludge as the supplementary fuel feedstock.

For indirect co-combustion the supplementary fuel has separate size reduction, milling and drying facilities and can then either be mixed with pulverised coal or transported separately to the boilers.

A schematic of the concept in operation at Gelderland is shown in Figure 14.6.

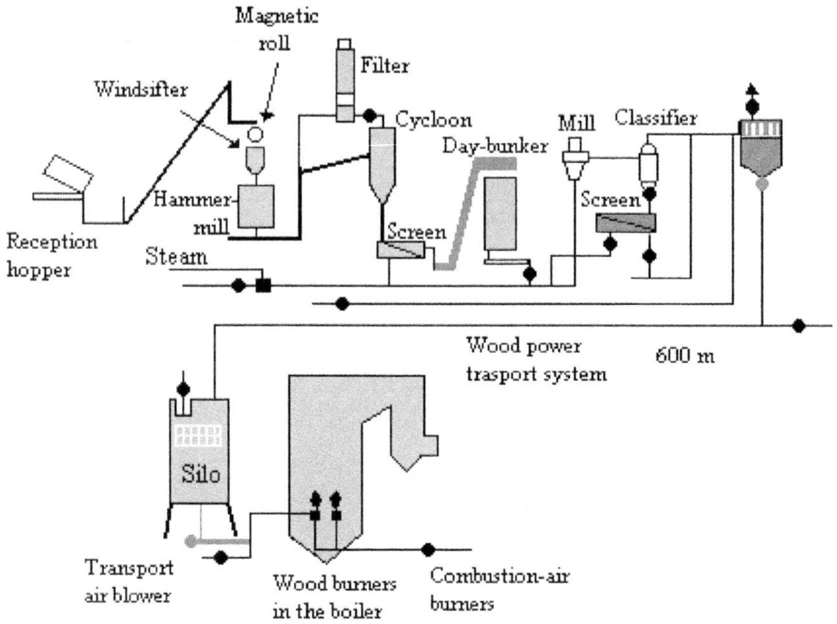

Figure 14.6 - Indirect co-combustion by separate size reduction, dry feeding and combustion at the Dutch Gelderland coal fired plant (EPON)

The main disadvantages are high energy consumption in size reducing and milling and additional investment. An alternative has been proposed by Fortnum in Finland and is outlined in Figure 14.7. [26] They claim it is possible to achieve up to 30% biomass in coal.

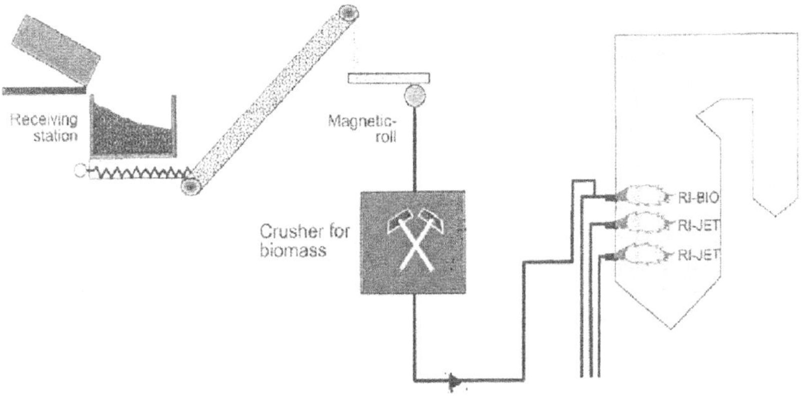

Figure 14.7 - Indirect co-combustion at Fortnum (Finland) [26]

In the Lurgi concept (Figure 8) the biomass fuel is initially gasified in an atmospheric circulating fluidized bed gasifier at 850°C following size-reduction and drying. The fuel gas produced is reduced in temperature cleaned using a scrubber and baghouse filter before combustion in the coal boiler.

Biomass co-firing in coal power plants represents the easiest way, in the near term, for power production from biomass. Co-firing, as compared to other processes, like "Direct combustion" and "Gasification" doesn't represent a production expansion option, but the substitution of fossil fuel with renewables for existing power plants.

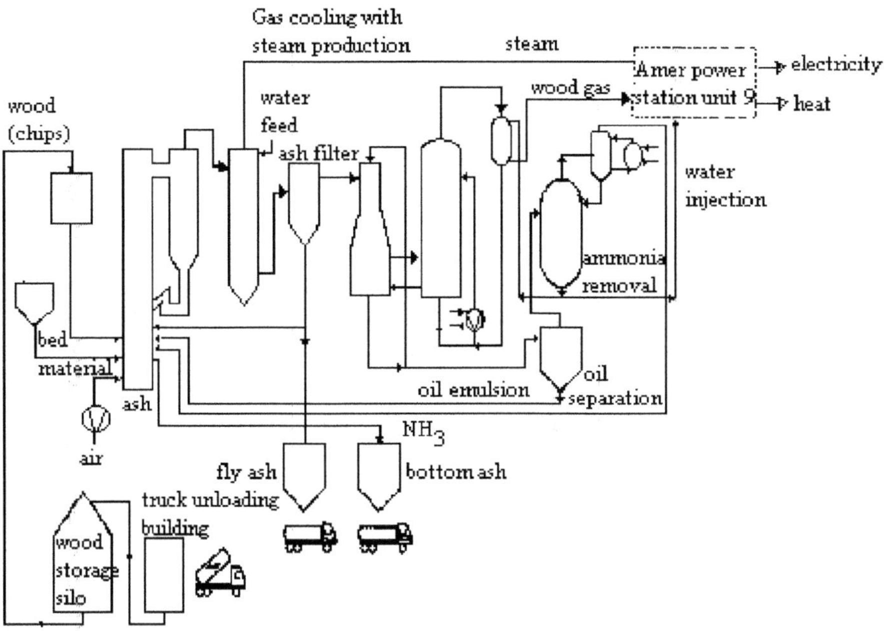

Figure 14.8 - Lurgi indirect combustion by upstream gasification at Amer-9 coal fired power plant (EPZ)

Co-firing technology has been successfully demonstrated with all the principal coal fueled steam generators. Plant modification for the coal power plants to operate in co-firing mode deals with fuel handling, storage, and feed system.

The principal advantages of the co-firing technology are:

- use of lignocellulosic biomass, (both from cultivation and from industrial refuse) with reduced CO_2 emission due to their renewable nature;
- improvement of flue gas emission quality, with particular reference to the SOx content, given the low sulfur content in biomass;
- use of waste heat for the drying of biomass;

- retrofitting of existing coal power plants, with low investment cost.

The principal disadvantages are:

- reduced percentage of substitution of biomass for coal (maximum 10-15%), in order to avoid exceeding reduction of efficiency;
- large production of ash;
- emission of alkaline metals and chlorine, with associated problems of corrosion.

Dealing with the economics, it has been calculated that for a 100MW power plant with 80% loading, the capital investment is in the range 180-250 US$/KW.

7.7 Biomass/MSW cofiring

In recent years many studies on stand-alone power plants fuelled by renewable fuels (biomass and MSW) have appeared, however the unfavorable scales, affecting negatively both the efficiency and the investment costs of the plants, added to social and environmental constrains, have prevented a positive matching between the potential availability of renewable fuels and the actual diffusion of bioelectricity through specifically dedicated plants. The access to a large scale power plant, with higher efficiency and lower specific cost than stand alone plants, could be a key factor in an economically successful exploitation of renewable fuels. [27] Of the many possible cofiring options, the integration between biomass gasification and natural gas combined cycles looks the most attractive.

Gasified biomass cofiring. The optimal size of biomass fuel gas generation plant lies in the range 10-50 MW, while the typical size for a combined cycle power plant can be in the order of 150-300MW. The integration of the two shows several advantages:

- favorable economics (optimal plant size, requirement of only one site, off-site and services sharing)
- the gas turbine specifications are easily met by dilution (avoiding the expensive purification of the low BTU syngas from biomass)
- the power generation plant is fully flexible (the gasifier mass flow remains within the gas turbine mass flow limits, with no mandatory constant year-round biomass supply)
- high environmental performances.

Gasified MSW cofiring. Due to the intrinsically variability and polluting potential of the charge, MSW gasification plants have to undertake far more energy and economically demanding operations than biomass operated plants would normally require, in order to fulfil the high value environmental goals they have conceived for. Nevertheless, as in the biomass case, the offer the opportunity of being integrated with conventional natural gas combined cycles, simply by adding the syngas produced by

MSW gasification to the natural gas stream fed to the gas turbine. The advantages of such operation are basically the same as in the previous case, the principal being the following:

- increased power generation section size (energetically and economically more favorable)
- low upgrading requirement for gas turbine combustion system (dilution with natural gas)
- potential savings in the post-gasification purification line (for the same reasons as above)

8. ADVANCED THERMAL CONVERSION TECHNOLOGIES

Advanced thermal conversion technologies (ATCT) for wastes, namely gasification and pyrolysis [28], are currently in a transition between research and demonstration phases, although a few technology developers claim that they can offer plants on a fully commercial scale. A large number of companies have been involved in the development of these technologies. Worldwide, about forty advanced thermal conversion technologies for waste are in operation. The scales range from small laboratory plants to about 50 kt/y demonstration plants. Some of those plants are at the commercial phase, and others close to achieving a commercial scale demonstration plant. The majority of the developed plants operate commercially at less than 100kt/y.

ATCT generally require a reasonably homogeneous feedstock, compared to mass burn systems. As a result, the most common feedstocks used with ATCT are sorted MSW (refuse derived fuel) and tires. Dealing with ATCT emissions, although a limited number of data is available, it seems likely that many of the technologies being developed do meet emission standards and have the potential to meet even tighter limits. The energy output of ATCT is determined by the feed and processing parameters. Net energy outputs are reported to be in the range of 260-1000 kWh of electricity per ton of MSW derived feedstock (some figures refer to unsorted MSW and some to RDF). The lower side of the range is for plant designed to minimize environmental emissions, including ash vitrification and wastewater treatment, while the upper side of the range is for plants utilizing combined cycle technologies. Higher energy output (3000 kWh/t) is reported for materials with higher calorific values such as rubber and plastic. As comparison, the electrical output of a typical mass burn combustion plant is 450-550 kWh/t. While the capital costs for ATCT, in the range 210-340 Euro/t/y, are substantially coincident with those of mass burn combustion plants (240-280 Euro/t/y, the operating costs are higher for ATCT (30-70 Euro/t), than for mass combustion plant (21-29 Euro/t). Such comparisons are only indicative, being derived by process figures produced in very different conditions (countries, environmental requirements, etc.).

A description of the principal ATCT follows.

8.1 Gasification

Gasification for power production involves the conversion of biomass in an atmosphere of steam and/or air to produce a low to medium calorific value gas. If air is present the ratio of oxygen to biomass is around 0.3. A large number of variables affect gasifier design. In air blown or directly heated gasifiers the heat to drive the process is generated internally within the gasifier. When air is used the resulting product gas is diluted with nitrogen and has a calorific value of ~ 4.9-5.9 MJ Nm^{-3} at 273 K, this can be increased to 12.8-13.8 MJ Nm^{-3} by using oxygen instead of air. [9] The cost of oxygen is high ($US 40-60 Mg^{-1}) and its use is generally restricted to applications involving the production of synthesis gas for downstream conversion.

The majority of gasification power applications have a gas turbine-based generation cycle with compression ratios of 10-20 so turbine inlet pressures are 1.0-2.5 MPa with a temperature of around 825°C. A pressurized gasifier will produce gas at a pressure suitable for direct turbine application and provide the highest overall process efficiency but will require high pressure feed systems and will produce in addition to the fuel gas, tars, chars and volatile alkali vapors which will have to be removed before the gas stream enters the turbine. Hot gas clean up technology is therefore required to maintain high efficiency, this will involve cracking of the tars either thermally (>900°C) or catalytically (825°C) followed by cooling to reduce alkali vapors (540-650°C) and passing through a ceramic filter to remove solids. The limits for alkali are 20 ppb in the turbine combustor exit gas but tars in the fuel gas are considered to be more problematic than alkalis for turbines.

If low or atmospheric pressure gasification is used the cleaned product gas has to be compressed to the appropriate pressure for turbine application. The tars are cracked and the temperature of the product gas reduced below the residual tar dew point before compression. The characteristics of the gas which constrain its suitability for compression include the amount of condensable tars and the water vapor content. The normal compressor gas inlet temperature is below 93°C.

The four primary gasification reactors i.e. moving bed reactors, bubbling fluid-bed, circulating fluid bed and entrained flow reactors have been used for biomass gasification, but biomass feedstocks are different to those of coal, (more reactive, low ash and low energy density).

8.2 Integrated gasification

One promising technology is the gasification of biomass. [29] Gasification involves transforming solid biomass into a gas that consists mainly of nitrogen (50%), carbon monoxide (20%), and hydrogen (15%). The process results in a combustible gas that can be used in a gas turbine, for example, to generate electricity. Also the gas can be separated from unwanted constituents prior to combustion. Specific enabling areas of research and development include improving hot gas cleanup to remove alkaline compounds, identifying the source of turbine blade deposits, and finding methods to remove particulate. In the USA, gasifier projects in Vermont and Hawaii are part of a major DOE initiative to demonstrate biomass gasification for electricity production.

Several processes have been proposed over the past ten years for advanced biomass gasification. The first plant to demonstrate the biomass integrated gasification combined cycle (BIGCC), built in 1993 a Vaernamo, Sweden, produced 6 MW of power an 9 MW of heat. The system includes a pressurized circulating fluidized bed gasifier, a gas turbine, and a steam turbine. The total efficiency of said plant is 82-83%, and the electrical efficiency is 33%. The project is a joint venture of Foster Wheeler and Sweden's largest privately owned electric utility, Sydcraft, Malmo.

Termiska Processer AB (TPS) also in Sweden has proposed a system that is based on low-pressure gasification. The TPS scheme involves gasification at 1.8 bar, followed by a series of gas conditioning steps prior to the gas turbine: cracking of tars to non condensable gases, cooling, baghouse filtration, scrubbing, compression, and re-heating.

The processes related to this technology are:

- pyrolysis (biomass heating to 600°C and conversion to gaseous hydrocarbons, H_2, CO, CO_2, H_2O and char);
- char gasification with steam and/or combustion.

Gasifiers may be direct (all the operations occur in a single system) or indirect (combustion occurs in a dedicated system). They, moreover, can be operated at high or low pressure. Biomass gasification systems most commonly proposed adopt a combined cycle called BGCC (biomass gasification combined cycle). The principal features of the technology are as follows:

- process with the best efficiency;
- important role of the pyrolytic step (biomass has a content in volatile compounds of 86% from dry mass, as compared to 30% produced from coal);
- syngas generated by direct gasifiers has an heat value significantly lower of the one produced by indirect gasifiers (5.6-7.5 Kj/Nm^3 vs. 13-18.7 Kj/Nm^3);

- in the case of low pressure gasifiers, gas cooling, cleaning and recompressing waste about 10% of the electric efficiency, on the other hand, high pressure and high temperature gas cleaning require more expensive apparatus; as result, the electric power produced has similar cost with both technologies;
- biomass have a low sulfur content and, as consequence, gas desulfurization is a simpler task;
- the most promising technology (high pressure BGCC) operating on large scale, could be commercially available into next five years;
- the electric energy production cost is 8-9 US cents/kWh (value close to the lower one for direct combustion):
- the technological improvements expected for the future (2010) would allow the production of larger power plants from the present 50-70 MWe to 100-110 MWe, with substantial increase in efficiency and associated reduction of both investment and operational costs.

In Table 14.11 are reported the most important UE initiatives in the field of biomass gasification.

Table 14.11 - Biomass gasification projects in Europe

No.	Name	Contractor	Capacity (MWe)	Biomass	Status
1	Vaernamo	Sydkraft	6	Wood	Operative
2	Energy Farm	Bioelettrica	12	Wood/SRF	Under Construction
3	ARBRE	ARBRE Energy	8	Wood/SRF	Under Construction
4	Greve		6,2	RDF	Operative
5	North Holland	ENW	30	Wood/RDF	Project
6	N. Umbria	N. Umbria	5	Sludge	Project
7	Biocomb	Draukraft	3,5	Wood/CoGas	Operative
8	Lahti	Lahdel	22 +/-	Various/CoGas	Operative
9	AMER	EPZ	20	Various/CoGas	Project

SRF: Short rotation forestry; RDF: Refuse derived fuel; CoGas: Co-gasification with coal

8.3 Moving bed gasification

The use of moving bed reactors for the gasification of biomass in Europe has been reviewed by Beennackers. [30]

<u>*Countercurrent (up draft gasifiers)*</u>

- countercurrent (up draft gasifiers) are efficient for heating applications below 10 MWt because the gas exits the reactor at relatively low temperatures. They can handle high moisture biomass feedstocks (up to 50% H_2O without pre-drying), and are not fuel size specific, they can also handle feedstocks with high ash content.

- Extensive tar clean up is required making this type of reactor unsuitable for power applications, all current applications are for district heating. Performance characteristics and costs are shown in Table 14.12.

Table 14.12 - Performance characteristics of countercurrent gasifiers

Average performance	
Operation time	8000 h/a
Availability	95-97%
Personnel (heating plant in total)	3-4
Heat generation	
Specific investment costs	350 kECU/MW
Specific operation costs	17 ECU/MWh
Specific heat generation costs	20 ECU/MWh

- The reactors show high efficiency even at part load operation, are flexible in terms of fuel type with little need for flue gas cleaning. Table 14.13 shows fuel type and size of some commercially operating up-draft gasifiers.

Table 14.13 - Commercially available up-draft gasifiers

Supplier	location	Fuel	Capacity
Bioneer (Fi)	10 sites	wood/peat	6 MWt
Voland (Dk)	Harbore and Kyndby	wood/peat	4MWt & 1.2 MWe
Daneco (It)	Villasatina	RDF	3 MWt & 0.6 MWe
	Isola d'Elba		6MWt

- It has been demonstrated by Voland (Denmark) that the tar content can be reduced by injection of secondary air at the top of the gasifier to prevent blocking of the pipes to the burner. If the gas is cleaned it can then be fed to an engine for small-scale power generation. In Italy Daneco use a wet scrubber followed by an activated carbon filter to clean the gas, which is then fed to a diesel engine. However, the overall biomass to electricity efficiency (20%) is low.

<u>*Co Current (Down draft gasifiers)*</u>

- Downdraft gasifiers produce less tar and are more suitable for power generation applications from biomass. Efficient operation require specific fuel size (e.g. 5x7x10 cm for Chevet gasifiers) and moisture content (~20%). The tars that are produced are more stable which makes clean up difficult and the reactors can not be scaled up beyond 1 MWe due to the throat design.
- the difficulty with scale up arises from the throat design, which plays an essential role in tar reduction. The air is introduced above the throat which creates high temperatures (>1000°C) in the combustion zone. A well-designed throat has uniform temperature through out the entire throat cross section area to yield complete cracking of the tars. With

scale up unavoidable cold spots appear with an increased tar content appearing in the producer gas.

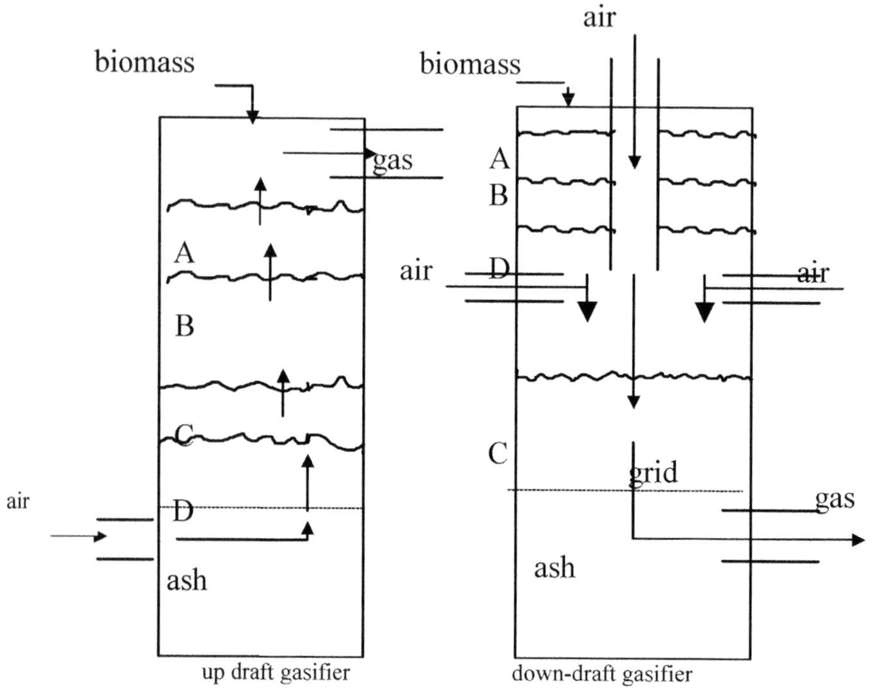

A :drying zone ; **B :** pyrolysis zone ;

C : reduction zone ; **D** :oxidation zone.

Figure 14.9 - Moving bed gasifiers

- Several hundred small scale down draft units have been installed in developing countries. Most units have a wet gas cleaning system and use a diesel engine to drive a generator of typically 20-100 kWe capacity. Their use in Western Europe is restricted due to stricter environmental controls for flue gases and they can never be operated unattended.
- Innovative throat design is required if down draft gasifiers are to be scaled up to a size sufficient for efficient gas clean up and power production. Several European companies now offer units with capacities in excess of 1 MWt (Table 14.14). The increased size is possible because extra air is taken in via the grid and the gas off-take is some distance above the grid.
- In Ireland since 1993 NIHPBS Loughall operates a down draft wood fired gasifier providing 120 kWt for space heating and 100 kWe using a diesel engine.

Table 14.14 - Commercial down-draft fixed bed gasifiers

	Location	Fuel	Capacity
Martezo (Fr)	Hogild	wood	135kWe
Chevet (Fr)	developing countries	wood, coconut husk	20-135 kWe
Wamsler (De)	Munich	wood	600-1500 kWt
Schelde (NL)	Vlissingen	sludge	IMWt
Melima (CH)	Endingen	wood	10 kWe
NIHPBS (UK)	Enniskillen (NL)	wood	100kWe + 200 kWt
Fluidyne (NZ)/ Terry Adams (UK)	Bristol (UK)	wood	30 kWe

8.4 Fluid bed gasification

Most advanced power production processes based on gasification of biomass use fluid bed gasifiers. In Table 14.15 are reported several options for advanced power production based on biomass pyrolysis and gasification. [31]

Table 14.15 - Summary of power production concepts. Condensing power production, flue gas dryer, except where otherwise specified. STIG = steam injected gas turbine [31]

Concept	Mode	Nominal capacity (MWe)	Sensitivities
Pressurized gasification	Combined cycle	30, 60	Flue gas/ steam dryer cogeneration
Atmospheric gasification	Combined cycle	5, 30, 60	cogeneration
Pressurized gasification	STIG	5, 30, 60	Flue gas/ steam dryer cogeneration
Atmospheric gasification	Diesel	5, 25	
Pyrolysis	Diesel	5, 25	cogeneration
Pyrolysis	Combined cycle	5, 30, 60	

They concluded that pressurized gasification IGCC has the highest efficiency, (approximately 45%), but the technical difficulties were greater than with atmospheric IGCC systems. STIG power generation systems are non-competitive due to poor efficiency while gasifiers coupled with diesel engines are non-competitive due to high investment cost.

The potential advantages of advanced power cycles utilizing gasification systems include increased energy conversion, which results in reduced feedstock consumption, reduced environmental impact and possibility of co-firing with coal and other fossil fuels. In terms of capital cost advanced gasification technologies result in reduced volumes of gas in comparison with direct combustion flue gas that must be subject to clean up resulting in lower equipment costs.

The European Union has initiated a plan to implement commercial gasification projects under the THERMIE program. The criteria for these projects included a dedicated biomass supply system based on high yield energy crops such as miscanthus, demonstration of advanced energy conversion systems and assurance of environmental benefits. Other demonstration biomass gasification pilot projects are on going in Brazil and the USA.

Table 14.16 - Biomass gasification plants

Title	Vermont	BGF	ARGRE	Biocycle	Energy farm
Location	Varnamo Sweden	Hawaii	Yorkshire UK	Denmark	Pisa, Italy
Process	Bioflow	RENUGAS	TPS	U-GAS/ RENUGAS	Lurgi
Proposers	Sydkraft Foster Wheeler		Niro (DK) ESB (Irl) TPS		ENEL (IT) Lurgi (De)
Gasifier	AtmCFB	PFB	AtmCFB	PFB	AtmCFB
Net Electric output MWe	6	5.0	8.0	7.2	11.9
Net thermal output MWt	9		0	6.78	0
Electrical efficiency %	32	30-35	30.6	39.8	33

The Hawaii Biomass Gasification project [32] is part of an US-DOE initiative to demonstrate high efficiency biomass gasification systems. The objective of the project is to scale up the 2 MWt (10 tons/day) Institute of gas Technology RENUGAS® pressurized air blown fluidized bed gasification plant to a 20 MWt using 100 tons/day of bagasse, a waste material from the cane sugar industry. Phase 1 has been completed with the gasifier (without hot gas clean up) being operated for over 100 hours at up to 50 tons/day and pressures up to 150 psi. Integrated hot gas clean up of the product gas will make it suitable for combined cycle power generation. A schematic of the process is shown in Fig 14.10.

The first complete IGCC power plant using biomass as a fuel has been operated in Varnamo in Sweden. Wood is gasified in an air blown pressurized circulating fluidized bed and generates 6 MWe using a gas turbine as well as 9 MWt for district heating in the local community (Fig 14.11). The feedstock is pre-dried to < 20% H_2O and screw fed to the gasifier, which operates at a temperature of 950-1000°C and a pressure of 20 bar. The gas is hot filtered at 400°C using a ceramic filter before being combusted and expanded through a gas turbine. Generating 4 MWe of electricity.

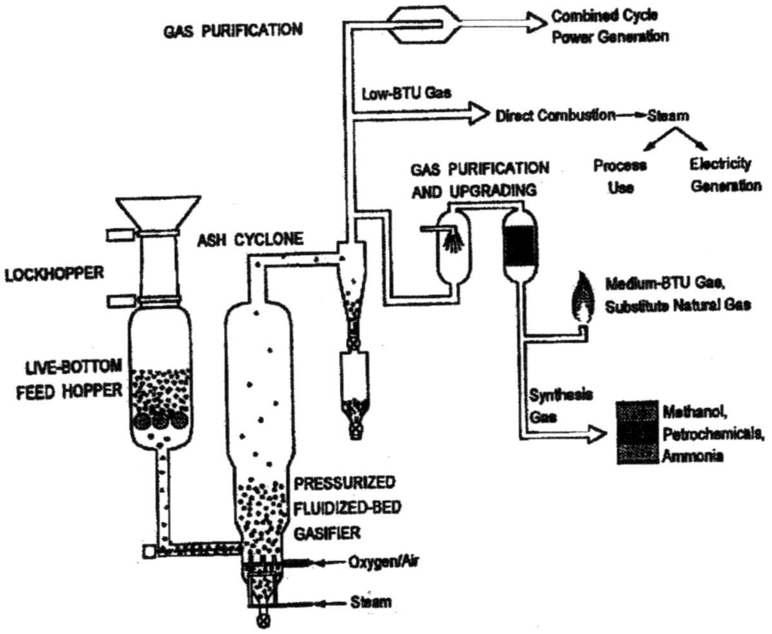

Figure 14.10 -The RENUGAS process.

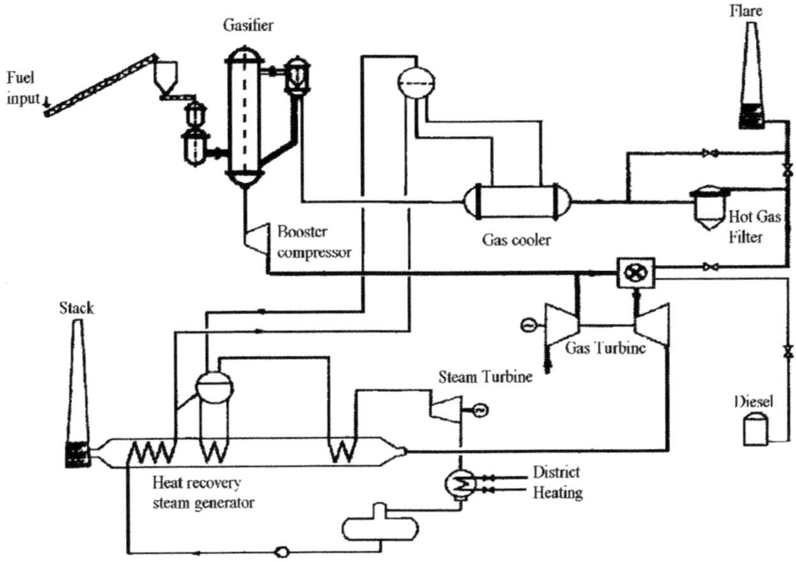

Figure 14.11 - The Varnamo demonstration plant

The Italian Energy Farm project which is due for commissioning in November 2000 is equipped with an atmospheric air blown circulating fluidized bed gasifier (Lurgi), a 10.9MWe single shaft gas turbine and a heat recovery steam generator which will provide steam to a 5MWe condensing steam-turbine. [33] Atmospheric rather than pressurized gasification was chosen to simplify the biomass feeding system and the gas clean up even though efficiency will be reduced due to the requirement to pressurize the fuel gas rather than just the air. Gas clean up will involve cyclones and bag filters followed by water scrubbing.

The UK ARBRE project will also use low pressure circulating fluidized bed gasification followed by tar destruction in a second circulating fluid bed cracker. The technology is known as the Thermiska processor TPS where dolomite in the second bed is used to catalyze the conversion of tars to gases and low molecular weight vapors.

9. FURTHER TECHNOLOGICAL EVOLUTION FOR THE BIOMASS TO ENERGY CONVERSION

For the direct combustion two technological improvements have bee presented: *Suspension Burning* that utilizes dried wood dust, and *Whole Tree Energy Process* where the whole tree is operated directly in the power plant.

In the case of gasification, *second generation turbines*, able to operate at 1250°C, are under development; hybrid systems such as *Fuel Cell/Gas Turbine* are also under development for the high efficient electricity production in remote regions.

9.1 Externally fired turbines

The biomass externally fired gas turbine system provides another effective way of adapting solid fuel such as biomass and coal in a gas turbine, thereby increasing overall plant efficiency, while lowering emissions. The main features of this turbine are an atmospheric combustor and a high temperature heat exchanger, replacing the conventional combustion system in an open gas turbine cycle. In this system, the biomass is directly fired in a biomass boiler. The discharge air from the compressor is heated in a high temperature heat exchanger in the boiler. Thus, only clean air is sent to the expander, so the turbine's gas path is never exposed to the corrosive elements in the fuel. The high temperature heat exchanger is the key piece of equipment that needs to be developed to achieve a high turbine inlet temperature.

9.2 Flue gas condensation

An important consideration for the future use of biomass fired power plants is the treatment of biomass flue gas. Biomass combustion flue gases have high moisture content. When the flue gas is cooled to a temperature below the dew point, water vapor starts to condense. By using flue gas condensation, sensible and latent heat can be recovered for district heating or other heat consuming processes; this increases the heat generation from a cogeneration plant by more than 30%.

Flue gas condensation not only recovers heat, but also captures dust and hazardous pollutants from flue gas at the same time. Most dioxins, chlorine, mercury, and dust are removed, and sulfur oxides are separated out to some extent. Another feature of the flue gas condensation is water recovery, which helps solving the problem of water consumption in evaporative gas turbines.

The technology can be integrated into the various biomass-based systems already discussed. In Sweden, for example, flue gas condensation is in use in more than 60 units for heat generation or for both heat and power generation. At least 40 of these units are biomass fueled plants mainly for district heating purposes, with a fuel input of 5 to 10 MW, corresponding to about 2000 GWh of additional heat produced per year.

9.3 Hydrogen by hydrothermal gasification of biomass

An innovative process, under study at the Forschungzentrum Karlsruhe, Germany [34,35], is the hydrothermal gasification of wet biomass. Wet biomass, such as algae and crops with high content of humidity, wet wastes and sewage sludge, could take advantage of this gasification technology, avoiding drying, a high energy consuming step, usually necessary in conventional gasification. In addition the new technology allows for the obtaining of hydrogen and carbon dioxide in one single process step, instead of producing synthesis gas, where carbon monoxide has to be converted into hydrogen and carbon dioxide in a second process step (water gas shift). In preliminary experiments, the best conditions, finalized to the optimization of the conversion to hydrogen and the minimization of CO and CH_4 production, have been found at temperatures around 600°C and pressure at about 250 bar. The development of the technology in order to become economically applicable, has to overcome problems such as: reliable high pressure feeding systems for slurries, fouling problems of the heat exchanger caused by salt precipitates, corrosion and embrittlement of construction materials. The above technology could represent a valuable and fast alternative to anaerobic digestion for the energetic conversion of wet biomass.

10. COMPARISON OF THE PRINCIPAL TECHNOLOGIES

The comparison between the principal presented technologies (direct combustion, co-firing, gasification) is summarized in Table 14.17.

Table 14.17 - Comparison of the principal technologies for bioelectricity production from lignocellulosic biomass

	Co-firing				Direct combustion				Gasification			
	1997	2005	2010	2020	1997	2005	2010	2020	1997	2005	2010	2020
Plant size, Mwe	100	150	200	300	50	100	150	184	75	75	100	110
Co-firing, % coal	10	15	15	15	-	-	-	-	-	-	-	-
Biomass humidity	25	25	25	25	50	50	50	50	24[3]	24[3]	24[3]	24[3]
Efficiency, %	32,7[4]	32,5[4]	32,5[4]	32,5[4]	23	27,7	27,7	33,9	36,0	37,0	37,0	41,5
Capital investment $/kW	271,6	240,5	230,3	216,7	1965	1510	1346	1115	2102	1650	1464	1258
Operating costs[1] US cents/kWh	1,62[2]	1,63	1,63[2]	1,63[2]	5,5[5]	4,74[5]	4,74[5]	3,87[5]	3,98[3]	3,55[3]	3,55[3]	3,29[3]

(1) Biomass cost: 2,5 $/GJ;
(2) Incremental costs due to the use of biomass (coal cost, 39 $/t);
(3) Drying of wood (humidity from 24 to 10%) (further drying implies1.5% points of electric efficiency reduction for each 10% points of removed humidity);
(4) Biomass efficiency (coal efficiency, 32.9%)

The principal comments are as follows:

- the comparison of *co-firing* and direct combustion shows for the first technology significantly lower costs, indicating it as a good option for the short term;
- the biomass percentage allowed for coal substitution is quite low (10-15%) and is not in accordance with an intensive use of biomass;
- large biomass power plants are well suited with the *direct combustion* and *gasification* technologies;
- the high present cost of electricity obtained with the last two technologies is expected to decrease, thanks to the increase in efficiency and plant scale-up;
- the best improvement is forecast (2020) for biomass gasification (cost of electricity 5.5 US cents/kWh ; 110 MWe power plant; electric efficiency 41.5%);
- the above cost is close to the one reported for the production of electric energy from coal (4.9 US cents/kWh; 500 MWe power plant; 40%

efficiency), but is still higher than the one obtained from a gas combine cycle plant (3.5 US cents/kWh; 500Mwe power plant; 52% efficiency).

11. OTHER THERMAL TECHNOLOGIES

11.1 Pyrolysis

Pyrolysis is the thermal degradation of a material in the complete absence of an oxidizing agent (air or oxygen). The process occurs at temperatures in the range 400-800°C. When applied to waste materials the action of the heat breaks the complex molecules into simpler ones. This results in the production of gas, liquid and char. These products can have several uses, depending on the nature of the feedstock, however, for waste based feedstock the most likely use is as fuel for energy generation.

The relative proportion of the pyrolysis products will depend on the temperature the material is subjected to, the time for which it is exposed to that temperature, and the nature of the material itself. Long exposure to low temperatures will maximize the production of char, whereas flash pyrolysis gives up to 80% liquid (by weight). Flash pyrolysis involves short exposure (< 1 sec) to temperatures around 500°C. Rapid quenching is necessary to freeze the reactions and condense gaseous species before simple molecules are formed which are normally gaseous under ambient conditions. If a gas is the principal product, it is likely to have a higher calorific value (15-30 MJ/NM^3) than that produced by gasification (due to partial oxidation of gaseous species).

The development of such technology has been reviewed by T. Bridgwater of the Aston University UK. [36]

11.2 Fast pyrolysis

Fast pyrolysis is a high temperature process in which biomass is rapidly heated in absence of oxygen. As a result it decomposes to generate mostly vapors and aerosol and some charcoal.

After cooling and condensation a dark brown mobile liquid is formed which has heating value about half of that of conventional fuel oil. While it is related to the traditional pyrolysis process to make charcoal, fast pyrolysis is an advanced process, which is carefully controlled to give high yields of liquid.

Features. The essential features of a fast pyrolysis process are:

- very high heating and heat transfer rates, which usually requires a finely ground biomass feed.
- carefully controlled pyrolysis reaction temperature of around 500°C in the vapor phase, with short vapor residence times of typically less than two seconds.
- Rapid cooling of the pyrolysis vapors to give the bio-oil product.

Main product, bio-oil is obtained in yields of up to 80% (wt) on dry fed, together with product char and gas which are used within the process, so there are no waste streams.

Fast pyrolysis system. A fast pyrolysis system consists of an integrated series of operations starting with a roughly prepared feedstock such as whole tree chips from short rotation coppice, energy crops such as miscanthus or sorghum, or agricultural residues such as straw. It is always necessary to provide for reception and handling and some storage of feed material.

Unless a naturally dry material such as straw is available, drying is usually essential, as all the feed water is included in the liquid product. Low-grade process heat would usually be employed.

Particles have to be very small to fulfil the requirements of rapid heating and to achieve high liquid yields. Feed specifications range from less than 200 microns for the rotating cone reactor to less then 2 mm for the fluid beds and less than 6 mm for the transported or circulating fluid beds. Ablative reactors can utilize whole tree chips, as the mechanism of heat transfer is different.

A wide range of reactor configurations have been investigated that show considerable diversity and innovation in meeting the basic requirements of fast pyrolysis. The best method is not yet established with most processes giving between 65-75% liquids based on dry wood input.

Some fine char is inevitably carried over from cyclones. Almost all of the ash in the biomass is retained in the char, so successful char removal gives successful ash removal. Char separation, however is difficult, but may be not necessary for all applications.

The time and temperature profile between formation of pyrolysis vapors and their quenching influence the composition and quality of the liquid product. High temperatures will continue to crack the vapors and the longer the vapors are at higher temperatures, the higher the extent of cracking. Vapor residence times of a few hundred milliseconds are necessary for optimum yields of chemicals and food additives, while fuels can tolerate vapor residence times of up to around 2 seconds. Longer residence times result in significant reductions in organic yields from cracking reactions.

The collection of liquids has long been a major difficulty in the operation of fast pyrolysis due to the nature of liquid product, which is mostly in the form of aerosols rather than true vapor.

While a wide range of reactor configurations has been operated, fluid beds are the most popular configurations, due to their ease of operation and ready scale-up.

Reactors:

- *Bubbling beds.* The technology has been selected for further development by several companies, such as Union Fenosa (Spain) who operates a 200 kg/h pilot unit, Dynamotive (Canada) with a 50 kg/h unit, and Wellman (UK) with a 200 kg/h unit. Circulating fluid beds and transport bed reactors have been developed to commercial status and are used in the USA for food flavoring and related products.
- *Ablative pyrolysis.* It is interesting as much larger particle sizes can be employed than in other systems and the process is limited by the rate of heat supply to the reactor rather than the rate of heat absorption by the pyrolysis biomass. Much of the pioneering work on this technology has been carried out by NREL (USA), by CNRS at Nancy (F), and by Aston University at Birmingham (UK).
- *Entrained flow reactor* has been developed at Georgia Tech Research Institute but due to difficulties encountered in achieving a good heat transfer the process is no longer operational.
- *Rotating cone reactor.* It has been invented at the University of Twente (NL) and operates effectively as a transported bed reactor, but with transport effected by centrifugal forces, rather than gas. The technology has been developed by BGT (NL) in a 200 kg/h pilot plant.
- *Vacuum pyrolysis.* The rate of heating is very low compared to the other systems described above, but the effect of fast pyrolysis, in terms of liquid product yield and quality, is achieved by removing the vapors as soon as they are formed by operating under vacuum.

11.3 Bio-oil

The pyrolysis liquid bio-oil is a dark brown free flowing liquid. Depending upon the starting feedstock and the mode of fast pyrolysis, the color can appear as dark red-brown, dark green or black, depending on the chemical composition and the presence of carbon particles.

The liquid contains several hundred different chemicals, ranging from low molecular weight and volatile formaldehyde and acetic acid, to complex high molecular weight phenols and anhydrosugars.

The liquid contains varying quantities of water, which forms a stable single-phase mixture, ranging from 15 % wt to a upper limit of about 40 % wt water. Pyrolysis liquid can tolerate the addition of some water, also if it is not considered as water soluble. It is also not miscible with petroleum derived fuels. The density of the liquid (1.2 kg/l) is higher than the one of light fuel oil (0.85 kg/l). This means that the liquid has about 42% of the energy content of fuel oil on a weight basis and of 61% on volumetric basis. The viscosity of bio-oil can vary from 25 cS to 1000 cS, depending on the water content. The liquid is chemically unstable, and this effect increases with the temperature. Heating the liquid to 100°C to try to remove water or distill off light fractions, causes a rapid reaction to occur, with production of a char residue of around 50 % wt of the original liquid.

In Table 18 are reported the typical properties and characteristics of a wood derived pyrolysis oil.

Table 14.18 -Typical characteristics of bio-oil

Physical property		Typical value
Mixture Content		15-30%
pH		2.5
Specific gravity		1.20
Elemental analysis on dry basis:	C	56.4%
	H	6.2%
	O (by difference)	37.3%
	N	0.1%
	Ash	0.1%
HHV as produced		16-19 MJ/kg
Viscosity (at 40°C and 25% water)		40-100 cp
Solids (char)		0.5%
Distillation		Max 50%, as liquid degrades

Bio-oil can substitute for fuel oil or diesel in many static applications including boilers, furnaces, engines and turbines for electricity generation. With reference to the last application Orenda (Canada) has modified and successfully run on bio-oil a 2.5 MWe gas turbine. There are also ranges of chemicals that can be extracted or derived, including food flavoring, specialties, resins, agri-chemicals, fertilizers etc.

Up-grading bio-oil to transportation fuels is technically feasible, but currently not economic.

Current status. Europe has only relatively recently become involved in direct production of liquid fuels from biomass. Examples of the realization of the technology are: the 200 kg/h Union Fenosa pilot plant in Spain (37,38); the 200 kg/h Egemin plant in Belgium (39); the 15 t/d Ensyn RTP3 pilot plant purchased by ENEL. [40]

In North America a number of commercial and demonstration plants have been operating at a scale up to 2000 kg/h. Ensyn (Canada) are marketing commercial fast pyrolysis plants of up to 10t/h throughput. [41]

12. BIOLOGICAL CONVERSION

Among the various biochemical processes available for energy production, anaerobic digestion represents the best opportunity for the valorization of high moisture content feedstocks, such as liquid wastes. The biogas obtained is an average calorific power gas, which can be used as such or mixed with other fuels, for energy production.

The biomass fermentation for the production of ethanol is a well-established technology. In recent years it has been subjected to renewed interest, thanks to co-current factors, such as the need for oxygenated gasoline additives consequent to the banning of MTBE, the surplus of some agriculture crops in US, the development of more efficient fermentation processes. A significant process improvement has been obtained by genetic engineering modification of *Saccharomyces* yeast, making them able to convert to ethanol, not only glucose but also xilose contained in the cellulosic biomass, with a consequent 30% improvement of biomass-to-ethanol conversion.

The bio-production of hydrogen is at its infancy. It is based on the use of phototrophic bacteria able to convert to hydrogen the volatile organic acids, in turn obtained by biomass acidogenic fermentation. This process, although at present able to produce a modest hydrogen output, has a high potential, because it allows the conversion of waste to energy by means of sunlight.

Follows a brief survey of consolidated and innovative technologies and a more detailed insight to anaerobic digestion.

12.1 Anaerobic digestion and biogas

Anaerobic digestion involves the breakdown of organic waste by bacteria in an oxygen-free environment. It is generally used as a waste treatment process but also produces a methane-rich biogas, which can be used to generate heat or electricity.

The anaerobic digestion equipment consists of a heated digester tank, a gas holder to store the biogas and a gas-burning engine/generator set, if electricity is to be produced. The organic waste is broken down in the tank with conversion up to 60% into biogas. The biogas has a calorific value typically between 50 and 70% that of natural gas and can be combusted directly in modified natural gas boilers or used to run internal combustion engines. The digestate residue (liquid and solid) can find application as fertilizer and soil conditioner, respectively. [1]

EU companies have considerable knowledge and experience in anaerobic digestion and are world leader in this technology. They can offer the full range of products and services and are likely to be well placed to export to any country.

Anaerobic digestion has become an established and proven technology for treatment of solid waste, coming from municipal solid waste, market or other industrial organics. In Europe are operating more than 50 plants able to treat at least 10% of organic solid waste coming from municipalities, with a minimum capacity of 3000 t/y. The treatment capacity in Europe has increased from 122.000 t/y in 1990 to 1.037.00 t/y by the year 2000 (+750%). The average capacity ranges from 8.700 t/y of Swiss plants to 30.000-50.000 t/y of plants constructed in France, Belgium and the Netherlands. The total average capacity has increased from 15.600 t/y in 1997 to 21.700 t/y in 2001.

A lot of research has been carried out to improve the process. The following are some of the most important steps. [42]

- Digestion plants were initially operated at mesophilic temperatures. Thermophilic operation developed afterwards, proved to be a reliable and accepted mode of fermentation. Benefits of the latter technology rely on a higher gas production and easier pathogen species killing.
- While in the 90's most of treatment capacity for solid waste was provided by wet digestion systems, in following years dry fermentation grew up, reaching a half of the total capacity by year 2000.
- One-phase or two-phase digestion. Two-phase digestion has been proposed as an important technological improvement, but has not been able to substantiate its claimed advantages. The added benefits in increasing the rate of hydrolysis and methanation have not been proven, and high digestion rates have been obtained also with one-phase systems. The added investment cost and operating complexity has limited the diffusion of the two-phase technology.
- Anaerobic digestion takes advantage, compared to aerobic composting, of the energy production. Although anaerobic digestion requires an additional average 15 kWh/t of waste treated in comparison to aerobic composting, the biogas generated yields a surplus energy of 150-250 kWh/t.

An important advantage of anaerobic digestion has been demonstrated to be the high flexibility in treating different types of waste streams, ranging from wet to dry and from clean organics to grey waste. In Table 19 are reported three plants treating significantly different organic waste. [42]

Although the anaerobic process with the today plants is more expensive than the aerobic composting plant (5-15 Euro/t), for large scale plants (> 50.000 t/y) costs tend to be similar.

A further sensitive parameter in favor of the anaerobic, as compared to the aerobic process, is given by the production of odors.

The weight of volatile organics produced per ton of waste amounts to 588g for the aerobic composting versus 3g for the anaerobic process. [43]

Biogas, obtained by anaerobic digestion of biomass represents, among the various processes available for energy production, the most developed technology for the valorization of high moisture content feedstocks, such as liquid wastes. The biogas obtained is an average calorific power gas, which can be used as such or mixed with other fuels, for energy production.

In order to allow an anaerobic digestion technology expansion several areas need to be addressed. Increase in reliability, reduction of capital costs, increase the value of process products, need to take account of the environmental benefits in the costs, better economic and environmental information on the successful schemes.

Table 14.19 - Comparison of waste composition and process parameters of three European plants [42]

Plant	Salzburg (Austria)	Brecht (Belgium)	Bassum (Germany)
Capacity	20.000	20.000	13.500
Kitchen waste	80	15	Grey
Garden waste	20	75	Waste
Paper waste	-	10	-
Dry matter substrate	31%	40%	57%
Volatile solids substrate (on dry matter)	70%	55%	51%
Dry matter content in the digested residue	22%	31%	41%
Nm^3 biogas/t waste	135	103	147

12.2 The biological treatment of the organic fraction of MSW

Many efforts are in progress to upgrade the organic fraction of MSW using biotechnology. In accordance with a review produced by the International Solid Waste Association [44] on the status and trends of the waste management in Europe, the most important technological approaches are composting and biomethanation. Unfortunately, the diffusion of these technologies is rather scarce: composting occurs at large scale only in the Netherlands, and biomethanation in performed only in small-scale plants. The limited diffusion of composting is caused by many reasons, including the high content in heavy metals, the bad smelling of plants, and the poor quality of the final product. Promising alternatives developed in the EniTecnologie Labs. [45,46] refer to hydrolytic-fermentative processes of the polymeric materials contained in the organic fraction of MSW, in order to obtain different acidic syrups that can be converted directly in gaseous energetic vectors, such as methane or hydrogen, or extracted and reacted with alcohol to obtain octane enhancers and Diesel substitutes. The above alternative approaches are depicted in Fig. 14.12.

Acids converted to methane or liquid fuels are volatile fatty acids, such as acetic, propionic and butyric acid. The above mentioned new technology could be advantageously coupled with new gasification processes that have the potential to increase energy recovery (gasification can give 30-35% energy conversion in terms of electricity, in comparison to 15-20% of incinerators using conventional cycles). The best option for the utilization of syngas originated by refuse could be its co-firing with natural gas. In this way the high demand of purification of syngas, due to its high corrosive action on turbines, could be lowered thanks to dilution with natural gas.

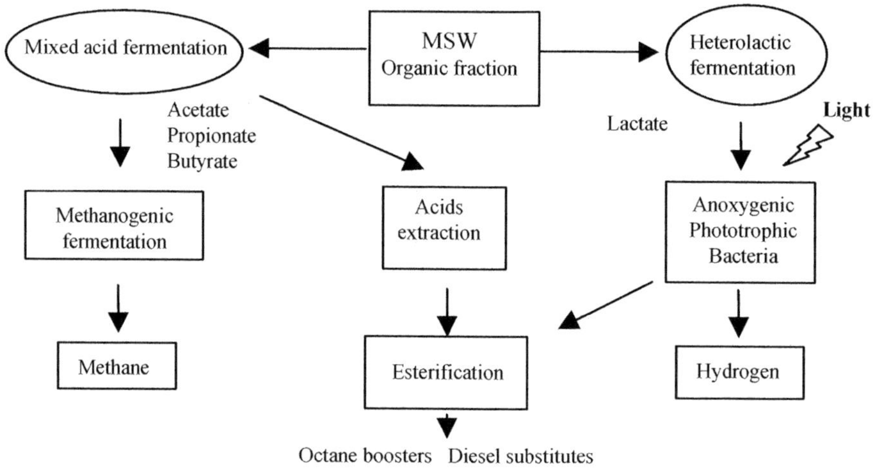

Figure 14.12 - New technologies for the upgrading/disposal of the organic fraction of municipal solid wastes

The bio-production of hydrogen is another innovative technology at its infancy. It is based on the use of phototrophic bacteria able to convert to hydrogen the volatile organic acids, in turn obtained by biomass acidogenic fermentation.

Anoxygenic phototrophic bacteria convert lactic acid, obtained through heterolactic fermentation, to hydrogen. [47] This last option, developed in the EniTecnologie Labs. under the Japanese RITE project, looks very attractive, trying to convert waste to energy by means of sunlight. The energy recovery from the organic fraction is at present still limited, ranging around 1.5%, but one of the most important advantages of this approach is linked to the downstream processing of materials originated by acidogenesis, that allows a good separation of inert, with consequent great improvement of the compost quality.

A promising alternative to the production of gaseous fuels (hydrogen and methane), is given by the production of different syrups containing volatile fatty acids, such as acetic, propionic and butyric acid, that can be extracted and reacted with alcohol to obtain octane enhancers and Diesel substitutes

(45). The energy recovery in the case of methyl esters production is 8%, somewhat higher than the one of hydrogen.

13. LIQUID BIOFUELS

Liquid biofuels are transport fuels, principally biodiesel and bioethanol/ETBE, two key liquid fuels processed from agricultural crops and other renewable feedstocks. Two minor liquid biofuels are biomethanol/MTBE and the previously treated bio-oil or pyrolysis oil.

Liquid biofuels are manufactured using well-understood and proven chemical processes. At present biodiesel is mostly produced from oilseed rape and sunflower oil. Bioethanol/ETBE is processed from wheat, sugar beet and sweet sorghum. Biomethanol can be produced from lignocellulosic material, bio-oil or pyrolysis oil is the result of advanced conversion of a biomass feedstock. The principal cost element in biofuel production is feedstock, so research is currently concentrated on the identification of more effective raw materials.

Biofuels (ethanol, ETBE, biodiesel) can be utilized as such or mixed to fossil derived fuels. In the case of ethanol, the most promising utilization is for the production of ETBE, the last being considered as a valid substitute for MTBE, after the alarm about the polluting activity of the last. In the case of biodiesel, its use in mixture with diesel presents a maximum of technical and economical advantages.

The market of liquid biofuels (biodiesel and bioethanol/ETBE) became important after the '70s oil crises as many countries sought indigenous supplies of liquid fuels. The 1995 EU biofuel production was of 910,000 tons, whereas the world production reached 15 M tons (mainly biodiesel). The production expected by 2010 is 4.9 M tons for EU (mostly bioethanol) and 19.5 M tons worldwide (1). Biofuels now are 2-3 times more expensive than their fossil counterparts, then policies taking into account the hidden environmental costs associated with the oil industry with consequent tax reduction for biofuels could make them competitive.

The European market needs to develop both in scale and user support, before considering competing with the large oil companies that dominate the transport fuel market. Brazil and USA have made major advances in bioethanol conversion from lignocellulosic material. The predicted expansion of bioethanol/ETBE in EU by 2010 is based on this process.

The main barrier to the development of liquid biofuel technology in the EU is uncertainty about the long-term supply of feedstock. Because of financial incentives, farmers use only non-set-aside land for food crops, whereas the acreage of set-aside land available for biofuel is being reduced.

13.1 Bioethanol

The biomass fermentation for the production of ethanol is a well-established technology. In recent years it has been subjected to renewed interest, thanks to co-current factors, such as the need for oxygenated gasoline additives consequent to the banning of MTBE, the surplus of some agriculture crops in US, the development of more efficient fermentation processes. A significant process improvement has been obtained by genetic engineering modification of *Saccharomyces* yeast, making them able to convert to ethanol, not only glucose but also xilose contained in the cellulosic biomass, with a consequent 30% improvement of biomass-to-ethanol conversion.

Bioethanol is produced by fermentation of biomass feedstock. The principal biomass sources are: *Residues* (Agricultural Production, Forestry Production, Agricultural Industry), *Agricultural crops surplus, Ad hoc cultivation* (mostly sugar cane, wheat and corn, and in lesser extent beetroot, sorghum and topinambur), *Municipal Solid Waste.*

The production cost of ethanol depends on the feedstock utilized. If a sugar or starch containing feedstock is used, the cost of starting materials is 60-70% of the overall process cost, whereas in the case of ligno-cellulose containing materials their cost amounts only to 20-30% of the process cost.

Obstacles to be overcome for the production of ethanol from ligno-cellulosic biomass are linked to the difficulty in obtaining high hydrolytic yields of sugars from cellulose, and to the utilization of emicellulose.

Hydrolysis may be performed in acid medium or enzymatically. Acid hydrolysis has been improved reaching (Madison Process) 60% yield in ethanol. One of the principal drawbacks of the chemical hydrolysis is the low fermentability of the hydrolyzate mixture. [48]

Enzymatic hydrolysis is, in principle more interesting, obtaining through a microbiological-biochemical process performed in mild conditions, the depolimerization of cellulose, obtaining an hydrolysate easily fermentable (49). The principal drawback of this approach is the low accessibility by the enzymatic attack to the cellulosic material. With the purpose to overcome this problem many pre-treatment procedures have been developed. Pre-treatment has the function of breaking the cellular structure of biomass, of depolymerizing cellulose, and of separating cellulose from lignine.

Pre-treatment procedures include mechanical methods, thermo-mechanical methods, chemical methods and biological methods. Mechanical and biological methods are considered expensive, the first for the high energetic need and the second for the slowness of the process. A steam-based process (200-220°C for 3-10 min) is being considered as the most preferred method. The following fundamental step is the selection of microorganisms with high cellulose and emicellulose hydrolytic activity. Bacteria useful in this purpose are of the *Bacillus, Cellulomonas* and

Clostridium species, whereas, in the case of fungi, *Thricoderma, Penicillium,* and *Aspergillus* species have been tested, among may others. Moreover, many microorganisms of bacteria and fungi species have been studied for the degradation of lignine. A further key step is the recovery and hydrolysis of the emicellulose fraction. The enzymatic approach is preferred to the acid hydrolysis, although some problems need to be overcome, mostly caused by the decrease of the enzymatic activity originated by the final product inhibition. In this purpose many efforts have been performed on the engineering side, with innovative membrane reactors, able of continuous separation of the produced glucose. By the industrial point of view, the biomass to ethanol conversion is performed in accordance to the following processes: Separate Hydrolysis and Fermentation (SHF) and Simultaneous Saccharification and Fermentation (SSF). SSF seems today a more promising system then SHF, because of the reduced plant and operation costs, and for the minimization of inhibition problems. Some difficulties are still originated by the difference in optimum temperature between hydrolysis (45-50°C) and fermentation (28-30°C). Other problems are originated by the difficult separation of the biocatalyst from the lignine residue. [50]

The cost of production of bioethanol is progressively reduced from 1979 (4.5 $/gallon) to 1994 (0.5 $/gallon) thanks to technology improvements [51,52] and to the contribution of genetic engineering that has produced bacteria able to allow pentose sugars fermentation. [53] An economic feasibility study of the bioethanol production from ligno-cellulosic biomass has put in evidence that the process is not profitable if based only on the selling of ethanol, but it can be by the selling also of the co-product furfurol.

The most important bioethanol production comes from Brazil, USA and South Africa, with a 10, 5 and 0.5 Mt/y production, respectively. [54]

While in Brazil, bioethanol is utilized as 100% fuel, in the other countries it is generally used as a 10-20% blend in gasoline. In USA there is a large effort for the improvement of technologies that could allow, by 2010 a reduction of bioethanol production cost of 30%, reaching competitiveness with fossil fuels. At EU level, there are a production plant in France (50,000 t/y) and one in Spain (80,000t/y). In both cases the production is utilized in the ETBE synthesis. In Sweden the construction of a demonstration plant for the production of ethanol from acid hydrolysis of lignocellulosic residues is in progress. Also l'Institut Francais du Pétrole has performed studies at pilot scale for the conversion of lignocellulosic residue to ethanol. [55] In Italy, in the period 1994-96, only a few tenth thousand tons of ETBE have been produced, utilizing the ethanol obtained from agriculture production surplus.

The recent decision of the California Governor to ban MTBE from gasoline, could modify substantially said forecast. The MTBE present production of about 20 Mt/y, could be drastically reduced in favor of ETBE, whose environmental impact is reduced in comparison to MTBE, due mainly to its lower water solubility.

13.2 Biodiesel

The most important chemical process for bioenergy production is the transesterification of vegetable oils to bio-diesel. The technology is mature and there are no major technical barriers to be overcome. The economic barriers that have prevented till now from a large utilization of such a fuel will be overcome by the increase of oil price and/or by environmental legislation imposing for urban areas the compulsory utilization of bio-diesel, given its beneficial effect in terms of sulfur and particulate emissions.

The principal environmental advantages given by biodiesel are: the absence of toxicity, the biodegradability and the negligible sulfur content. The principal disadvantages are: high cost and low cetane value.

In USA all the activities dealing with biodiesel are co-ordinated by the National Biodiesel Board, with a budget of $ 3.5 M/y. In Europe, at present, biodiesel is the largest marketed biofuel (600-700 Kt/y) and is produced in France, Germany, Austria, Belgium and Italy. [22] The biodiesel market remains at low level, mostly because of its high production cost, still three times higher than conventional diesel fuel. The most optimistic forecast sees a 30-35% reduction on production cost, not easily achievable due to the high cost of raw material. In Italy the biodiesel production reaches 100 Kt/y, corresponding to 0.8% of the automotive diesel market. Referring to the forecast biodiesel production by 2012, amounting to 500 Kt/y, said figure would correspond to 3.2% of the actual automotive diesel market.

In the case the use of biodiesel is compulsory regulated as the diesel fuel for public transport in urban agglomerates greater than 100,000 inhabitants, and as a blend component for regular diesel, the above forecast figures could be overcome.

14. CONCLUSIONS

The present survey has put evidence on the outstanding importance of biomass as renewable energy source. The present R&D efforts are focused on the increasing of competitiveness of three energetic products: bioethanol, biodiesel, and bioelectricity.

Bioethanol. The technological efforts performed till now, finalized to the introduction of ethanol into the liquid fuels market, have not been completely successful.

At present, scientific research seems, however, to allow the overcoming of the economic barriers. By the help of genetic engineering, it has been possible to modify microorganisms in order to utilize, as carbon source, not only exose, but also pentose sugars, contained in biomass. The improvement of the hydrolytic process of the lignocellulosic material will further contribute, for the medium term, for ethanol to reach competitiveness with gasoline.

Biodiesel. The development of this fuel seems mostly driven by political, instead of technological tasks. Its present production cost is non-competitive with traditional diesel fuel, and this figure is forecast unchanged for the medium term. Niche utilization will be possible as city diesel, due to its low sulfur content.

Bioelectricity. Production costs of electricity from biomass are the ones closest to competitiveness with fossil fuels. In particular cases, such as co-firing of waste from wood industry and, in general, whatever biomass source costing less than 3-4$/GJ, allows electricity production at competitive costs (~2.2 US cents/kWh). Said technology, however, is able to burn only limited percentage (10-15%) of biomass mixed to coal. Technologies able of more intensive use of biomass, such as combustion and gasification, are at the moment more expensive (9.2 and 7.6 US cents/kWh, respectively). Forecast figures (2020) for the two technologies are more appealing (6.0 and 5.5 US cents/kWh, respectively), placing them between the kWh production cost from natural gas (>3.7 US cents) and the one from coal (>7 US cents).

The exploitation of the energetic conversion of biomass in Europe by 2010 could increase from 3.5% to 12% the contribution to the total energetic consumption given by said source of renewable energy, corresponding to an increase of the energetic output of 90 Mtoe. The contribution that could be made by biogas exploitation from livestock production, agro-industrial effluents, sewage treatment and landfill by 2010 is estimated at 15 Mtoe. As far as solid residues are concerned, it is estimated that 30 Mtoe could be mobilized annually by 2010 for the power and the heating and industrial process heat market. Energy crops also need to be considered with the aim of reaching the objective. 45 Mtoe is the amount estimated, of this 18 Mtoe could be in the form of liquid biofuels and 27 Mtoe as biomass for heat and/or power.

REFERENCES

1. The Atlas Project (1997). "Energy Technology: the next steps" European Commission - Joule Thermie.
2. E.U. Von Weizsacker, A.B. Lovins, L.H. Lovins Factor 4, Doubling Wealth Halving Resource Use, London, Earthscan, 1997.
3. OECD Paris Benign Energy? The environmental implications of renewables, 1998.
4. Eurostat 1997.
5. European Commission - Common Position CE 9/98; January 19, 1998.
6. Italian Interministerial Committee for the Economic Planning. "Guide lines for national policies and measures for reduction of greenhouse gas emissions". CIPE Decision N° 137; October 19, 1999 (Italian).

7. European Commission. "Energy for the future: Renewable Enrgy Sources (RES)"Energia per il futuro: le fonti energetiche rinnovabili"; Green Paper for a Community Strategy. Communication 576 def.; Brussels, November 20, 1996.
8. "New Renewable Energy Resources", World Energy Council, 1997.
9. Bain, R.L., R.P. Overend, K.R. Craig. "Biomass-fired power generation" Fuel Processing Technology, 54, 1-16, 1998.
10. Hall, D.O., & J. I. House "Biomass energy in Western Europe to 2050" Land Use Policy 12 (1) , 37-48, 1995.
11. Hall, D.O. "Biomass energy in industrialized countries- a view of the future" Forest Ecology and Management, 91, 17-45, 1997.
12. European Commission, Energy for the future: Renewable sources of energy (Rets) White paper for a Community Strategy and Action Plan. Communication from the commission FOR/086/97, Brussels), 1997.
13. Kuiper L.C., R. Sikkema & J.A. Stolp. "Establishment needs for short rotation forestry in the EU to meet the goals of the commission's white paper on renewable energy" Biomass and Bioenergy 15(6), 451-456, 1998.
14. Diamantidis, N.D., Koukios, E.G. "Agricultural crops and residues as feedstock for non-food products in Western Europe" Industrial Crops and Products 11, 97-106, 2000.
15. Panoutsou, C., "Energy potential of agricultural residues in EU" in: Kopetz, H., et al. (Eds.) Biomass for Energy and Industry, Proceedings of 10th European Conference, 8-11th June 1998, Wurzburg, Germany, 863-866, 1998.
16. El Bassam, N. "Energy Plant Species- Their use and impact on environment and development" James & James, London, 1998.
17. Lako, P., Gilien D.J. "European Biomass Scenarios and the Need for the Import of Biomass" ECN Petten, 1997.
18. Price R. "Energy from wastes" Financial Times Energy Publishing, 1996.
19. Beone, G., Luciani, G., Luciani, M. ENEA, "Data Bank of Italian Incineration Plants"; ENEA, 1995 (Italian).
20. Developing a new generation of sustainable technologies: Report of Workshop of the Renewable energy Working Party of the IEA, Paris, 2000.
21. ENEA. Renewable Energy Sources. Green Paper; July16, 1998 (Italian).
22. Italian Ministery of Agriculture and Forestry. "Renewable Energy from Biomass"; National Programme, June 24, 1998 (Italian).
23. Kruse, A., Abeln, J., Dinjus, E., Kluth, M., Pertich, G., Schacht, M., Sandri, E., Schmieder H. "Gasification of biomass and model compounds in hot compressed water". Forschungzentrum Karlsruhe, 1999.
24. Yan, J., Alvfors, P., Eidensten, L., Svedberg, G. "A future for biomass" American Society of Mechanical Engineers, MeMagazine, October 1997.
25. Beekes, M.L. et al., "Co-combustion of biomass in pulverized coal-fired Boilers in the Netherlands", 1998.
26. Van Ree, R., Korbee, R., et al., "Operational experiences of (in) direct co-combustion in coal fired and gas fired power plant in Europe" Powergen Europe, Brussels,2001.
27. De Vita, L. et al. "Cofiring: an opportunity for the diffusion of renewable fuels" 1st World Conference and Exhibition on Biomass for Energy and Industry. Seville, Spain, 5-9 June, 2000.
28. IEA Bioenergy/Caddet "Advanced thermal conversion technologies for energy from solid wastes", 1998.
29. Babu, S.P. "Thermal gasification of biomass technology developments". Biomass and Bioenergy 9(1-5), 271-285, 1995.
30. Beennackers A.A.C.M. "Biomass gasification in moving beds, a review of European technology" Renewable Energy 16, 1180-1186, 1999.
31. Solantausta, Y., Bridgewater, A.T. & Beckman. D. "Feasibility of power production with pyrolysis and gasification systems" Biomass and Bioenergy, 9, 257-269, 1995.

32. Lau, F.S. "The Hawaii project" Biomass and Bioenergy. 15(3), 233-238, 1998.
33. De Lange, H.J., Barbucci, P. "The Thermie Energy Farm Project" Renewable Energy 16, 1004-1006, 1999.
34. Kruse, A., Abeln, J., Boukris, N., Dinjus, E., Kluth, M., Petrich, G., Sandri, E., Schacht, M., Smieder, H., "Proceedings of the International GVC-Meeting "High Pressure Chemical Engineering" FZKA 6271, Karlsruhe, 111, 1999.
35. Schmieder, H., Abeln, J., Boukris, J., Dinjus, E., Kruse, A., Kluth, M., Petrich, G., Sandri, E., Schacht, M., "Hydrothermal Gasification of Biomass and Organic Wastes" Proc.5th Conf. On Supercrit. Fluids and their Applications, 13-16 June, Garda, Italy, 347-354, 1999.
36. Bridgwater, T., "A guide to fast pyrolysis of Biomass for Fuels and Chemicals" Pyrolysis Network Guide N°1, 1999.
37. Cuevas, A., Ceinoso, C., Scott, D.S. "Pyrolysis oil production and its perspectives" Proc. Power production from biomass II, March 1995.
38. Cuevas, A., Proc. EEC Joule Contractors Meeting, Athens, June 1993.
39. Maniatis, K., Bayens, J., Roggeman, G., Peeters, H. "Flash pyrolysis of biomass in an entrained bed reactor" in: 6th EC Conference on Biomass for Energy, Industry and Environment, Athens, Greece, 22-26 April 1991, OR 09.02.
40. Trebbi, G., Rossi, C., Pedrelli, G., "Plans for the production and utilization of bio-oil from biomass fast pyrolysis" In: Bridgewater, A.V., Boocock, D.G.B. Eds., Development in Thermochemical Biomass Conversion Vol. 1. Chapman and Hall, 378-87, 1997.
41. Smith, S.L., Graham, R.G., Freel, B. "The development of commercial scale rapid thermal processing of biomass" in: D.L. Klass, ed. First Biomass conference of the Americas - Energy, Environment, Agriculture, and Industry, Vol. 2, CO, USA, NREL, 1194-1200, 1993.
42. De Baere, L., "Anaerobic digestion of solid waste: State of the art" 2nd Intl. Symposium on Anaerobic Digestion of Solid Wastes. Barcelona, July 15-17, 1999.
43. De Bo, I., et al. "Optimization of a sampling method for volatile organic compounds and its application to composting plants" Engineering Thesis, University of Ghent, 1997.
44. Rogalski,W., Charlton, J. "Status and trends for biological treatment of organic waste in Europe" ISWA Working Group on Biological Treatment of Waste, 1995
45. D.Addario, E. et al. (1993) "The acidogenic digestion of the organic fraction of municipal solid waste for the production of liquid fuels" Water Sci. Technol., 27, 183-192, 1993.
46. D'Addario, E., et al. "Upgrading/disposal of municipal solid waste through the production of volatile fatty acids" Environmental Technol., 16, 1115-1124, 1995.
47. Fascetti, E., et al. "Photosynthetic hydrogen evolution with volatile organic acids derived from the fermentation of source selected municipal solid wastes" Int. J. Hydrogen Energy 23, 753-760, 1998.
48. Larson, S., Palmquist, E., Hahn-Hagerdale, B., Stanberg, K., Zacchi, G., Nilvenbrant, N. "The generation of fermentation inhibitors during dilute acid hydrolysis of softwood", Enzyme Microb. Technol., 24, 151, 1999.
49. Avella, R. "Research Activities on the Energetic Use of Biomass"; ENEA Casaccia, Roma, 2000 (Italian).
50. Stemberg, K., Bollok, M., Raczey, K., Galbe, M., Zacchi, G., "Effect of substrate and cellulose concentration on simultaneous saccharification and fermentation of steam-pretreated softwood for ethanol production" Biotechnol. Bioeng., 68, 204, 2000.
51. Coombs, J. "Bioconversion assessment study", European Commission, Directorate General XII, Science Research and Development, 1996.
52. Von Sivers, M., Zacchi, M., Olsson, L., Hahn-Hagerdal, B., "Cost analysis of ethanol production from willow using recombinant Escherichia coli"; Biotechnol. Prog., 10, 555-560, 1994.

53. Finkelstein, M., Glassner, D., "Bioethanol -The future is now" 2nd European Motor Biofuels Forum, Brussels, 10-13 October, 1999.
54. Ondrey, G., Armesto, C. "A Renewed Boost for Ethanol" Chemical Engineering, 35-39, February 1999.
55. Ballerini, D., Desmarquest, J.C., Pourquie, J. "Ethanol Production from Lignocellulosics: Large Scale Experimentation and Economics" Bioresource Technology, 50, 17-23, 1994.

SECTION VI

ASSESSMENT OF CO_2 UTILIZATION AND INDUSTRIAL PERSPECTIVE

This Section summarizes the activity performed within WP6 of the BRRT-CT98-5089 RUCADI Project, "Recovery andUtilization of Carbon Dioxide" by the following Partners:

Antonella Caroppo, *Nowatech Ass.*, Italy
Angeliki A.Lemonidou, *CPERI*, Greece
Marcella Narracci, *Nowatech Ass.*, Italy
John McMullan, *University of Ulster*, UK
Dimitri Mignard, *University of Edinburgh*, UK
Marco Ricci, *PolimeriEuropa*, Italy
Mortasa Sahibzada, *University of Edinburgh*, UK
Julia Valla, *CPERI*, Greece

Chapter 15

METHANOL PRODUCTION FROM NATURAL GAS

Assessment of CO_2 utilization in natural gas reforming

Angeliki A. Lemonidou, Julia Valla and Iacopos A. Vasalos
CPERI – Dept. Chemical Engineering Aristotele "Univ. of Thessaloniki" PO Box 1517, 54006 University City Thessaloniki – Greece
lemonido@eng.auth.gr

1. INTRODUCTION

Carbon dioxide (CO_2) is widely recognized as a major greenhouse gas contributing to global warming. This greenhouse gas is produced in large quantities worldwide by the heavy industry, such as steel production, chemical and petrochemical manufacturing, cement production etc., including power generation from fossil fuel. In the past decades, the continuous and rapid development of such industries has caused considerable concern. Awakening to the CO_2-mediated global warming problem has led to several international forums on climate change called by the United Nations. As a result all developed countries are called to regulate the CO_2 emissions in order to eliminate the long term risk associated with climate change. Such a stipulation may adversely affect the development of a large number of industries in the near future. [1, 2]

To mitigate the global warming problem, the removal of CO_2 from the industrial flue gases seems to be necessary. Carbon dioxide recovered from flue gases can be sequestered in oceans or spent gas and oil wells in an attempt to mitigate atmospheric CO_2 increase. [3, 4] An alternative and more effective way is the recovered gas to be used for the production of chemicals, fuels and other useful products. The motivation for the production of chemicals lies in the fact that CO_2 is a cheap, totally renewable feedstock that can be used for the synthesis of materials like polymers and the synthesis of intermediates and chemicals via new routes that are

M. Aresta (ed.), Carbon Dioxide Recovery and Utilization, 379-394.

competitive with the current established methods. [5] Even though the impact of chemical utilization of CO_2 on the global carbon balance is small, by no means can it be considered insignificant. However, the utilization of CO_2 as a feedstock for fuels and chemicals has to fulfill economic, technical and environmental prerequisites in order to be feasible. [4]

One of the most promising and feasible routes for the use of CO_2 is the synthesis of methanol. Methanol, a commodity chemical with a global demand of around 36 MMT in 2000 is commercially manufactured via indirect route based on natural gas. [6] It is actually used as a chemical feedstock, mainly for the manufacturing of formaldehyde and methyl tert-butyl ether (MTBE), a gasoline octane booster. The demand for MTBE is declining steadily due to environmental concerns. However, a much larger potential for the methanol market is in the fuels sector. The conversion of remote natural gas into easily transportable methanol is a challenge for the gas industry. According to the Shell Briefing Service an annual demand of 2000MMT of gas for transportation is foreseen, as well as another 4000MMT of coal equivalent for power generation. [7] Stimulated research efforts over the last few years have been directed towards the application of methanol as a fuel for fuel cells. The methanol reforming system is an attractive option for on board hydrogen generation at low temperature with very promising results. [8]

Methanol is produced commercially from synthesis gas ($CO + H_2$) via the following basic reaction over zinc/copper based catalyst:

$$2H_2 + CO \leftrightarrows CH_3OH \quad (\Delta H = -91 \text{ kJ/mol}) \tag{1}$$

The overall methanol synthesis is exothermic and equilibrium limited, being favored by low temperature and high pressure. Commercial reactors are fixed bed, operating in the range of 230 to 270 °C and 50 to 100 atm. [6]

The mixture of H_2 and CO (syngas) is industrially produced by steam reforming of natural gas, a mature and most established process, over Ni based catalysts. [9] The main stoichiometric equation that describes the steam reforming of methane is:

$$CH_4 + H_2O \leftrightarrows CO + 3H_2 \quad (\Delta H = 206.3 \text{ kJ/mol}) \tag{2}$$

The water gas shift reaction is always active in methane reforming environment:

$$CO + H_2O \leftrightarrows CO_2 + H_2 \quad (\Delta H = -41 \text{ kJ/mol}) \tag{3}$$

Reaction (2) is highly endothermic and is typically carried out at high pressure in the range of 20 to 30 atm. Since equilibrium is not favored by high pressure, excess steam is utilized for high methane conversion. Reaction (3) is known as slightly exothermic and its equilibrium is shifted to the left at the high steam reforming temperatures. However, when large excess steam is used, the reaction leads to production of CO_2 at the expense of CO. The syngas produced has a surplus of hydrogen, which results in recycling a large quantity of not converted gas in the methanol section. The H_2/CO ratio can be corrected by adding CO_2 to the reformer feed. CO_2 reacts with CH_4 (dry reforming) producing syngas with $H_2/CO=1$ according to the reaction:

$$CH_4 + CO_2 \leftrightarrows 2CO + 2H_2 \quad (\Delta H = 247.3 \text{ kJ/mol}) \qquad (4)$$

Dry reforming is about 20% more endothermic than steam reforming, and the major problem that limits the industrial application of this process is the high formation of carbon. [10]

The combination of dry reforming with steam reforming for the production of methanol can accomplish important objectives: (i) to produce syngas with the desired H_2/CO ratio, which results to improved natural gas conversion to methanol and to reduced overall plant energy requirements per ton of product, (ii) to mitigate the carbon formation that is significant for dry reforming and (iii) to provide an approach for the reduction of CO_2 emitted from the natural gas combustion in the reformer.

Various technologies have been developed for the removal of CO_2 from concentrated CO_2 sources based on absorption by chemical solvents, physical absorption, cryogenic separation, membrane separation and CO_2 fixation by biological or chemical methods. Among these methods the absorption of CO_2 by chemical solvents appears to offer an interesting and practical alternative. In fact CO_2 absorption by alkylamines (MEA) has been the most popular and effective one and many investigators have performed extensive research in the past decades. [11] According to Halmann and Steinberg [12], the energy required for CO_2 recovery by MEA absorption is 0.6 kWh/kg, which is 30% less than this required by refrigeration or membrane separation.

The objective of this work was to study the effect of CO_2 recovered from reformer flue gases on the capacity, energy efficiency and economy of the methanol process. The methodology is composed of the following:

- Equilibrium analysis of synthesis gas and methanol production
- Development of process flow scheme for the production of methanol from natural gas
- Technical assessment of the process using ASPEN/SP simulator
- Cost assessment of methanol production process.

2. EQUILIBRIUM ANALYSIS

The approach to equilibrium is an important factor for any syngas production system. Even though reforming is a catalytic process, product distribution at the reactor exit is not kinetically controlled by the catalyst but is determined by thermodynamic equilibrium. According to reaction (1), methanol synthesis requires mixtures of CO and H_2, satisfying the balance ratio $R=H_2/CO=2$. The feasibility of using feed mixtures consisting of CH_4, with H_2O and CO_2 was analyzed in relation with the production of syngas for methanol. It was found that under certain conditions the syngas produced exhibited a balance ratio for methanol synthesis that lies above the carbon limit curve. [13] Extensive equilibrium studies were performed with Equil software [14] in order to determine the effect of feed composition H_2O/CH_4 (HC) and CO_2/CH_4 (CC), temperature and pressure on methane conversion, R ratio and methanol production.

The results showed (Figure15.1) that equilibrium methane conversion at 950°C and 25 atm is an increasing function of both HC and CC ratios.

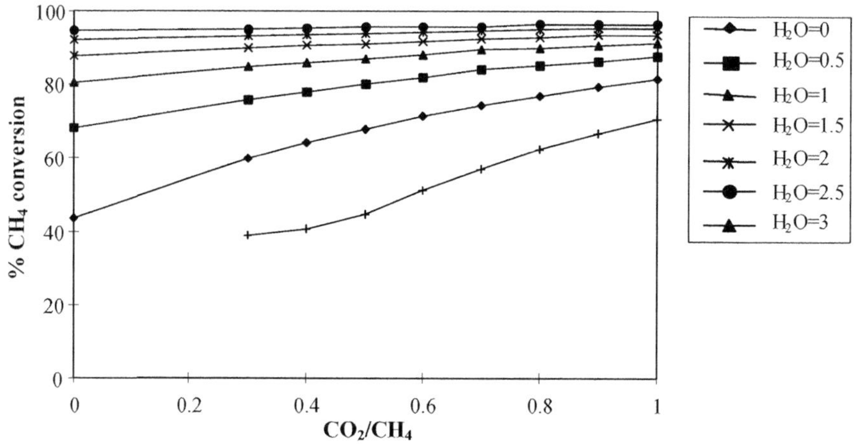

Figure 15.1 - Methane conversion as a function of H_2O/CH_4 and CO_2/CH_4 ratios (T=950 °C and P=25atm)

At low HC ratios the addition of CO_2 in the feed results in higher methane conversions while at high HC ratios the presence of CO_2 in the feed does not affect conversion considerably. The presence of increased amounts of CO_2 in the reacting mixture is in general beneficial for methane conversion. The higher the CC ratio the higher is the methane conversion. The stoichiometric ratio R is significantly affected by the feed composition. For constant CC the increase in steam concentration results in higher H_2/CO values. For a wide range of CC ratios (0.3 – 1) the value of R=2 can be attained at 950°C and 25 atm (Figure 15.2).

Apart from the results obtained at temperature 950 °C and pressure 25 atm, extensive calculations were performed using different reaction temperatures and pressures. It was observed that CH_4 and CO_2 conversion is slightly affected by varying the pressure, while it increases significantly with increasing the temperature.

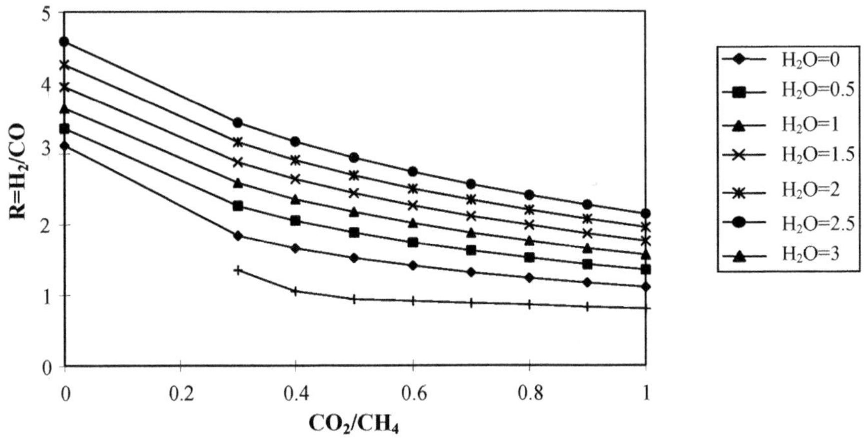

Figure 15.2 - Syngas ratio R as a function of CO_2/CH_4 and H_2O/CH_4 ratios (T=950 °C and P=25 atm)

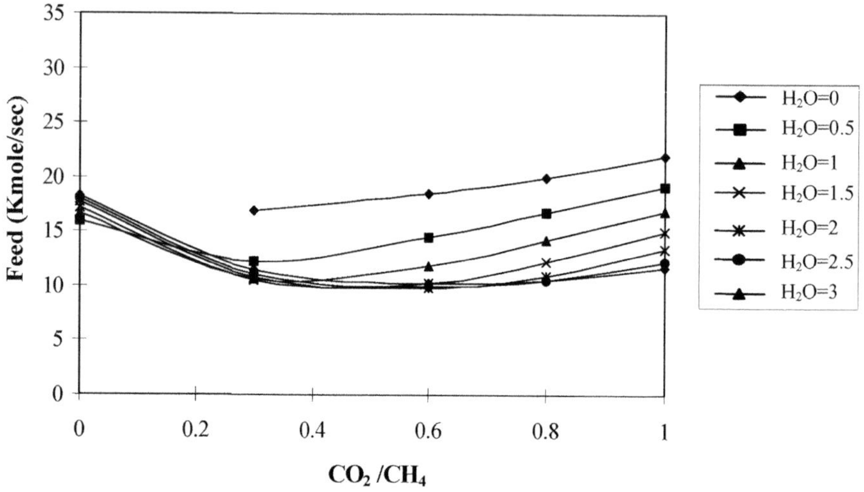

Figure15. 3 - Methanol reactor feed as a function of CO_2/CH_4 and H_2O/CH_4 ratios (T=270 °C, P=100atm)

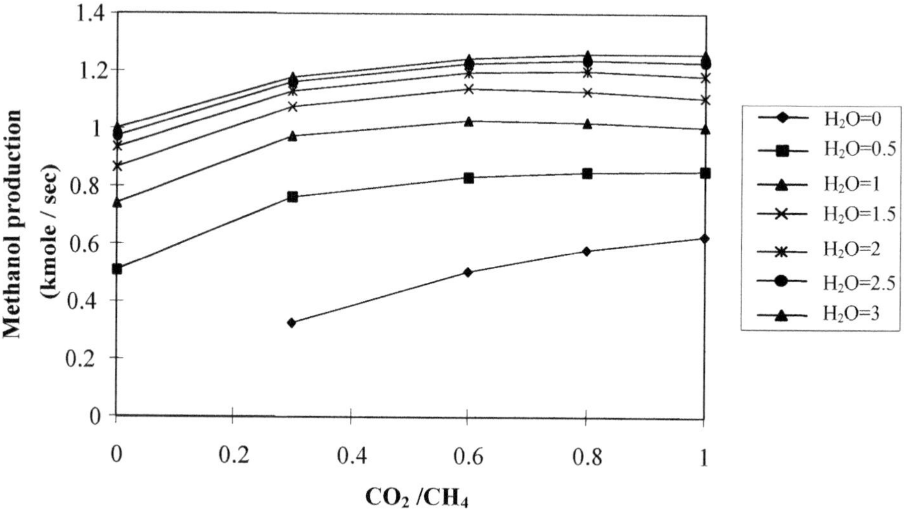

Figure 15.4 - Methanol production as a function of $CO_2/CH_4$and H_2O/CH_4 ratios (T=270 °C, P=100atm)

Furthermore the effect of feed composition on methanol reactor feed and on methanol production was examined. The conditions selected for the equilibrium calculations in the methanol reactor were 270°C and 100 atm, which are the conditions most often used in -industrial methanol reactors. [6] The flow to the reactor is minimized (minimum recycle) in the range CO_2/CH_4=0.3-0.9, while for CO_2/CH_4=0 (steam reforming) the feed flow is quite high due to the increased recycle (Figure 15.3). The effect of CO_2 addition to the reformer feed is illustrated in Figure 15.4. Methanol production rate is positively affected by the presence of CO_2. The production rate is an increasing function of HC due to the increased methane conversion in the syngas reactor. It was also observed that for CO_2/CH_4>0.6 and H_2O/CH_4>1.5 the amount of methanol is relative insensitive to CO_2/CH_4 (Figure 15.4).

Based on these results it was concluded that, for a temperature of 950°C and a pressure of 25 atm (these conditions are commonly applied in industrial steam reformers), the upstream addition of carbon dioxide in steam reformer has beneficial effects both on methanol yield and methanol reactor feed. The optimum feed ratios, satisfying the balance R=2, which were considered for the study, were: H_2O/CH_4=1.5 and CO_2/CH_4=0.6. Under these conditions the thermodynamic potential for carbon formation is zero in agreement with the data of equilibrium chart presented by Dibbern et al. [10]

The following cases were considered for the techno-economical assessment:

- Steam reforming (base case) H_2O/CH_4=2.5.
- Steam-CO_2 reforming H_2O/CH_4=1.5 & CO_2/CH_4=0.6.

3. DEVELOPMENT OF PROCESS FLOW DIAGRAM

In order to evaluate the steam reforming and the steam-CO_2 reforming processes, a common conceptual design was developed for the production of methanol from natural gas. The concept is illustrated by the process flow scheme of natural gas to methanol in Figure 15.5. The design basis common for the two processes is presented in Table 15.1. The natural gas feed used contained 94,5% methane and was available at a temperature of 25°C and a pressure of 15 atm.

The simulation process diagram is divided into 4 main sections.

3.1 Syngas section

Natural gas available at 25 °C and 15 atm is compressed to 30 atm and preheated in the natural gas preheater against the hot reactor effluent. It is then mixed with hydrogen and fed to the desulfuriser, where the sulfur components are converted to hydrogen sulfide. The sulfur free gas (less than 0.5 ppm S) is led to the syngas reactor. The endothermic steam and carbon dioxide reforming reactions take place in a tubular fixed bed reactor.

Table 15.1 - Design basis common to synthesis gas generation processes

Design capacity	**2,500 tn/day methanol, 99.5% fuel grade**
Natural gas feed rate	2,050,000 Nm^3/day
Natural gas feed	% vol
CH_4	94.5
CO_2	0.5
N_2	0.8
C_2H_6	2.7
C_3H_8	1.5
Syngas reactor	Temperature: 950 °C, Pressure: 25 atm
Methanol reactor	Temperature: 270 °C, Pressure: 100 atm

Preheated natural gas is mixed with process steam and, in the case of the steam-CO_2 process, with the recovered CO_2. The steam/methane and carbon dioxide/methane ratios are fixed according to the thermodynamic equilibrium studies. The feedstock mixture is preheated to 600°C by the furnace flue gases. The reformer exit conditions were set at 25atm and 950°C. The reformer fuel comprises natural gas and purge gas from the methanol recycle stream. The purge ratio was set at 7.5% in both cases. It is assumed that the efficiency of the radiant section is 50% and the heat flux to this section 745,000 KJ/m^2. The temperature of the flue gases is set at 150°C. The heat carried by the flue gases is used for preheating the reactor feed and fuel, producing high, medium steam and preheating the boiler feed-water.

The hot reactor effluents at 950°C are cooled down to 35°C in a series of heat exchangers and boilers. The cooled gases are compressed in a multistage steam driven compressor to 100 atm with interstage cooling and water separation.

3.2 Methanol section

Compressed syngas mixed with recycle gas is preheated to 250°C by methanol reactor product gas and is fed to the reactor. The exothermic reaction takes place in catalyst filled tubes at 270°C and the generated heat is removed producing medium pressure steam. The reactor product is cooled down to 38°C in a series of a feed preheater, an air cooler and a water cooler. The condensed liquid consisting mainly of methanol and water is separated at 96 atm in the methanol separator, expanded to 4 atm and fed to the distillation column. The gas product leaving the separator, comprising mainly unreacted H_2, CO and CO_2, as well as inert gases CH_4 and N_2, is recycled to the reactor after compression at 100 atm. In order to keep the inert gases at low fractions, part of this stream is purged and is used as fuel in the reformer. Raw methanol (about 20% H_2O content) from the expansion vessel is fed to the distillation column, where methanol product is drawn at 99.5 % vol purity. Methanol is cooled down to 38°C and fed to storage.

3.3. MEA unit

The monoethanolamine absorption was selected as the appropriate process for the recovery of CO_2 from the low-pressure flue gases.

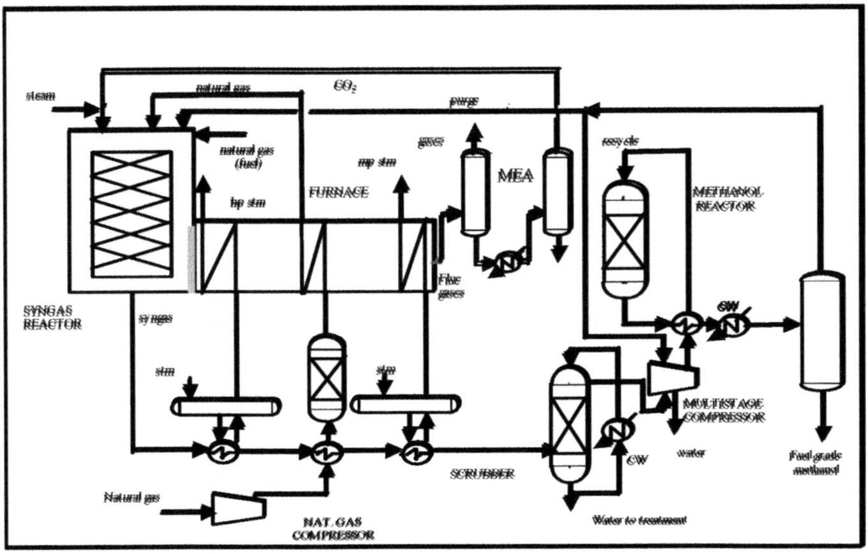

Figure 15.5 - Flow diagram of natural gas to methanol process

The process includes MEA extraction, compression at 30 atm and dehydration. The stream delivered has a composition of ~100% CO_2. Dehydration of the gas to water concentrations of <50 ppm is necessary to prevent corrosion of the pipeline by carbonic acid. The main parts of this section comprise the absorber and stripper columns, together with the associated piping, heat exchange and separation equipment. [11,15]

3.4 Utilities

The plant utilities considered in the assessment are high pressure steam for driving compressors, pumps, blowers etc., medium pressure as process steam, low pressure steam for distillation reboiler and off-plant utilities, cooling water and electricity.

4. MATERIAL AND ENERGY BALANCE

The advanced process simulator ASPEN/SP was used to simulate a methanol plant based on steam and dry reforming to syngas and its further conversion to methanol. In order to assess the advantages and the disadvantages of CO_2 addition in the methanol production process the steam-CO_2 reforming was compared with the conventional steam reforming process. Methanol capacity for the conventional steam reforming was set at 2,500 tn/day fuel grade.

The reformer for this process was simulated as an equilibrium reactor using the RGIBBS model of ASPEN/SP. The heat duty for the reformer was supplied from the combustion of natural gas and purge stream which contains 90% H_2. The methanol reactor was simulated as an isothermal reactor using the REQUIL model of ASPEN/SP.

The results of the material balance for the steam reforming process (SR) and the steam-CO_2 reforming process (SCR), are shown in Table 15.2. The conversion of CH_4 obtained in the steam reformer at 950°C and 25 atm was 91% while in the mixed reforming it was slightly lower, 87%. The concentration of unconverted CH_4 in the reformer exit gas was 2.2% (on dry basis) for the steam reformer and 3% for the steam-CO_2 reformer. The inerts concentration of the syngas is a very important factor for the methanol plant, since the methanol synthesis occurs in a loop with significant recycle. Therefore, any CH_4 that is not converted in the reformer will be accumulated in the methanol synthesis loop thus reducing the partial pressure of active components. [16] Due to the high steam/methane ratio used, the syngas produced had a very high H_2 content (R=4.11) in the conventional reformer. The syngas ratio attained in the presence of CO_2 is much closer to the stoichiometrically needed for methanol production (R=1.98).

The addition of CO_2 in the reformer feed has a pronounced effect in methanol production. The amount of methanol produced is 18% higher compared to the steam reforming. In addition, the flow of recycle is drastically reduced in steam-CO_2 reforming (50% less). This minimum recycle results in smaller methanol reactor volume and consequently lower capital investment. The amount of the required natural gas fuel varies in the two cases because purge gas is also used as fuel in the furnace of the reformer.

Table15. 2- Material balance for steam (SR) and steam-CO_2 reforming (SCR)

	SR	SCR
Reformer feed		
Natural gas feed, Nm^3/h	85 200	85 200
Steam, Nm^3/h	201 600	120 960
CO_2, Nm^3/h	-	48 384
Reformer		
CH_4 conversion, %	91	87
H_2/CO	4.11	1.98
Methanol reactor		
Syngas feed, Nm^3/h	286 800	401 100
Recycle, Nm^3/h	1 115 250	494 325
Purge Fuel, Nm^3/h	90 317	40 320
Methanol, t/h	104	156
Methanol, t/day	2 496	2 945
Furnace		
Natural Gas Fuel, Nm^3/h	7 741	22 982
MEA unit		
Flue gas flow rate, Nm^3/h		860 000
CO_2, Nm^3/h		64 500
Total CO_2 recovered, t/day		2 280

Energy balances of methanol processes were also calculated using the ASPEN/SP simulator. The differences between the two processes are not significant. The results are summarized in Table 15.3. In steam-CO_2 reforming the furnace unit was designed to produce more energy, because the syngas production via CO_2 reforming is more endothermic. The energy duty that is required for the recompression of the recycled unconverted syngas stream is higher in the methanol reactor when steam reforming is used for syngas production. This is expected because of the inappropriate H_2/CO ratio that results to higher unconverted syngas flowrates. The amount of heat released in the methanol reactor is higher in the case of steam-CO_2 reforming due to the increased methanol production. Thus, more heat is transferred for the production of middle pressure steam.

The energy and the utility consumption per ton of methanol produced are presented in Table 15.4. The energy consumption was based on the net calorific value of natural gas. The slightly lower energy consumption of the steam-CO_2 reforming (31 GJ/t) compared to conventional steam reforming (31.9 GJ/t) is due to the substitution of natural gas with CO_2. The same difference in energy consumption between the two processes was also reported in the literature. [16] An important factor, which contributes to the large capital cost of methane conversion processes, is thermal efficiency. [17]

Table15.3- Energy balance for steam (SR) and steam-CO_2 reforming (SCR)

	SR, MW	SCR, MW
Natgas Compression	2.84	2.84
Furnace Duty	-502	-509
Syngas Reaction Duty	274	278
Heat Recovery	-154	-152
Syngas Compression	29	28
Recycle Compression	2.00	1.00
Methanol Reactor Duty	-86	-115

Thermal efficiency is defined as the ratio of the lower heating value of the liquid product to that of the feedstock. The conventional methanol scheme, which is based on steam reforming, operates with an energy consumption of 31.9 GJ/tn of methanol, which corresponds to a thermal efficiency of 62.8%. In the case of steam-CO_2 reforming the energy consumption of 31 GJ/tn corresponds to a thermal efficiency of 64%. Lange's data on energy consumption and thermal efficiency for methanol production via conventional steam reforming are in close agreement with the calculated values. [7]

Table 15.4 - Energy and utility consumption

Consumption/tn MeOH	SR	SCR
Energy (GJ)	31.9	31
CO_2 (kg)	-	760
Electric Power (kWh)	14	16
Boiler Feed Water (kg)	2 478	2 010
Cooling Water (m^3)	94	65

5. COST ASSESSMENT

The methanol production cost is affected both by the feedstock cost and the initial plant investment. It is well known that about 60% of the total investment cost in modern methanol plants, is attributed to the synthesis gas generation section. Thus, new and improved processes for the production of synthesis gas are of major technological and economic importance. [18-20]

The methodology applied for the cost estimation comprised capital and production cost calculations and profitability evaluation. The capital cost calculations, were based on study estimate (factor estimate) using the major equipment cost and existing historical data for 1995. [7]

$M.E. = M.E._{base} * [(\text{Section Capacity})/(\text{Base Capacity})^n*[(\text{Index } 2001)/(\text{Index } 1995)]$

n= scaling exponent.

The Chemical Engineering Plant Cost Indexes for the year 2001 were used under the assumption that the plant would be erected in 2001 in Europe.

The economic analysis of the MEA unit for the CO_2 recovery was based on a more general method named "Power factor applied to plant – capacity ratio" [21]:

$I_F = I_{Fbase}$*[(Section Capacity/(Base Capacity)]$^{0.7}$*[(Index 2001)/(Index 1997)]

A design basis was selected for recovering the CO_2 from flue gases emitted by a typical 500 MW power plant. [15,22]

The economic analysis showed that the total capital investment required for the methanol plant based on steam-CO_2 reforming is 3.4% lower than investment for the plant based on steam reforming. However, with the addition of the MEA unit the overall investment increases approximately 15% (Figure 15.6). Methanol price is strongly dependent on the particular syngas generation process and the cost of added CO_2 (Figure 15.7). When CO_2 is readily available without cost, a cost reduction of 10% is feasible. This saving in product value comes mainly from the improved methanol yield in the steam-CO_2 process scheme. When CO_2 is recovered by using the MEA unit, then the price of methanol becomes 5.5% higher than the base case. These outcomes fairly agree with the economic evaluation of Haldor-Topsoe about the technical and economic advancements of CO_2 addition in the steam reformer in the new methanol plant located at Bandar Immam in the Persian Gulf. [16]

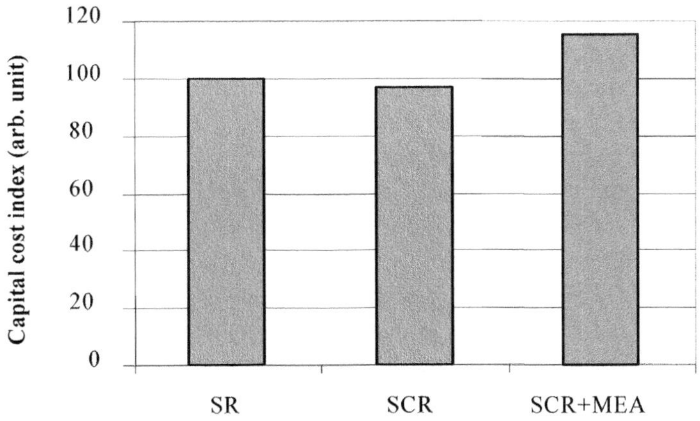

Figure 15.6 - Estimated relative capital cost of natural gas to methanol processes (SR:steam reforming, SCR:steam-CO_2 reforming, SCR+MEA: steam-CO_2 reforming plus MEA unit)

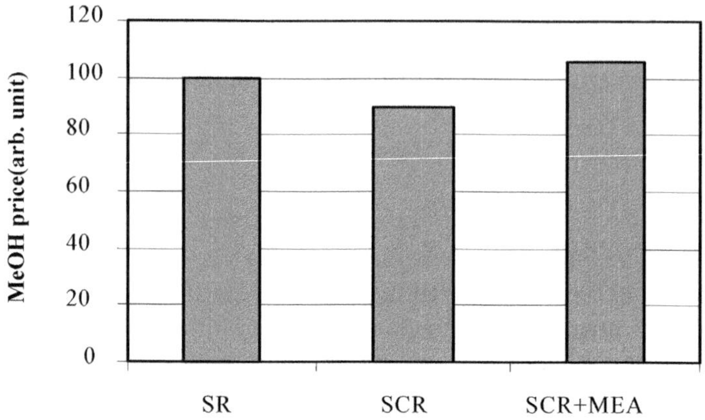

Figure 15.7- Estimated relative methanol price of natural gas to methanol processes (SR:steam reforming, SCR:steam-CO_2 reforming, SCR+MEA: steam-CO_2 reforming plus MEA unit)

6. CONCLUSIONS

The current worldwide emphasis on reducing greenhouse gas emissions provides an opportunity to revisit how energy is produced and used, consistent with the need for sustainable development. Both the scales of the problem and the efforts needed for its solution are extremely large. The real challenge is to find ways to reduce the effects of emissions without requiring expensive restrictions on industrial or economic growth and to find feasible ways to enhance emissions management technology.

The present study focused on the removal and utilization of CO_2 greenhouse gas produced by a methanol industrial process. According to the results approximately 3000 tn/day CO_2 are generated and disposed to the atmosphere per 2500 tn/day methanol product by a conventional steam reforming process of natural gas. However, the thermodynamic equilibrium analysis and the material balance calculations showed that the addition of CO_2 (CO_2/CH_4=0.6/1.0) upstream in steam reformer produce the optimum syngas ratio for methanol.

Considering the harmful effects of CO_2 to the environment and taking into account that the product capacity is significantly increased by 18%, when CO_2 is added to the process, with slightly higher energy efficiency (3%), the recovery of CO_2 from the flue gases seems to be a reliable approach both in terms of energy generation and CO_2 emission reduction.

The economic evaluation of this alternative methanol process based on syngas generation from steam-CO_2 reforming showed that the required total capital investment will be approximately 15% higher than the conventional steam reforming process, while the product price will be about 5.5% higher. However, it should be emphasized that the scenario of using the recovered CO_2 in the process results in 75% less CO_2 emissions than the conventional system.

The decision whether the alternative process is viable can be easily drawn if we consider the environmental damage due to the increasing CO_2 emissions and the harmful effects of a possible climate change. The recent increase in environmental concern forces industry and society to make serious attempts to decrease CO_2 emissions and incur the extra expense in order to avoid any negative environmental impact.

REFERENCES

1. Riemer, P., "Greenhouse gas mitigation technologies-an overview of the CO_2 capture, storage and future activities of the IEA greenhouse R&D programme" Energy Convers. Mgmt. 37, 665-670, 1996.
2. Steinberg, M. "Efficient natural gas technologies: A response to global warming" Chemtech 29, 31-36, 1999.
3. Grimston, M.C., Karakoussis, V., Fouquet, R., van der Horst, R., Pearson, P., Leach, M., "The European and global potential of carbon dioxide sequestration in tackling climate change" Climate Policy 1, 155-171, 2001.
4. Hanisch, C. "Exploring options for CO_2 capture and management" Environ. Sci. Technol 33(3), 66-70, 1999.
5. Aresta, M., Dibenedetto, A., Tommasi, I. "Developing innovative synthetic technologies of industrial relevance based on carbon dioxide as raw material" Energy & Fuels 15, 269-273, 2001.
6. Tijm, P.J.A., Waller, F.J., Brown, D.M. "Methanol technology developments for the new millennium" Appl. Catal. A 221, 275-282, 2001.
7. Lange, J-P. "Prespectives for manufacturing methanol at fuel value" Ind. Eng. Chem. Res., 36, 4282-4290, 1997.
8. Mizsey, P., Newson, E.M. "Comparison of different vehicle power trains" J. Power Sources, 102, 205-209, 2001.
9. Rostrup-Nielsen J.R., "Catalytic Steam Reforming" in Catalysis Science and Technology, J.R. Anderson and M.Boudart, Ed. Berlin: Springer-Verlag, 1984.
10. Dibbern, H.C., Olesen, P., Rostrup-Nielsen, J.R., Tottrup, R.B., Udengaard, N.R. "Hydrocarbon Processing" 65(1), 71-74, 1986.
11. Leci, D.L. "Development requirements for absorption processes for effective CO_2 capture from pilot plants" Energy Convers. Mgmt. 38, 45-50, 1997.
12. Halmann, M.M., Steinberg, M., "Greenhouse gas-carbon dioxide mitigation", New York Lewis Publishers, 1998.

13. Tjatjopoulos, G.J., Vasalos, I.A. "Feasibility analysis of ternary feed mixtures of methane with oxygen, steam, and carbon dioxide for the production of methanol synthesis gas", Ind. Eng. Chem. Res., 37, 1410-1421, 1998.
14. Tjatjopoulos, G.J., Vasalos, I.A. "Equil software", CPERI, Thessaloniki Greece, 1994.
15. Kadam, K.L. "Power plant flue gas as a source of CO_2 for microalgae cultivation: economic impact of different process options" Energy Convers. Mgmt. 38, 505-510, 1997.
16. Holm-Larsen, H. "CO_2 reforming for large scale methanol plants-an actual case" in Studies in Surface Science and Catalysis, J.J. Spivey, E Iglesia and T.H. Fleisch, Ed. Amsterdam: Elsevier Science B.V. 2001.
17. Lange, J.P., Tijm, P.J.A. "Processes for converting methane to liquid fuels: economic screening through energy management" Chem. Eng. Sci. 51, 2379-2387, 1996.
18. Dave, N., Foulds, G.A. "Comparative assessment of catalytic partial oxidation and steam reforming for the production of methanol from natural gas" Ind. Eng. Chem. Res., 34, 1037-1043, 1995.
19. Michel, S. "Reduce methanol manufacturing costs" Hydrocarbon Processing, 79(3), 116C-116I, 2000.
20. Aasberg-Petersen, K., Bak Hansen, J.H., Christensen, T.S., Dybkjaer, I., Seier-Christensen, P., Sub-Nielsen, C., Winter Mandsen, S.E.L., Rostrup-Nielsen, J.R. "Technologies for large-scale gas conversion" Appl. Catal. A, 221, 379-387, 2001.
21. Remer, D., Chai, L.H., "Estimate cost of scaled up process plants" Chem. Eng. 97, 138-143, 1990.
22. Steinberg, M. "Methanol as an agent for CO_2 mitigation" Energy Convers. Mgmt. 38, 423-430, 1997.

Chapter 16

CARBON DIOXIDE AS A BUILDING BLOCK FOR ORGANIC INTERMEDIATES: AN INDUSTRIAL PERSPECTIVE

Marco Ricci
Polimeri Europa S.p.A., Istituto Guido Donegani, Via Fauser 4, Novara-Italy
marco.ricci@polimerieuropa.com

1. INTRODUCTION

In this Chapter, current industrial processes will be described, in which carbon dioxide is used as a C_1 building block for the syntheses of organic intermediates. So, other interesting processes which also use carbon dioxide, but not as a synthetic building block (*e.g.*, as a cheap neutralising agent in the textile industry, or as a solvent, or as a blowing agent for plastic foams, or in the production of inorganic carbonates), will not be further mentioned.

Exploiting carbon dioxide as a raw material offers a well known paradox. On one side, there are several reasons for producing chemicals from it: *i)* carbon dioxide is widely recognised as a very cheap, hardly toxic C_1 building block which could, in principle, replace toxic reagents like phosgene; *ii)* carbon dioxide is a renewable feedstock, unlike oil or coal; *iii)* producing chemicals from carbon dioxide would help in controlling its atmospheric concentration. On the other hand, only very few processes use it as a raw material. The explanation of this paradox is the great thermal stability of carbon dioxide itself, witnessed by its very negative $\Delta G°_f$ (-394.6 kJ/mole) [1]. Because of this stability, relatively few processes are known which are thermodinamically allowed, *i.e.* with $\Delta G < 0$.

Nevertheless, a large proportion of all the recovered carbon dioxide is used in the synthesis of methanol and urea, quite often in the same place where its production occurs. Much smaller amounts of other chemicals

M. Aresta (ed.), Carbon Dioxide Recovery and Utilization, 395-402.

(salicylic acid, other aromatic hydroxycarboxylic acids, cyclic organic carbonates) are also produced using carbon dioxide as a raw material.

2. METHANOL

Most of the methanol produced world-wide is derived from synthesis gas which, in turn, is obtained by the steam reforming of natural gas (methane). So:

$$CH_4 + H_2O \leftrightarrows CO + 3\ H_2 \tag{1}$$

$$CO + 2\ H_2 \leftrightarrows CH_3OH \tag{2}$$

The steam reforming allows to produce a synthesis gas with a H_2/CO ratio close to 3, whereas a H_2/CO ratio of 2 is optimal for the methanol synthesis. For this reason, in order to reach the best stoichiometry for methanol synthesis, carbon dioxide is added, when available, to the synthesis gas at the exit of the steam reformer, to exploit the reaction:

$$CO_2 + 3\ H_2 \leftrightarrows CH_3OH + H_2O \tag{3}$$

which is still an exothermic reaction (although to a lesser extent compared with carbon monoxide hydrogenation: $\Delta H_{300\ K}$ = - 49,16 and - 90,77 kJ/mol, respectively [2]), due to the formation of a very stable water molecule. So, ideally, the overall stoichiometry of methanol synthesis is:

$$3\ CO + CO_2 + 9\ H_2 \leftrightarrows 4\ CH_3OH + H_2O \tag{4}$$

Worldwide annual production of methanol is around 20 x 10^6 t [2, 3]. However, it has been estimated that only 25% of the maximum possible amount of carbon dioxide is actually employed [3]. So, a consumption of ca. 2 x 10^6 t/year of carbon dioxide would result for the methanol synthesis.

3. UREA

Urea is the most important fertiliser in the world. Almost two-fifths of the world's population rely, particularly in less developed countries, on urea for food supply. World urea demand has been estimated to be, in 1997, ca. 89 x 10^6 t/y [4]. Urea prices are very different, *e.g.*, in U.S.A. and Europe: assuming, however, an average price of 130 €/t ([5]; exchange rate: 1 € = 1

$), the current value of the urea production can be estimated to be ca. 11.5 x 10^9 €. With 330 plants around the world, urea production is very spread. However, more than 54% of global capacity is located in South East Asia.

Urea production occurs *via* a two steps reaction of carbon dioxide and ammonia [4]. In the first step, the reagents combine to give ammonium carbamate:

$$2\,NH_3 + CO_2 \leftrightarrows H_2N\text{-}COO^-NH_4^+ \qquad (5)$$

which, upon dehydration, affords urea in the second step:

$$H_2N\text{-}COO^-NH_4^+ \leftrightarrows H_2NCONH_2 + H_2O \qquad (6)$$

In the actual practice, the two steps occur virtually simultaneously. Typical reactions conditions are 185-190 °C and 150-220 atm. Yields are essentially quantitative: some biuret ($H_2NCONHCONH_2$) also forms, but it is not removed from the product in which, however, it cannot exceed 1-2%. Several technologies are available upon license, which differ for reagents ratio in the feed, reagents recovery and recycling, and product finishing (granulated *vs.* prilled urea). They also differ a little about specific consumption, mainly of utilities, rather than of raw materials.

Formally, urea production needs a huge (65 x 10^6 t/year) carbon dioxide consumption. However, urea plants are typically parts of integrated facilities, also comprehensive of ammonia plants. The overall reaction in producing ammonia *via* natural gas (methane) steam reforming is [6]:

$$CH_4 + 0.3035\,O_2 + 1.131\,N_2 + 1.393\,H_2O \rightarrow CO_2 + 2.262\,NH_3 \qquad (7)$$

or, in different terms:

$$CH_4 + 1.4345\ \text{air} + 1.393\ H_2O \rightarrow CO_2 + 2.262\ NH_3 \qquad (8)$$

So, carbon dioxide is itself a product of the steam reforming step and its amount is quite balanced, for urea synthesis, with that of ammonia. As a consequence, the whole process accounts for the net consumption of relatively small amounts of carbon dioxide: the theoretical need is 0,131 mol per mol of urea, corresponding to 85 kg of carbon dioxide per t of urea. Based on these figures, a net consumption of 7,5 x 10^6 t/year would result.

4. AROMATIC HYDROXYCARBOXYLIC ACIDS

The sodium salt of salicylic acid (2-hydroxybenzoic acid) is produced, on industrial scale, by the the Kolbe-Schmitt synthesis, starting from carbon dioxide and dry sodium phenoxide:

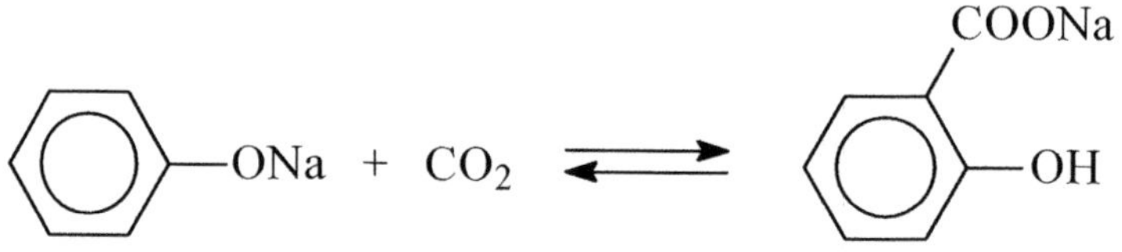

(9)

The process is usually run in batchwise manner, in a stream of carbon dioxide, at 150-160 °C, and under a pressure of 5 bar. The yield is about 90% [7].

The world capacity of salicylic acid is approximately 60.000 t/year, mainly used (60%) in the manufacture of the analgesic drug aspirin (2-acetoxybenzoic acid). Other applications, however, are growing, including as an intermediate in the production of drugs different than aspirin, of agrochemicals, of dyes, and as a component of phenolic resins [8]. Assuming an average price of 3,5 €/kg for salicylic acid ([9], exchange rate: 1 € = 1 $) this relatively small market has the significant value of about 210 x 10^6 €.

If potassium (rather than sodium) phenoxide is reacted with carbon dioxide (220 °C; 5,4 bar), formation of 4-hydroxybenzoic acid occurs in 80% yield [10]:

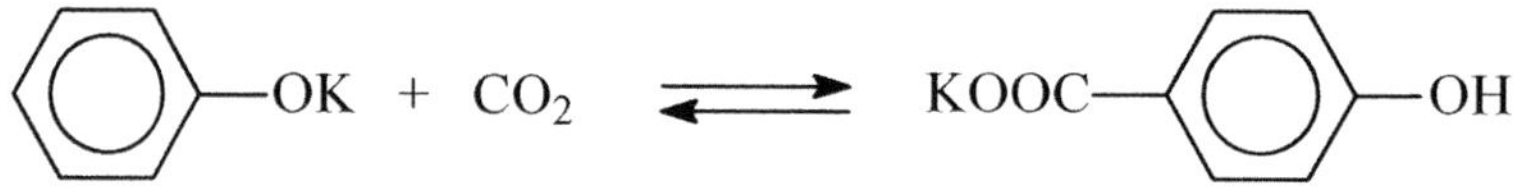

(10)

4-Hydroxybenzoic acid is a useful intermediate in the syntheses of food preservatives, dyes, pharmaceuticals, and pesticides. It has been also employed as a monomer for polyesters and for liquid crystals polymers.

The carboxylation of phenoxides can be extended to that of the alkaline salts of naphthols. Different hydroxynaphthoic acids are obtained, depending upon the raw material (1- or 2-naphthol), the alkali used, and the reaction conditions adopted [11]. Thus, treatment of the dried sodium salt of 1-naphthol with carbon dioxide at 125-135 °C affords 1-hydroxy-2-naphthoic acid, whereas salts of 2-naphthol give (115-130 °C; 3-4,5 bar) 2-hydroxy-1-naphthoic acid. In both cases, the yield is about 80%. However, 3-hydroxy-2-naphthoic acid (BON, β-oxynaphthoic, acid) acid is formed from the dry sodium salt of 2-naphthol if the reaction with carbon dioxide is run above 200 °C (typically, at 220-260 °C and 15 bar). In this case, the 2-naphthol conversion is about 50% and the acid yields are about 80-95% based on consumed 2-naphthol. Finally, substantial yields of 6-hydroxy-2-naphthoic acid are obtained by reacting carbon dioxide at 170-230 °C with the potassium salt of 2-naphthol. All these hydroxynaphthoic acids are used as intermediates in the synthesis of dyes, *e.g.* of azo and triphenylmethane dyes. 6-Hydroxy-2-naphthoic acid has been also used as a raw material for thermoplastic polyesters. Typical productions are of few thousands t/year, with the exception of the BON acid, which is one of the most important naphthalene-derived colorant intermediates: its production is probably close to 30.000 t/year.

5. CYCLIC ORGANIC CARBONATES

Few cyclic organic carbonates can be found on the market, although in relatively small amounts (probably, less than 5.000 t/year). The most representative among them are ethylenecarbonate and propylenecarbonate. They are produced either by reaction of phosgene with ethylene- and propylene-glycol (respectively), or exploiting the reaction of carbon dioxide with epoxides:

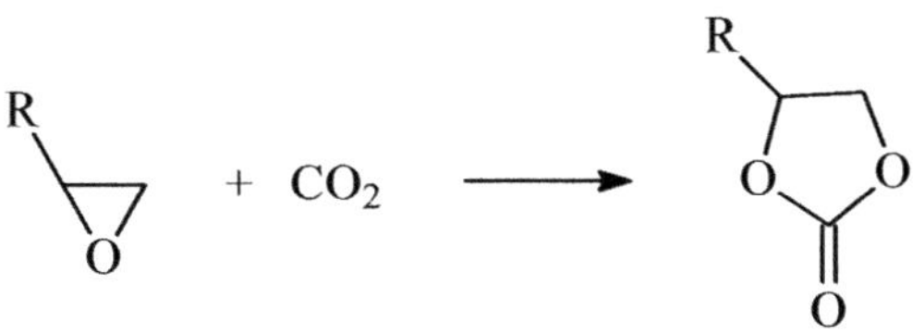

$R = H, CH_3$ (11)

Quaternary ammonium salts are suitable catalysts for the latter reaction.

6. OVERLOOK AND PERSPECTIVES

The manufacture of urea is, by far, the largest exploitation of carbon dioxide as a synthetic building block. The second biggest consumption is provided by the methanol synthesis. Much smaller amounts of carbon dioxide are used to produce other chemicals. So, we can estimate that, overall, about 70 x 10^6 t/year of carbon dioxide are consumed for making industrial organic chemicals. Most of this carbon dioxide, however, is actually produced on purpose for urea plants and the net consumption is probably close to 15% of this figure.

However, several other products could be, in principle, prepared using carbon dioxide as a raw material (*e.g.*, organic carbonates and carbamates; isocyanates; carboxylic acids, esters, and amides; lactones; substituted ureas), provided that suitable synthetic methods will be developed.

From an industrial point of view, one of the most important among such products is the dimethyl carbonate (DMC). Its current production is nearly 130.000 t/y, most of which is for captive use in the phosgene-free synthesis of polycarbonates *via* a double transesterification process (of DMC with phenol to get the diphenyl carbonate, and then of the latter with bisphenol A to get polycarbonate). According to available information, two lines of 50.000 t DMC/y each are run by General Electric (USA) in Spain, to produce polycarbonates, and the realisation of a third 50.000 t/y line has been already approved by the company board. Moreover, 12.000 t/y are also produced by a joint-venture General Electric-Mitsui in Japan to the same scope. In all these cases, the Polimeri Europa (Italy, formerly EniChem) phosgene-free DMC synthesis is used (starting from carbon monoxide, methanol and oxygen), upon licence.

In addition, 16-17.000 t/y of DMC are also available on the free market. Leader in the field is Polimeri Europa (8.000 t/y, with its own technology [12]), but the French company SNPE (ca. 4.000 t/y, *via* phosgene) and the Japanese company UBE (4-5.000 t/y, *via* methylnitrite) are also very active and are currently expanding their production facilities.

The average price of DMC is close to 1 €/kg. Thus, the value of its current production for the free market is about 16-17 $x10^6$ €.

Potential for further DMC development is very high, either as an oxygenated solvent with low toxicity or as a valuable oxygenated fuel additive: should the latter use be implemented, DMC market would rise to tens Mt/y. From a strictly chemical point of view, the availability of a low cost synthesis of DMC (possibly using carbon dioxide as a raw material) would enhance the profitability of phosgene-free routes to polycarbonates (already on stream), and aromatic and aliphatic isocyanates synthesis, (currently under development: see [13] for leading references). About these products, few economic figures follow.

The global polycarbonate capacity (most *via* the very toxic phosgene) is around 2,2 x 10^6 t/y. With an average price of 2,7 €/kg, the value of the market can be estimated to be close to 6 x 10^9 €. Most of the capacity (35%) is located in Western Europe, whereas leading companies are General Electric Plastics (which represents ca. 35% of the market), and Bayer (Germany, with ca. 27% of the market) [14].

The global demand for toluene diisocyanate (TDI) is put at around 1,2 x 10^6 t. Assuming an average price of 1,9 €/kg, the value of this market is ca.

2,3 x 10^9 €. Western Europe plays a pivotal role in this business, accounting for 28% of the worldwide capacity. Moreover, two out of the four leader companies operating in the field are European, namely Bayer (1st position, ca. 22% of the capacity), and BASF (Germany, 4th position, ca. 12% of the capacity). At the 2nd position with ca. 15% of the capacity is Dow Chemical (USA), which has recently taken over the activities of EniChem. Lyondell (USA) represents ca. 14% of the market, after having taken over the activities of ARCO which, in turn, had previously bought Olin activities [15].

On its turn, the global demand for methylenediphenyl diisocyanate (MDI) is put at 2,2 x 10^6 t/y and is expected to continue growing at around 8-11%/y anywhere. Assuming an average price of 1,3 €/kg, the value of the market is ca. 2,9 x 10^9 €. European plants account for more than 40% of the overall capacity. Leader in the field is Dow Chemicals which represents ca. 25% of the market. Two European companies follow: Bayer, with ca. 19% of the market, and BASF, with ca. 18% of the market [16].

The production of isocyanates other than TDI and MDI is, by far, less important. For instance, as far as aliphatic isocyanates are concerned, the overall production has been, in 1993, of ca. 55.300 t. Again, Western Europe plays an important role, with about 50% of the overall capacity [17], and Bayer is leader in the field [18].

REFERENCES

1. Reid R.C., Prausnitz J.M., Polig B.E., The properties of gases and liquids, 4th ed., McGraw-Hill, 668, 1988.
2. Fiedler E., Grossmann G., Kersebohm B., Weiss G., Witte C., Methanol, in Ullmann's Encyclopedia of Industrial Chemistry, Fifth Edition, VCH, A 16, 465-486, 1990.
3. Smalberg F., Moulijn J., van Bekkum H., Global carbon dioxide emission and consumption, Green Chemistry 2 (6), G97-G100, 2000.
4. Meessen J.H., Petersen H., Urea, in Ullmann's Encyclopedia of Industrial Chemistry, Fifth Edition, VCH, A 27, 333-365, 1996.
5. Burridge E. (Ed.), Urea, European Chemical News, 10-16 February, 16, 2003.
6. Bakemeier H., Huberich T., Krabetz R., Liebe W., Schunck M., Mayer D., Ammonia, in Ullmann's Encyclopedia of Industrial Chemistry, Fifth Edition, VCH, A 2, 143-242, 1985.
7. Boullard O., Leblanc H., Besson B., Salicylic Acid, in Ullmann's Encyclopedia of Industrial Chemistry, Fifth Edition, VCH, A 23, 477-483, 1993.
8. Thomas M.R., Salicylic acid and related compounds, in Kirk-Othmer Encyclopedia of Chemical Technology, Fourth Edition, Wiley, 21, 601-626, 1997.
9. Anonimous, Chemical Prices, Chemical Market Report, February 24, 38, 2003.

10. Ritzer E., Sundermann R., Hydroxycarboxylic Acids, Aromatic, in Ullmann's Encyclopedia of Industrial Chemistry, Fifth Edition, VCH, A 13, 519-526, 1989.
11. Booth G., Naphthalene Derivatives, in Ullmann's Encyclopedia of Industrial Chemistry, Fifth Edition, VCH, A 17, 9-57, 1991.
12. Romano U., Dimethyl carbonate and its production technology, Chim. Ind. (Milan), 75, 303-306, 1993.
13. (a) Romano U., Tesei R., Method for preparing aromatic urethans, U.S. Patent 4 100 351, 1978; Chem. Abs., 88, 37459, 1978; (b) Gurgiolo A.E., Preparation of carbamates from aromatic amines and organic carbonates, U.S. Patent 4 268 683, 1981; Chem. Abs., 95, 168832, 1981; (c) Fu Z.H., Ono Y., Synthesis of methyl N-phenyl carbamate by methoxycarbonylation of aniline with dimethyl carbonate using Pb compounds as catalysts, J. Mol. Catal., 91, 399-405, 1994; (d) Cesti P., TDI senza fosgene, Chim. Ind. (Milan), 81, 324, 1999; (e) Cesti P., Bosetti A., Mizia F., Notari M., Ricci M., Rivetti F., Romano, U., Integrated process for the preparation of aromatic isocyanates and procedures for effecting the relative intermediate phases, Eur. Patent Appl. EP 01 01056, 2001.
14. Burridge E. (Ed.), Polycarbonate, European Chemical News, 24-30 June 14, 2002.
15. Burridge E. (Ed.), TDI, European Chemical News, 22-28 July 14, 2002.
16. Burridge E. (Ed.), MDI, European Chemical News, 15-21 July 15, 2002.
17. Richter R.H., Priester Jr. R.D., Isocyanates, Organic, in Kirk-Othmer Encyclopedia of Chemical Technology, Fourth Edition, Wiley, 14, 902-934, 1995.
18. Anonimous, Chemical Market Report, September 1, 11, 1997.

Index

GPSR Compliance

The European Union's (EU) General Product Safety Regulation (GPSR) is a set of rules that requires consumer products to be safe and our obligations to ensure this.

If you have any concerns about our products, you can contact us on ProductSafety@springernature.com

In case Publisher is established outside the EU, the EU authorized representative is:

Springer Nature Customer Service Center GmbH
Europaplatz 3
69115 Heidelberg, Germany

Batch number: 10755284

Printed by Printforce, the Netherlands